OXFORD MASTER SERIES IN CONDENSED MATTER PHYSICS

OXFORD MASTER SERIES IN PHYSICS

The Oxford Master Series is designed for final-year undergraduate and beginning graduate students in physics and related disciplines. It has been driven by a perceived gap in the literature today. While basic undergraduate physics texts often show little or no connection with the huge explosion of research over the last two decades, more advanced and specialized texts tend to be rather daunting for students. In this series, all topics and their consequences are treated at a simple level, while pointers to recent developments are provided at various stages. The emphasis in on clear physical principles like symmetry, quantum mechanics, and electromagnetism which underlie the whole of physics. At the same time, the subjects are related to real measurements and to the experimental techniques and devices currently used by physicists in academe and industry. Books in this series are written as course books, and include ample tutorial material, examples, illustrations, revision points, and problem sets. They can likewise be used as preparation for students starting a doctorate in physics and related fields, or for recent graduates starting research in one of these fields in industry.

CONDENSED MATTER PHYSICS

1. M. T. Dove: *Structure and dynamics: an atomic view of materials*
2. J. Singleton: *Band theory and electronic properties of solids*
3. A. M. Fox: *Optical properties of solids*
4. S. J. Blundell: *Magnetism in condensed matter*
5. J. F. Annett: *Superconductivity, superfluids, and condensates*
6. R. A. L. Jones: *Soft condensed matter*

ATOMIC, OPTICAL, AND LASER PHYSICS

7. C. J. Foot: *Atomic physics*
8. G. A. Brooker: *Modern classical optics*
9. S. M. Hooker, C. E. Webb: *Laser physics*

PARTICLE PHYSICS, ASTROPHYSICS, AND COSMOLOGY

10. D. H. Perkins: *Particle astrophysics*
11. T. P. Cheng: *Relativity and cosmology*

Superconductivity, Superfluids, and Condensates

JAMES F. ANNETT

Department of Physics
University of Bristol

OXFORD
UNIVERSITY PRESS

OXFORD
UNIVERSITY PRESS

Great Clarendon Street, Oxford, OX2 6DP,
United Kingdom

Oxford University Press is a department of the University of Oxford.
It furthers the University's objective of excellence in research, scholarship,
and education by publishing worldwide. Oxford is a registered trade mark of
Oxford University Press in the UK and in certain other countries

First Edition published in 2004
Reprinted 2005, 2008, 2009, 2010, 2011 (twice), 2012, 2013

Published in the United States of America by Oxford University Press
198 Madison Avenue, New York, NY 10016, United States of America

British Library Cataloguing in Publication Data
Data available

Library of Congress Cataloging in Publication Data
Data available

ISBN 978–0–19–850756–7

Preface

Ever since their original discovery nearly 100 years ago, superconductors and superfluids have led to an incredible number of unexpected and surprising new phenomena. The theories which eventually explained superconductivity in metals and superfluid ^{4}He count among the greatest achievements in theoretical many-body physics, and have had profound implications in many other areas, such as in the construction of the "Higgs mechanism" and the standard model of particle physics.

Even now there is no sign that the pace of progress is slowing down. Indeed recent years have seen renewed interest in the field in following the 1986 discovery of cuprate high temperature superconductivity and the 1995 announcement of Bose–Einstein condensation (BEC) in ultra-cold atomic gases. These breakthroughs have tremendously widened the scope of the area of "low temperature physics" from 165 K (only about -100°C, a cold day at the North Pole) the highest confirmed superconducting transition temperature ever recorded, to the realm of nano-Kelvin in laser trapped condensates of atomic gases. Furthermore an incredibly wide range of materials is now known to be superconducting. The field is no longer confined to the study of the metallic elements and their alloys, but now includes the study of complex oxides, carbon-based materials (such as fullerene C_{60}), organic conductors, rare earth based compounds (heavy fermion materials), and materials based on sulphur and boron (MgB_2 superconductivity was discovered in 2001). Commercial applications of superconducting technology are also increasing, albeit slowly. The LHC ring currently (in 2003) being installed at the CERN particle physics center is possible only because of considerable recent advances in superconducting magnet technology. But even this uses "traditional" superconducting materials. In principle, even more powerful magnets could be built using novel high temperature superconducting materials, although these materials are difficult to work with and there are many technical problems still to be overcome.

The goal of this book is to provide a clear and concise first introduction to this subject. It is primarily intended for use by final year undergraduates and beginning postgraduates, whether in physics, chemistry, or materials science departments. Hopefully experienced scientists and others will also find it interesting and useful.

For the student, the concepts involved in superfluidity and superconductivity can be difficult subject to master. It requires the use of many different elements from thermodynamics, electromagnetism, quantum mechanics, and solid state physics. Theories of superconductivity, such as the Bardeen Cooper Schrieffer (BCS) theory, are also most naturally written in the mathematics of quantum field theory, a subject which is well beyond the usual undergraduate physics curriculum. This book attempts to minimize the use of these advanced mathematical techniques so as to make the subject more accessible to beginners.

Of course, those intending to study superconductivity at a more advanced level will need to go on to the more advanced books. But I believe most of the key concepts are fully understandable using standard undergraduate level quantum mechanics, statistical physics, and some solid state physics. Among the other books in the **Oxford Master Series in Condensed Matter**, the volumes *Band theory and electronic properties of solids* by John Singleton (2001), and *Magnetism in condensed matter* by Stephen Blundell (2001) contain the most relevant background material. This book assumes an initial knowledge of solid state physics at this level, and builds upon this (or equivalent level) foundation.

Of course, there are also many other books about superconductivity and superfluids. Indeed each chapter of this book contains suggestions for further reading and references to some of the excellent books and review articles that have been written about superconductivity. However, unlike many of these earlier books, this book is not intended to be a fully comprehensive reference, but merely an introduction. Also, by combinining superconductivity, superfluids and BEC within a single text, it is hoped to emphasize the many strong links and similarities between these very different physical systems. Modern topics, such as unconventional superconductivity, are also essential for students studying superconductivity nowadays and are introduced in this book.

The basic framework of the earlier chapters derives from lecture courses I have given in Bristol and at a number of summer and winter schools elsewhere over the past few years. The first three chapters introduce the key experimental facts and the basic theoretical framework. First, Chapter 1 introduces BEC and its experimental realization in ultra-cold atomic gases. The next chapter introduces superfluid ^{4}He and Chapter 3 discusses the basic phenomena of superconductivity. These chapters can be understood by anyone with a basic understanding of undergraduate solid state physics, quantum mechanics, electromagnetism, and thermodynamics. Chapter 4 develops the theory of superconductivity using the phenomenological Ginzburg–Landau theory developed by the Landau school in Moscow during the 1950s. This theory is still very useful today, since it is mathematically elegant and can describe many complex phenomena (such as the Abrikosov vortex lattice) within a simple and powerful framework. The next two chapters introduce the BCS theory of superconductivity. In order to keep the level accessible to undergraduates I have attempted to minimize the use of the mathematical machinery of quantum field theory, although inevitably some key concepts, such as Feynman diagrams, are necessary. The effort is split into two parts: Chapter 5 introducing the language of coherent states and quantum field operators, while Chapter 6 develops the BCS theory itself. These two chapters should be self-contained so that they are comprehensible whether or not the reader has had prior experience in quantum field theory techniques. The final chapter of the book covers some more specialized, but still very important, topics. The fascinating properties of superfluid ^{3}He are described in Chapter 7. This chapter also introduces unconventional Cooper pairing and is based on a series of review articles in which I discussed the evidence for or against unconventional pairing in the high temperature superconductors.

For a teacher considering this book for an undergraduate or graduate level course, it can be used in many ways depending on the appropriate level for the students. Rather than just starting at Chapter 1 and progressing in linear fashion, one could start at Chapter 3 to concentrate on the superconductivity parts alone. Chapters 3–6 would provide a sound introduction to superconductivity up to the level of the BCS theory. On the other hand, for a graduate level course one could start with Chapters 4 or 5 to get immediately to the many-body physics aspects. Chapter 7 could be considered to be research level or for specialists only, but on the other hand could be read as stand-alone reference by students or researchers wanting to get a quick background knowledge of superfluid ^{3}He or unconventional superconductivity.

The book does not attempt to cover comprehensively all areas of modern superconductivity. The more mathematically involved elements of BCS and other theories have been omitted. Several more advanced and comprehensive books exist, which have good coverage at a much more detailed level. To really master the BCS theory fully one should first learn the full language of many-particle quantum field theory. Topics relating to the applications of superconductivity are also only covered briefly in this book, but again there are more specialized books available.

Finally, I would like to dedicate this book to my friends, mentors, and colleagues who, over the years, have shown me how fascinating the world of condensed matter physics can be. These include Roger Haydock, Volker Heine, Richard Martin, Nigel Goldenfeld, Tony Leggett, Balazs Györffy, and many others too numerous to mention.

James F. Annett[1]
University of Bristol, March 2003

[1] I will be happy to receive any comments and corrections on this book by Email to james.annett@bristol.ac.uk

one could start in Chapter [illegible] supersymmetric [illegible] Chapters 1–4 [illegible] the level of the [illegible] theory [illegible] with Chapters 7 or 8 [illegible] aspects. Chapter [illegible] could [illegible] only, but on the other hand [illegible] could be [illegible] by students or researchers wanting to [illegible] background knowledge [illegible] unconventional superconductivity.

This book does not attempt [illegible] superconductors [illegible] other [illegible] books [illegible] many [illegible] of many particle [illegible] superconductivity [illegible] in this book [illegible]

Finally I would like to thank [illegible] Martin [illegible]

[illegible]
University of Bristol [illegible]

Contents

Bose–Einstein condensates

1.1 Introduction

Superconductivity, **superfluidity**, and **Bose–Einstein condensation** (BEC) are among the most fascinating phenomena in nature. Their strange and often surprising properties are direct consequences of quantum mechanics. This is why they only occur at low temperatures, and it is very difficult (but hopefully not impossible!) to find a room temperature superconductor. But, while most other quantum effects only appear in matter on the atomic or subatomic scale, superfluids and superconductors show the effects of quantum mechanics acting on the bulk properties of matter on a large scale. In essence they are **macroscopic quantum phenomena**.

In this book we shall discuss the three different types of macroscopic quantum states: superconductors, superfluids, and atomic Bose–Einstein condensates. As we shall see, these have a great deal in common with each other and can be described by similar theoretical ideas. The key discoveries have taken place over nearly a hundred years. Table 1.1 lists some of the key discoveries, starting in the early years of the twentieth century and still continuing rapidly today. The field of low temperature physics can be said to have its beginnings in 1908, where helium was first liquified at the laboratory of H. Kammerling Onnes in Leiden, The Netherlands. Very soon afterwards, superconductivity was discovered in the same laboratory. But the theory of superconductivity was not fully developed for until nearly forty years later, with the advent of the

Table 1.1 Some of the key discoveries in the history of superconductivity, superfluidity, and BEC

1908	Liquefaction of ^{4}He at 4.2 K
1911	Superconductivity discovered in Hg at 4.1 K
1925	Bose–Einstein condensation (BEC) predicted
1927	λ transition found in ^{4}He at 2.2 K
1933	Meissner–Ochsenfeld effect observed
1938	Demonstration of superfluidity in ^{4}He
1950	Ginzburg–Landau theory of superconductivity
1957	Bardeen Cooper Schrieffer (BCS) theory
1957	Abrikosov flux lattice
1962	Josephson effect
1963–4	Anderson–Higgs mechanism
1971	Superfluidity found in ^{3}He at 2.8 mK
1986	High temperature superconductivity discovered, 30–165 K
1995	BEC achieved in atomic gases, 0.5 μK

Bardeen Cooper Schrieffer (BCS) theory.[1] In the case of BEC it was the theory that came first, in the 1920s, while BEC was only finally realized experimentally as recently as 1995.

[1]Some offshoots of the development of superconductivity have had quite unexpected consequences in other fields of physics. The **Josephson effect** leads to a standard relationship between voltage, V, and frequency, ν: $V = (h/2e)\nu$, where h is Planck's constant and e is the electron charge. This provides the most accurate known method of measuring the combination of fundamental constants h/e and is used to determine the best values of these constants. A second surprising discovery listed in Table 1.1 is the **Anderson–Higgs mechanism**. Philip Anderson explained the expulsion of magnetic flux from superconductors in terms of *spontaneous breaking of gauge symmetry*. Applying essentially the same idea to elementary particle physics Peter Higgs was able to explain the origin of mass of elementary particles. The search for the related *Higgs boson* continues today at large accelerators such as CERN and Fermilab.

Despite this long history, research in these states of matter is still developing rapidly, and has been revolutionized with new discoveries in recent years. At one extreme we have a gradual progression to systems at lower and lower temperatures. Atomic BEC are now produced and studied at temperatures of nano-Kelvin. On the other hand high temperature superconductors have been discovered, which show superconductivity at much higher temperatures than had been previously believed possible. Currently the highest confirmed superconducting transition temperature, T_c, at room pressure is about 133 K, in the compound $HgBa_2Ca_2Cu_3O_{8+\delta}$. This transition temperature can be raised to a maximum of about 164 K when the material is subjected to high pressures of order 30 GPa, currently the highest confirmed value of T_c for any superconducting material. Superconductivity at such high temperatures almost certainly cannot be explained within the normal BCS theory of superconductivity, and the search for a new theory of superconductivity which can explain these remarkable materials is still one of the central unsolved problems of modern physics.

This book is organized as follows. In this chapter we start with the simplest of these three macroscopic quantum states, BEC. We shall first review the concept of a BEC, and then see how it was finally possible to realize this state experimentally in ultra-cold atomic gases using the modern techniques of laser cooling and trapping of atoms. The following two chapters introduce the basic phenomena associated with superfluidity and superconductivity. Chapters 4–6 develop the theories of these macroscopic quantum states, leading up to the full BCS theory. The final chapter goes into some more specialized areas: superfluidity in ^{3}He and superconductors with unconventional Cooper pairing.

1.2 Bose–Einstein statistics

In 1924 the Indian physicist S.N. Bose wrote to Einstein describing a new method to derive the Plank black-body radiation formula. At that time Einstein was already world-famous and had just won the Nobel prize for his quantum mechanical explanation of the photoelectric effect. Bose was a relatively unknown scientist working in Dacca (now Bangladesh), and his earlier letters to European journals had been ignored. But Einstein was impressed by the novel ideas in Bose's letter, and helped him to publish the results.[2] The new idea was to treat the electromagnetic waves of the black-body as a gas of **identical particles**. For the first time, this showed that the mysterious light quanta, introduced by Planck in 1900 and used by Einstein in his 1905 explanation of the photo-electric effect, could actually be thought of as particles of light, that is, what we now call photons. Einstein soon saw that the same method could be used not only for light, but also for an ideal gas of particles with mass. This was the first proper quantum mechanical generalization of the standard classical theory of the ideal gas developed by Boltzmann, Maxwell, and Gibbs. We know now that there are two distinct quantum ideal gases, corresponding to either Bose–Einstein or Fermi–Dirac statistics. The method of counting quantum states introduced by Bose and Einstein applies to **boson** particles, such as photons or ^{4}He atoms.

[2]For some more historical details see "The man who chopped up light" (Home and Griffin 1994).

The key idea is that for identical quantum particles, we can simply count the number of available quantum states using combinatorics. If we have N_s identical bose particles in M_s available quantum states then there are

$$W_s = \frac{(N_s + M_s - 1)!}{N_s!(M_s - 1)!}, \tag{1.1}$$

available ways that the particles can be distributed. To see how this factor arises, imagine each available quantum state as a box which can hold any number of identical balls, as sketched in Fig. 1.1. We can count the number of arrangements by seeing that the N_s balls and the $M_s - 1$ walls between boxes can be arranged in any order. Basically there are a total of $N_s + M_s - 1$ different objects arranged in a line, N_s of those are of one type (particles) while $M_s - 1$ of them are of another type (walls between boxes). If we had $N_s + M_s - 1$ distinguishable objects, we could arrange them in $(N_s + M_s - 1)!$ ways. But the N_s particles are indistinguishable as are the $M_s - 1$ walls, giving a reduction by a factor $N_s!(M_s - 1)!$, hence giving the total number of configurations in Eq. 1.1.

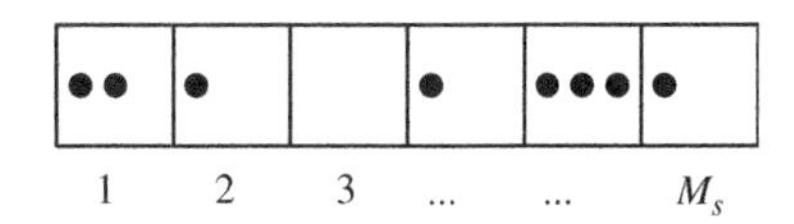

Fig. 1.1 N_s boson particles in M_s available quantum states. We can count the number of possible configurations by considering that the N_s identical particles and the $M_s - 1$ walls between boxes and can be arranged along a line in any order. For bosons each box can hold any number of particles, $0, 1, 2 \ldots$.

We now apply this combinatoric rule to the thermodynamics of an ideal gas of N boson particles occupying a volume V. Using periodic boundary conditions, any individual atom will be in a plane-wave quantum state,

$$\psi(\mathbf{r}) = \frac{1}{V^{1/2}} e^{i\mathbf{k}\cdot\mathbf{r}}, \tag{1.2}$$

where the allowed wave vectors are

$$\mathbf{k} = \left(\frac{2\pi n_x}{L_x}, \frac{2\pi n_y}{L_y}, \frac{2\pi n_z}{L_z}\right), \tag{1.3}$$

and where L_x, L_y, and L_z are the the lengths of the volume in each direction. The total volume is $V = L_x L_y L_z$, and therefore an infinitessimal volume $d^3k = dk_x dk_y dk_z$ of k-space contains

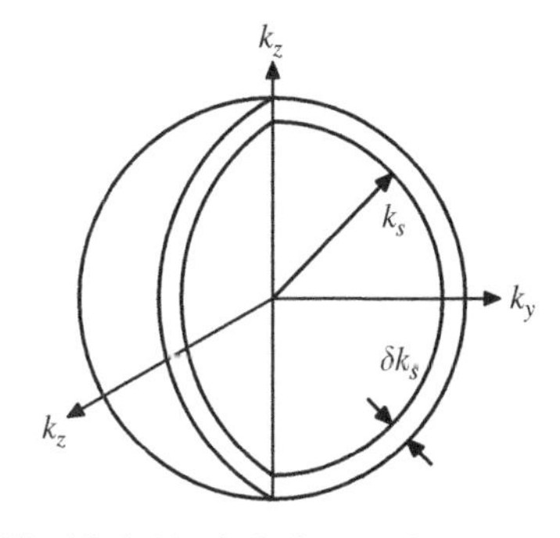

Fig. 1.2 A thin shell of states of wave vector between k_s and $k_s + \delta k_s$. The shell has volume $4\pi k_s^2 \delta k_s$ and so there are $4\pi k_s^2 \delta k_s V/(2\pi)^3$ quantum states in the shell.

$$\frac{V}{(2\pi)^3} d^3k \tag{1.4}$$

quantum states.[3]

[3] The volume *Band Theory and Electronic Properties of Solids*, by John Singleton (2001), in this **Oxford Master Series in Condensed Matter Physics** series explains this point more fully, especially in Appendix B.

Each of these single particle quantum states has energy

$$\epsilon_{\mathbf{k}} = \frac{\hbar^2 k^2}{2m}, \tag{1.5}$$

where m is the particle mass. We can therefore divide up the available single particle quantum states into a number of thin spherical shells of states, as shown in Fig. 1.2. By Eq. 1.4 a shell of radius k_s and thickness δk_s contains

$$M_s = 4\pi k_s^2 \delta k_s \frac{V}{(2\pi)^3} \tag{1.6}$$

single particle states. The number of available states between energy ϵ_s and $\epsilon_s + \delta\epsilon_s$ is therefore

$$\begin{aligned} M_s &= \frac{V m^{3/2} \epsilon^{1/2}}{\sqrt{2}\pi^2\hbar^3} \delta\epsilon_s, \\ &= V g(\epsilon_s) \delta\epsilon_s, \end{aligned} \tag{1.7}$$

where

$$g(\epsilon) = \frac{m^{3/2}}{\sqrt{2}\pi^2\hbar^3} \epsilon^{1/2} \tag{1.8}$$

is the density of states per unit volume, shown in Fig. 1.3.

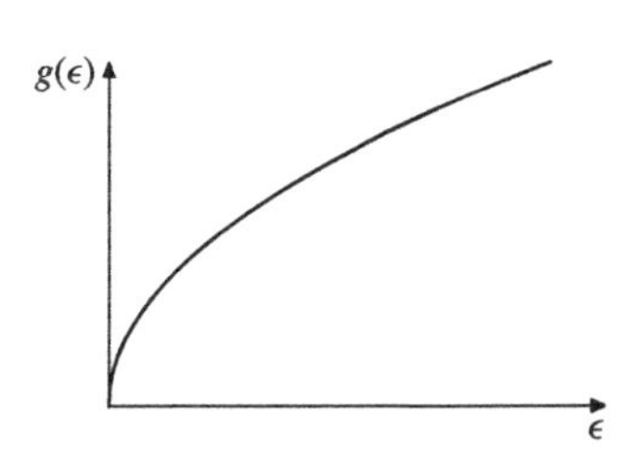

Fig. 1.3 The single particle density of states, $g(\epsilon)$, of a three dimensional gas of particles.

The fundamental principles of statistical mechanics tell us that the total entropy of the gas is $S = k_{\mathrm{B}} \ln W$, where k_{B} is Boltzmann's constant and W is the number of available microstates of a given total energy E. To determine W we must consider how the N atoms in the gas are distributed among the k-space shells of states of different energies. Suppose that there are N_s atoms in shell s. Since there are M_s quantum states in this shell, then we can calculate the total number of available quantum states for this shell using Eq. 1.1. The total number of available microstates for the whole gas is simply the product of the number of available states in each k-space shell,

$$W = \prod_s W_s = \prod_s \frac{(N_s + M_s - 1)!}{N_s!(M_s - 1)!}. \tag{1.9}$$

Using Stirling's approximation, $\ln N! \sim N \ln N - N$, and assuming that N_s, $M_s >> 1$, we have the entropy

$$S = k_{\mathrm{B}} \ln W = k_{\mathrm{B}} \sum_s [(N_s + M_s) \ln (N_s + M_s) - N_s \ln N_s - M_s \ln M_s]. \tag{1.10}$$

In thermal equilibrium the particles will distribute themselves so that the numbers of particles in each energy shell, N_s, are chosen so as to maximize this total entropy. This must be done varying N_s in such a way as to keep constant the total number of particles,

$$N = \sum_s N_s \tag{1.11}$$

and the total internal energy of the gas

$$U = \sum_s \epsilon_s N_s. \tag{1.12}$$

Therefore we must maximize the entropy, S, with the constraints of fixed N and U. Using the method of Lagrange multipliers, this implies that

$$\frac{\partial S}{\partial N_s} - k_{\mathrm{B}}\beta \frac{\partial U}{\partial N_s} + k_{\mathrm{B}}\beta\mu \frac{\partial N}{\partial N_s} = 0, \tag{1.13}$$

where the Lagrange multiplier constants have been defined as $k_{\mathrm{B}}\beta$ and $-k_{\mathrm{B}}\beta\mu$ for reasons which will be clear below. Carrying through the differentiation we find

$$\ln (N_s + M_s) - \ln N_s - \beta\epsilon_s + \beta\mu = 0. \tag{1.14}$$

Rearranging to find N_s we find the result first obtained by Bose and Einstein,

$$N_s = \frac{1}{e^{\beta(\epsilon_s - \mu)} - 1} M_s. \tag{1.15}$$

The average number of particles occupying any single quantum state is N_s/M_s, and therefore the average occupation number of any given single particle states of energy $\epsilon_{\mathbf{k}}$ is given by the **Bose–Einstein distribution**

$$f_{\mathrm{BE}}(\epsilon) = \frac{1}{e^{\beta(\epsilon - \mu)} - 1}. \tag{1.16}$$

In this formula we still have not properly identified the two constants, β and μ, which were introduced simply as Lagrange multipliers. But we can easily find their correct interpretation using the first law of thermodynamics for a gas of N particles,

$$dU = T\,dS - P\,dV + \mu\,dN, \tag{1.17}$$

where T is the temperature, P is the pressure, and μ is the chemical potential. Rearranging gives

$$dS = \frac{1}{T}(dU + P\,dV - \mu\,dN). \tag{1.18}$$

The entropy is given by $S = k_B \ln W$ calculated from Eq. 1.10 with the values of N_s taken from Eq. 1.15. Fortunately the differentiation is made easy using a shortcut from Eq. 1.13. We have

$$\begin{aligned} dS &= \sum_s \frac{\partial S}{\partial N_s} dN_s, \\ &= k_B\beta \sum_s \left(\frac{\partial U}{\partial N_s} - \mu\frac{\partial N}{\partial N_s}\right) dN_s \text{ from Eq. 1.13}, \\ &= k_B\beta(dU - \mu\,dN). \end{aligned} \tag{1.19}$$

Comparing with Eq. 1.18, we see that

$$\beta = \frac{1}{k_B T}, \tag{1.20}$$

and the constant μ which we introduced above is indeed just the chemical potential of the gas.

The method we have used above to derive the Bose–Einstein distribution formula makes use of the thermodynamics of a gas of fixed total particle number, N, and fixed total energy U. This is the **microcanonical ensemble**. This ensemble is appropriate for a system, such as a fixed total number of atoms, such as a gas in a magnetic trap. However, often we are interested in systems of an effectively infinite number of atoms. In this case we take the **thermodynamic limit** $V \to \infty$ in which the density of atoms, $n = N/V$, is held constant. In this case it is usually much more convenient to use the **grand canonical ensemble**, in which both the total energy and the particle number are allowed to fluctuate. The system is supposed to be in equilibrium with an external heat bath, maintaining a constant temperature T, and a particle bath, maintaining a constant chemical potential μ. If the N-body quantum states of N particles have energy $E_i^{(N)}$ for $i = 1, 2, \ldots$, then in the grand canonical ensemble each state occurs with probability

$$P^{(N)}(i) = \frac{1}{\mathcal{Z}} \exp\left[-\beta(E_i^{(N)} - \mu N)\right], \tag{1.21}$$

where the grand partition function is defined by

$$\mathcal{Z} = \sum_{N,i} \exp\left[-\beta(E_i^{(N)} - \mu N)\right]. \tag{1.22}$$

All thermodynamic quantities are then calculated from the grand potential

$$\Omega(T, V, \mu) = -k_B T \ln \mathcal{Z} \tag{1.23}$$

using

$$d\Omega = -SdT - PdV - Nd\mu. \tag{1.24}$$

It is quite straightforward to derive the Bose–Einstein distribution using this framework, rather than the microcanonical method used above.[4]

[4]For the derivation of the Bose–Einstein and Fermi–Dirac distribution functions by this method, see standard thermodynamics texts listed under further reading, or Appendix C in the volume *Band Theory and Electronic Properties of Solids* (Singleton 2001).

1.3 Bose–Einstein condensation

Unlike the classical idea gas, or the Fermi–Dirac gas, the Bose–Einstein ideal gas has a thermodynamic phase transition, **Bose–Einstein condensation**. In fact it is quite unique, since it is a phase transition occurring for non-interacting particles. The phase transition is driven by the particle **statistics** and not their interactions.

At the phase transition all the thermodynamic observables have an abrupt change in character. This defines the critical temperature, T_c. The term "condensation" is used here by analogy with the normal liquid–gas phase transition (such as in the van der Waals theory of gases), in which liquid drops condense out of the gas to form a saturated vapor. In the same way, below the critical temperature T_c in BEC "normal gas" particles coexist in equilibrium with "condensed" particles. But, unlike a liquid droplet in a gas, here the "condensed" particles are not separated in space from the normal particles. Instead, they are separated in **momentum space**. The condensed particles all occupy a single quantum state of zero momentum, while the normal particles all have finite momentum.

Using the Bose–Einstein distribution, 1.16, the total number of particles in the box is

$$N = \sum_{\mathbf{k}} \frac{1}{e^{\beta(\epsilon_{\mathbf{k}}-\mu)} - 1}. \tag{1.25}$$

In the thermodynamic limit, $V \to \infty$, the possible $\mathbf{k}$ values become a continuum and so we should normally expect to be able to replace the summation in Eq. 1.25 with an integration

$$\sum_{\mathbf{k}} \to \int \frac{V}{(2\pi)^3} d^3k.$$

If this is valid, then Eq. 1.25 becomes

$$N = \frac{V}{(2\pi)^3} \int \frac{1}{e^{\beta(\epsilon_{\mathbf{k}}-\mu)} - 1} d^3k, \tag{1.26}$$

and so the particle density is

$$n = \frac{1}{(2\pi)^3} \int \frac{1}{e^{\beta(\epsilon_{\mathbf{k}}-\mu)} - 1} d^3k, \tag{1.27}$$

or, in terms of the density of states per unit volume $g(\epsilon)$ from Eq. 1.8,

$$n = \int_0^\infty \frac{1}{e^{\beta(\epsilon-\mu)} - 1} g(\epsilon)\, d\epsilon. \tag{1.28}$$

This equation defines the particle density $n(T, \mu)$ as a function of the temperature and chemical potential. But, of course, usually we have a known particle density, n, and wish to find the corresponding chemical potential μ. Therefore we must view Eq. 1.28 as an equation which implicitly determines the chemical potential, $\mu(T, n)$, a function of temperature and the particle density n.

Rewriting Eq. 1.28 in terms of the dimensionless variables $z = e^{\beta\mu}$ (called the fugacity), and $x = \beta\epsilon$ gives

$$n = \frac{(mk_BT)^{3/2}}{\sqrt{2}\pi^2\hbar^3}\int_0^\infty \frac{ze^{-x}}{1-ze^{-x}}x^{1/2}\,dx. \tag{1.29}$$

To calculate this integral we can expand

$$\begin{aligned}\frac{ze^{-x}}{1-ze^{-x}} &= ze^{-x}\left(1+ze^{-x}+z^2e^{-2x}+\cdots\right),\\ &= \sum_{p=1}^{\infty} z^p e^{-px}.\end{aligned} \tag{1.30}$$

This expansion is clearly convergent provided that z is smaller than 1. Inserting this into Eq. 1.29 we can now carry out the integral over x using

$$\begin{aligned}\int_0^\infty e^{-px}x^{1/2}\,dx &= \frac{1}{p^{3/2}}\int_0^\infty e^{-y}y^{1/2}\,dy,\\ &= \frac{1}{p^{3/2}}\frac{\sqrt{\pi}}{2},\end{aligned}$$

where the dimensionless integral is a special case of the Gamma function,

$$\Gamma(t) = \int_0^\infty y^{t-1}e^{-y}\,dy \tag{1.31}$$

with the value $\Gamma(3/2) = \sqrt{\pi}/2$. Combining the numerical constants, the particle density is therefore given as a function of the fugacity, z, by

$$n = \left(\frac{mk_BT}{2\pi\hbar^2}\right)^{3/2} g_{3/2}(z), \tag{1.32}$$

where the function $g_{3/2}(z)$ is defined by the series

$$g_{3/2}(z) = \sum_{p=1}^{\infty}\frac{z^p}{p^{3/2}}. \tag{1.33}$$

In order to evaluate the particle density in Eq. 1.32, we must consider the shape of the function $g_{3/2}(z)$. Using the ratio test for convergence, one can easily show that the series Eq. 1.33 converges when $|z| < 1$, but diverges if $|z| > 1$. At $z = 1$ the series is just convergent,

$$g_{3/2}(1) = \sum_{p=1}^{\infty}\frac{1}{p^{3/2}} = \zeta\left(\frac{3}{2}\right) = 2.612, \tag{1.34}$$

where

$$\zeta(s) = \sum_{p=1}^{\infty}\frac{1}{p^s}$$

is the Riemann zeta function. On the other hand, the function has infinite derivative at $z = 1$, since

$$\frac{dg_{3/2}(z)}{dz} = \frac{1}{z}\sum_{p=1}^{\infty}\frac{z^p}{p^{1/2}}, \tag{1.35}$$

which diverges at $z = 1$. With these limiting values we can make a sketch of function $g_{3/2}(z)$ between $z = 0$ and $z = 1$, as shown in Fig. 1.4.

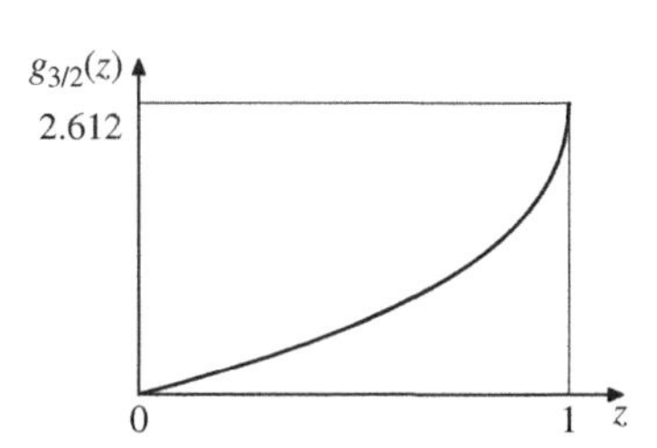

Fig. 1.4 The function $g_{3/2}(z)$ as defined in Eq. 1.33. At $z = 1$ the function is finite but its derivative is infinite.

Equation 1.32 gives the density, n in terms of $g_{3/2}(z)$. Turning it around, we can say that the value of z, and hence the chemical potential μ, is determined by

$$g_{3/2}\left(e^{\beta\mu}\right) = \left(\frac{2\pi\hbar^2}{mk_\mathrm{B}T}\right)^{3/2} n. \tag{1.36}$$

If we are at high temperature T or low density n, then the right-hand side of this equation is small, and we can use the small z expansion $g_{3/2}(z) \approx z + \cdots$ to obtain,

$$\mu \approx -\frac{3}{2}k_\mathrm{B}T \ln\left(\frac{mk_\mathrm{B}T}{2\pi\hbar^2 n^{2/3}}\right). \tag{1.37}$$

This gives a negative chemical potential, as sketched in Fig. 1.5.

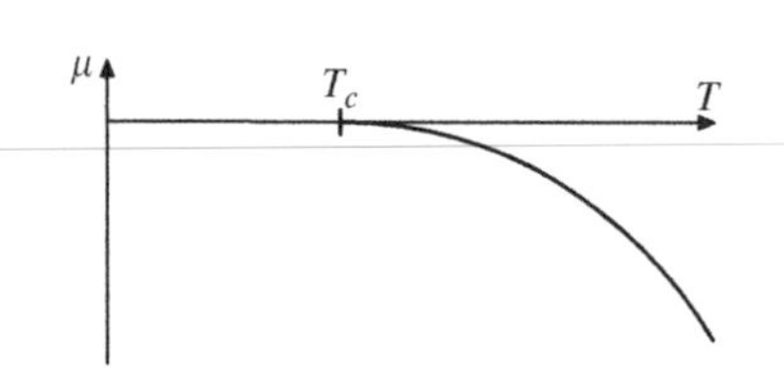

Fig. 1.5 Chemical potential, μ, of a Bose gas as a function of temperature, T. At $T = 0$ all the particles are in the condensate and $n_0 = n$. On the other hand, above the critical temperature T_c all the particles are in the normal component, and $n_0 = 0$.

However, on cooling the gas to lower temperatures the value of z gradually increases until it eventually equals one. At this point the chemical potential, μ, becomes zero. The temperature where this happens (for fixed density, n), defines the critical temperature, T_c,

$$T_c = \frac{2\pi\hbar^2}{k_\mathrm{B}m}\left(\frac{n}{2.612}\right)^{2/3}, \tag{1.38}$$

where $g_{3/2}(z)$ has reached its maximum finite value of 2.612. This T_c is the **BEC temperature**.

But what happens when we cool the gas below T_c? Einstein realized that as soon as the chemical potential μ becomes zero the number of particles in the lowest energy quantum state becomes infinite. More precisely we can say that out of a total of N particles in the gas, a macroscopic number N_0 occupy the one quantum state with $\epsilon_\mathbf{k} = 0$. By a "macroscopic number" we mean that N_0 is proportional to the system volume, so that there is a finite fraction of all of the particles, N_0/N, are in the one quantum state. Recall that we are working in the thermodynamic limit, $V \to \infty$. The Bose–Einstein distribution predicts an occupation of the $\epsilon_\mathbf{k} = 0$ state of

$$N_0 = \frac{1}{e^{-\beta\mu} - 1}. \tag{1.39}$$

Rewriting Eq. 1.39 we obtain

$$\mu = -k_\mathrm{B}T \ln\left(1 + \frac{1}{N_0}\right) \sim -k_\mathrm{B}T\frac{1}{N_0}. \tag{1.40}$$

If a finite fraction of the particles are in the ground state, then as $V \to \infty$ we will have $N_0 \to \infty$ and hence $\mu \to 0$. Therefore below the BEC temperature T_c, the chemical potential is effectively zero, as shown in Fig 1.5.

Below T_c we must take the $\mathbf{k} = 0$ point into account separately, and so we must replace Eq. 1.25 by

$$N = N_0 + \sum_{\mathbf{k}\neq 0}\frac{1}{e^{\beta\epsilon_\mathbf{k}} - 1}, \tag{1.41}$$

where the chemical potential μ is zero. If we again replace the $\mathbf{k}$ summation by an integral (excluding the one point $\mathbf{k} = 0$), the density of particles is now

$$n = n_0 + \frac{(mk_\mathrm{B}T)^{3/2}}{\sqrt{2}\pi^2\hbar^3}\int_0^\infty \frac{e^{-x}}{1 - e^{-x}} x^{1/2}\, dx, \tag{1.42}$$

The definite integral can be evaluated by the same methods as before, and is equal to $\Gamma(3/2)\zeta(3/2)$ and we finally obtain for $T < T_c$,

$$n = n_0 + 2.612\left(\frac{mk_\mathrm{B}T}{2\pi\hbar^2}\right)^{3/2}. \tag{1.43}$$

The particle density, n, can therefore be divided into a **condensate density**, n_0 and a remainder (or normal density) n_n,

$$n = n_0 + n_n. \tag{1.44}$$

The fraction of particles in the condensate can be written compactly

$$\frac{n_0}{n} = 1 - \left(\frac{T}{T_c}\right)^{3/2}, \tag{1.45}$$

as illustrated in Fig. 1.6. It is clear from this expression that at $T = 0$ all of the particles are in the ground state, and hence $n_0 = n$, but at higher temperatures n_0 gradually decreases. n_0 eventually equals zero at the critical temperature T_c, and is zero above T_c.

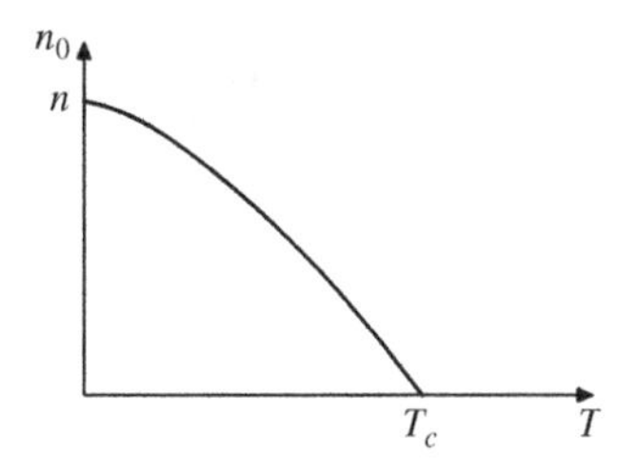

Fig. 1.6 BEC density, n_0, as a function of temperature, T.

Using these results, other thermodynamic quantities of the Bose gas can also be calculated exactly. For example, the total internal energy of the gas is

$$\begin{aligned} U &= V\int_0^\infty \frac{\epsilon}{e^{\beta(\epsilon-\mu)}-1}g(\epsilon)\,d\epsilon, \\ &= V(k_\mathrm{B}T)^{5/2}\frac{m^{3/2}}{\sqrt{2}\pi^2\hbar^3}\int_0^\infty \frac{ze^{-x}}{1-ze^{-x}}x^{3/2}\,dx. \end{aligned} \tag{1.46}$$

The average energy per particle can be calculated by dividing by the particle number obtained previously, giving for $T > T_c$

$$u = \frac{U}{N} = \frac{3}{2}k_\mathrm{B}T\frac{g_{5/2}(z)}{g_{3/2}(z)} \tag{1.47}$$

and for $T < T_c$

$$u = \frac{3}{2}k_\mathrm{B}\frac{T^{5/2}}{T_c^{3/2}}\frac{g_{5/2}(1)}{g_{3/2}(1)}. \tag{1.48}$$

Here the function $g_{5/2}(z)$ is defined by

$$g_{5/2}(z) = \sum_{p=1}^{\infty}\frac{z^p}{p^{5/2}} \tag{1.49}$$

and the numerical constant $g_{5/2}(1)$ equals $\zeta(5/2) = 1.342$.

In the limit where T is much larger than T_c we have a normal Bose gas, and

$$u \sim \frac{3}{2}k_\mathrm{B}T \tag{1.50}$$

(since both $g_{5/2}(z) \approx z$ and $g_{3/2}(z) \approx z$ when z is small). Obviously, this result is exactly the same as the energy per particle in a classical monatomic ideal gas. Physically it implies that the Bose–Einstein statistics of the particles becomes irrelevant at high temperatures $T \gg T_c$.

By examining the heat capacity of the gas, we can see that the temperature T_c represents a true thermodynamic phase transition. The heat capacity can

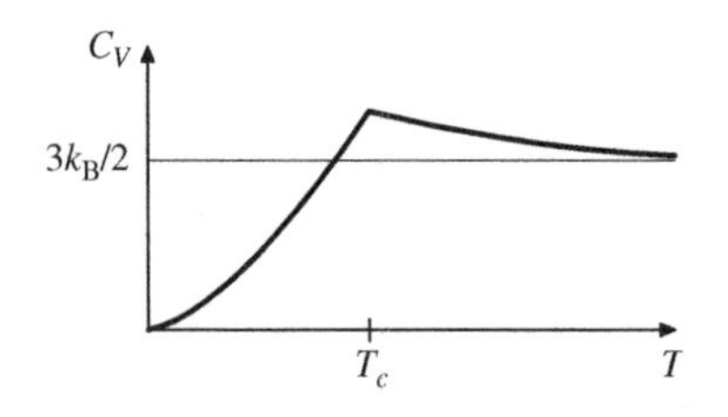

Fig. 1.7 Heat capacity of a Bose–Einstein ideal gas as a function of temperature, T. The cusp at T_c implies that BEC is a thermodynamic phase transition.

be obtained by differentiating the internal energy while keeping the density n constant,

$$C_V = \frac{\partial u}{\partial T} \tag{1.51}$$

per particle. We obtain $C_V \sim 3k_B/2$ for $T \gg T_c$ just as for a classical ideal gas and

$$C_V = \frac{15}{4} \frac{g_{5/2}(1)}{g_{3/2}(1)} \left(\frac{T}{T_c}\right)^{3/2} k_B \tag{1.52}$$

below T_c. This is sketched in Fig 1.7. At T_c the heat capacity has a cusp, or a discontinuity in slope. This implies that the free energy is not analytic at T_c, showing that BEC is indeed a thermodynamic phase transition.[5] Other thermodynamic quantities, such as the entropy or pressure can also be calculated. See Exercise (1.6) or texts on themodynamics, such as Huang (1987), for more details.

[5]The BEC phase transition is usually interpreted as a **first-order phase transition**. The two phases are a gas of normal particles with nonzero momentum, and a "condensate" of zero momentum particles. At any nonzero temperature below T_c we have a mixture of these two distinct phases. The proof that the full thermodynamic behavior is consistent with that expected for such a two-phase mixture is developed below in Exercise (1.6).

Finally, it is interesting to examine the origin of the BEC a little more carefully. Originally we replaced the whole sum over the discrete $\mathbf{k}$ plane wave states by a continuum integral. Then we realized that it is neccessary to treat the $\mathbf{k} = 0$ state specially, but the rest of the states were still represented by a continuum. Why is this justified? Consider the occupation number of the first states with finite wave number $\mathbf{k}$. In a cubic box of side length L, the very first low energy states have $k \approx 2\pi/L$, and hence $\epsilon_{\mathbf{k}} \approx h^2/mL^2 = V^{-2/3}h^2/m$. The occupation of these states is

$$N_1 = \frac{1}{e^{\beta(\epsilon_{\mathbf{k}}-\mu)} - 1} \sim \frac{1}{e^{\beta V^{-2/3}h^2/m} - 1} = O(V^{2/3})$$

where the notation $O(n)$ means "of order n" in the limit $V \to \infty$. We can see that, although the occupations of the finite $\mathbf{k}$ states grow with V, they grow much more slowly than the value of N_0. In fact $N_1/N_0 = O(V^{-1/3}) \to 0$ as $V \to \infty$. Therefore in the infinite size system limit, the occupations of any individual single particle plane wave state $\mathbf{k}$ is negligible compared with the one special state at $\mathbf{k} = 0$. In the thermodynamic limit we can quite correctly make the continuum approximation for all of the $\mathbf{k}$ states except $\mathbf{k} = 0$ with no significant error.

1.4 BEC in ultra-cold atomic gases

In the late 1930s, quite soon after Einstein's prediction of BEC, it was discovered that liquid ^{4}He becomes a superfluid below the lambda point at about 2.2 K. Since a ^{4}He atom contains two electrons, two protons, and two neutrons, it is, as a whole, a boson (i.e. it has total spin zero). Therefore it was natural to postulate a link between BEC and the superfluidity of ^{4}He. Interestingly, if one uses the density of ^{4}He of $\rho \approx 145$ kg m^{-3} and the atomic mass $m \approx 4m_p$, to find the particle density $n = \rho/m$ one obtains a value of T_c of about 3.1 K using Eq. 1.38, which is quite close to the superfluid transition of ^{4}He.

Unfortunately the BEC theory, as described earlier, was for the ideal-gas, and completely neglected any interactions between the particles. But in the case of liquid helium the particle density is fairly high and the particle interactions cannot be neglected. Therefore we cannot really view helium as a suitable test case for the concept of BEC. Indeed there are many differences between the properties of superfluid ^{4}He, as described in the next chapter, and the predictions for the ideal Bose gas.

In fact only in 1995 were actual physical examples of a BEC finally realized. These were provided not in helium, but in very dilute gases of alkali metal atoms. The techniques for trapping and cooling atoms in magnetic and laser traps had been developed and improved gradually over the preceding two decades. At first sight it may seem surprising that one can achieve the conditions of temperature and density such that BEC can occur in such systems. The densities of atoms in the traps are typically of order $10^{11}-10^{15}$ cm^{-3}, which is many orders of magnitude less than the atomic density of ^{4}He, which is about $n \sim 2 \times 10^{22}$ cm^{-3}. Furthermore the atomic masses of the alkalis are very much higher than for ^{4}He, especially for heavy alkali atoms such as ^{87}Rb. Using Eq. 1.38 we would expect T_c values perhaps $10^{-6}-10^{-8}$ times smaller than for the parameters of ^{4}He. In other words we expect T_c values of order 10 nK – 1 μK.

It is remarkable that the techniques for cooling and trapping atoms with lasers and magnetic traps can now achieve such incredibly low temperatures in the laboratory. Explaining in detail exactly how this is done would take us far from the main topics of this book. Here, we shall only give a brief outline of some of the fundamental principles involved.

First, how can we view a single large object, such as a rubidium atom as a boson? In quantum mechanics a particle will be a boson if it has an integer spin. Alkali metal atoms are in the first column of the periodic table, and so they have a single valence electron in the outermost s-orbital, for example 2s for lithium (Li), 3s for sodium (Na), 4s for potassium (K), or 5s for rubidium (Rb). The other electrons are in completely full shells of quantum states, and as a result they have a total orbital angular momentum and total spin of zero. The only other contribution to the total spin of the atom is the nuclear spin. If the nuclear isotope is one with an odd number of protons and neutrons it will have a net half-integer spin. For example, ^{7}Li, ^{23}Na, and ^{87}Rb all have $S = 3/2$ nuclei. In this case the total spin of the atom will be the sum of the nuclear spin and the valence electron spin, which will be an integer. Recall that in quantum mechanics adding two spins S_1 and S_2 leads to the possible values of the total spin,

$$S = |S_1 - S_2|, |S_1 - S_2| + 1, \ldots, S_1 + S_2 - 1, S_1 + S_2. \tag{1.53}$$

The spin $S = 3/2$ nucleus and the $S = 1/2$ valence electron spin combine to give states with a total spin of either $S = 2$ or $S = 1$. If we can prepare the gas so that only one of these types of states is present, then this will be a gas of particles each with an integer spin. Therefore we can view this as a gas of Bose particles. On the other hand, if atoms in both $S = 1$ and $S = 2$ quantum states are present in the gas, then this is effectively a mixture of two different species of bosons, since the two types of atoms are distinguishable from each other.

In order to see how these atoms can be trapped by a magnetic field we must consider the energy levels of the atom and how they are affected by a magnetic field. For definiteness, let us assume that the alkali atom has an $S = 3/2$ nucleus. It is helpful to first find the explicit spin wave functions for the different quantum states. First we find the states with maximum total spin, $S = 2$. For this total spin there are five different states, corresponding to z-components of total spin given by quantum numbers $M_S = 2, 1, 0, -1, 2$. The wave function corresponding to the maximum value, $M_S = 2$, can be represented as

$$|S = 2, M_S = 2\rangle = \left|\frac{3}{2}, \frac{1}{2}\right\rangle,$$

where we use the notation $|m_{s1}, m_{s2}\rangle$ to denote the state where the nucleus is in state m_{s1} and the electron is in state m_{s2}. The other M_s quantum states with total spin $S = 2$ can be found by acting with the spin lowering operator $\hat{S}^- = \hat{S}_1^- + \hat{S}_2^-$. Using the identity

$$\hat{S}^- |m\rangle = \sqrt{s(s+1) - m(m-1)}|m-1\rangle \tag{1.54}$$

successively we obtain the five quantum states $M_s = 2, 1, 0, -1, -2$,

$$\begin{aligned}
|S = 2, M_S = 2\rangle &= \left|\frac{3}{2}, \frac{1}{2}\right\rangle, \\
|S = 2, M_S = 1\rangle &= \frac{1}{2}\left(\sqrt{3}\left|\frac{1}{2}, \frac{1}{2}\right\rangle + \left|\frac{3}{2}, -\frac{1}{2}\right\rangle\right), \\
|S = 2, M_S = 0\rangle &= \frac{1}{\sqrt{2}}\left(\left|\frac{1}{2}, -\frac{1}{2}\right\rangle + \left|-\frac{1}{2}, \frac{1}{2}\right\rangle\right), \\
|S = 2, M_S = -1\rangle &= \frac{1}{2}\left(\sqrt{3}\left|-\frac{1}{2}, -\frac{1}{2}\right\rangle + \left|-\frac{3}{2}, \frac{1}{2}\right\rangle\right), \\
|S = 2, M_S = -2\rangle &= \left|-\frac{3}{2}, -\frac{1}{2}\right\rangle.
\end{aligned} \tag{1.55}$$

The three states with total spin $S = 1$ and $M_S = 1, 0, -1$ must be orthogonal to the corresponding $S = 2$ states, and this requirement determines them uniquely to be

$$\begin{aligned}
|S = 1, M_S = 1\rangle &= \frac{1}{2}\left(\left|\frac{1}{2}, \frac{1}{2}\right\rangle - \sqrt{3}\left|\frac{3}{2}, -\frac{1}{2}\right\rangle\right), \\
|S = 1, M_S = 0\rangle &= \frac{1}{\sqrt{2}}\left(\left|\frac{1}{2}, -\frac{1}{2}\right\rangle - \left|-\frac{1}{2}, \frac{1}{2}\right\rangle\right), \\
|S = 1, M_S = -1\rangle &= \frac{1}{2}\left(\left|-\frac{1}{2}, -\frac{1}{2}\right\rangle - \sqrt{3}\left|-\frac{3}{2}, \frac{1}{2}\right\rangle\right).
\end{aligned} \tag{1.56}$$

In zero magnetic field the $S = 2$ and $S = 1$ states have slightly different energy due to the weak **hyperfine interaction** between the nucleus and the outermost unpaired valence electron. In zero magnetic field all five of the $S = 2$ states are degenerate with each other, as are the three $S = 1$ states, as shown in Fig. 1.4. But a magnetic field leads to a Zeeman splitting of these degenerate states. To a good approximation, we can write the relevant effective Hamiltonian as,

$$\hat{H} = J\hat{\mathbf{S}}_1 \cdot \hat{\mathbf{S}}_2 + 2\mu_{\mathrm{B}}\hat{\mathbf{S}}_{2z}B_z, \tag{1.57}$$

where J is the hyperfine interaction between the nuclear and valence electron spins, $2\mu_{\mathrm{B}}$ is the magnetic moment of the valence electron ($\mu_{\mathrm{B}} = e\hbar/2m_e$, the Bohr magneton) , and B_z is the magnetic field which is assumed to be in the z direction. In writing this simplified Hamiltonian we have ignored the magnetic moment of the nucleus, which is very much smaller than that of the valence

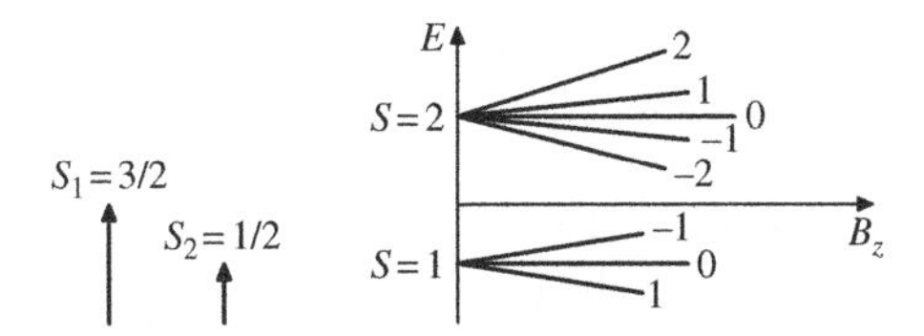

Fig. 1.8 Energy levels of a typical alkali metal atom with an $S = 3/2$ nucleus, as a function of a weak external magnetic field, B_z. At $B_z = 0$ the $S = 2$ and $S = 1$ total spin states are separated by the nuclear hyperfine coupling. The weak magnetic field Zeeman further splits these $S = 2$ and $S = 1$ hyperfine levels into their different M_S states as shown.

electron. It is not difficult to find the eigenstates of this Hamiltonian, especially in the limit of small magnetic fields, B_z. If $B_z = 0$ we can use the identity

$$\hat{\mathbf{S}}_1 \cdot \hat{\mathbf{S}}_2 = \frac{1}{2}\left((\hat{\mathbf{S}}_1 + \hat{\mathbf{S}}_2)^2 - \hat{\mathbf{S}}_1^2 - \hat{\mathbf{S}}_2^2\right) \tag{1.58}$$

to find the two energy levels

$$E_2 = +\frac{3}{4}J \qquad E_1 = -\frac{5}{4}J \tag{1.59}$$

corresponding to $S = 2$ and $S = 1$, respectively, as sketched in Fig. 1.4. In a small magnetic field we can use perturbation theory in the term $\hat{H}' = 2\mu_B \hat{\mathbf{S}}_{2z} B_z$ corresponding to the magnetic dipole moment of the valence electron. First order perturbation theory gives the energy shifts $\Delta E = \langle \hat{H}' \rangle$, which are linear in B_z, as shown in Fig. 1.4.

This magnetic field dependence of the quantum state energies is exploited in a **magnetic atom trap**. The trap is constructed by producing a region of space in which the magnetic field has a local minimum. At first sight it is surprising that it is possible to produce a local minimum in magnetic field, because the field must also obey Maxwell's equations for a region of free space

$$\nabla \cdot \mathbf{B} = 0, \tag{1.60}$$

$$\nabla \times \mathbf{B} = 0. \tag{1.61}$$

In fact it is indeed possible to both obey these and to have a local minimum in the field magnitude $|\mathbf{B}(\mathbf{r})|$.[6] Now, if we prepare an atom in a quantum state such as $S = 2$ $M_S = 2$ in Fig. 1.4, then it will lower its energy by moving toward a region of smaller magnetic field. It will therefore be attracted into the magnetic trap, which will appear to the atom as a local minimum in potential energy. This potential is only a local minimum, as shown in Fig. 1.9. Atoms which have too much kinetic energy, that is, are too "hot," will not be bound by the trap and will escape. While atoms which have less kinetic energy, that is, are "cold," will be bound by the local minimum in potential energy.

Comparing with the energy level diagram in Fig. 1.4 one can see that several different quantum states lower their energy by moving to a region of lower field, and so could be bound by a magnetic trap. In Fig. 1.4 the $S = 2$, $M_S = 2$, and $M_S = 1$ states and the $S = 1$, $M_S = -1$ state could be trapped. Indeed one could also have a mixture of atoms in these states bound together in the trap. But because the atoms in different quantum states would be distinguishable from one another, in the case of a mixture there would be effectively two different types of boson particles in the trap. In such a case the two distinct boson gases are in thermal equilibrium with each other.

[6] Interestingly one can prove that it is impossible to find a magnetic field $\mathbf{B}(\mathbf{r})$ for which $|\mathbf{B}(\mathbf{r})|$ has a local **maximum**. This result is known as Earnshaw's theorem. Fortunately the theorem does not rule out the existence of a local **mimimum**.

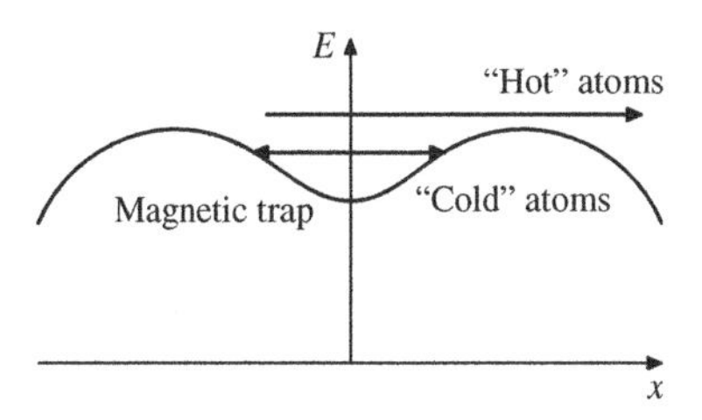

Fig. 1.9 A magnetic trap provides a local minimum in energy. Atoms which are too energetic can escape, while atoms with lower kinetic energy are trapped. Also atoms in a quantum state M_S whose energy decreases with $|\mathbf{B}|$ will see a local maximum in potential energy, not a minimum, and so will be expelled from the trap.

The effective potential energy of the trap in Fig. 1.9 also gives rise naturally to a simple mechanism for cooling the trapped gas. Individual atoms which have high kinetic energies will escape over the barrier and carry away their energy. The remaining atoms will be cooler on average. This is simply cooling by evaporation! By controlling the barrier height, one can control the cooling rate, and hence the final temperature of the system. In this way, temperatures below 1 μK can be achieved.

Unlike the ideal Bose gas we studied earlier in this chapter, the alkali atoms in a magnetic trap do in fact interact with each other. In fact the interactions can be quite strong, since the atoms strongly repel each other at short distances. Also, at large interatomic distances there is a van der Waals attraction force between the atoms. These interactions would eventually lead to the atoms in the trap binding strongly together into atomic clusters. But fortunately the rate at which this happens is very slow. The reason is that collisions between the atoms will be almost entirely **two-body** collisions between pairs of atoms. Since these are elastic collisions, no binding can take place. Binding would only be possible in a three-body collision in which one pair of atoms could form a bound state, while the excess kinetic energy is carried off by the third atom. The rate of three body collisions will be tiny when the density of atoms in the trap, n, is sufficiently low. Typically the length scale of the interatomic interaction is of order 0.2–0.3 nm, while the density of atoms in the magnetic trap is of order $n \sim 10^{11}{-} \sim 10^{15}\ \mathrm{cm}^{-3}$, corresponding to a typical interatomic spacing of $r_s \sim 50$–600 nm where r_s is defined by $n = 1/(4\pi r_s^3/3)$. Therefore the probability of three particles colliding simultaneously is low, and it is possible to maintain the atoms in the trap for a reasonably long time (seconds, or perhaps a few minutes) to perform experiments.

On the other hand the two-body collisions between particles are not entirely negligible. First, it is important to note that the two-body collisions do not allow transitions between the different hyperfine quantum states in Fig. 1.4. This will be the case for the $S = 2, M_S = 2$ or the $S = 1, M_S = -1$ states in Fig. 1.4. Therefore particles prepared in one of these low field seeking quantum states will remain in the same state. Second, the interparticle interactions contribute to the overall potential energy of the atoms in the trap, and cannot be neglected. Pairwise interparticle interactions are also necessary so that **thermal equilibrium** is established during the timescale of the experiment. Pairwise collisions lead to redistribution of energy and are necessary for the system to achieve thermal equilibrium.

We can treat the interatomic pair interactions approximately as follows. Since the interaction only acts of a very short length scale compared with the typical interparticle separations we can approximate the pair interaction by a Dirac delta-function

$$V(\mathbf{r}_1 - \mathbf{r}_2) \approx g\delta(\mathbf{r}_1 - \mathbf{r}_2). \tag{1.62}$$

The interaction can therefore be characterized by a single constant, g. Using scattering theory this can also be expressed as a two-body s-wave scattering length, a_s, defined by

$$g = \frac{4\pi a_s \hbar^2}{m}. \tag{1.63}$$

Usually g and a_s are positive, corresponding to a net repulsive interaction. On average the effects of this interactions can be represented as an additional potential felt by each particle, resulting from the average interaction with the other particles. This mean-field contribution to the potential can be written

$$V_{\text{eff}}(\mathbf{r}) = gn(\mathbf{r}), \tag{1.64}$$

where $n(\mathbf{r})$ is the density of atoms at point $\mathbf{r}$ in the trap. In this approximation the atoms in the magnetic trap obey an effective Schödinger equation

$$\left(-\frac{\hbar^2}{2m}\nabla^2 + V_{\text{trap}}(\mathbf{r}) + V_{\text{eff}}(\mathbf{r})\right)\psi_i(\mathbf{r}) = \epsilon_i\psi_i(\mathbf{r}), \tag{1.65}$$

where $V_{\text{trap}}(\mathbf{r})$ is the effective potential of the magnetic trap, as in Fig. 1.9, including both the magnetic field energy and gravity. Equation 1.65 is effectively a nonlinear Schrödinger equation, since the potential depends on the particle density which in turn depends on the wave functions via the Bose–Einstein distribution

$$n(\mathbf{r}) = \sum_i \frac{1}{e^{\beta(\epsilon_i-\mu)} - 1}|\psi_i(\mathbf{r})|^2. \tag{1.66}$$

As usual, the chemical potential μ is determined from the constraint of the constant total number of atoms in the trap, N,

$$N = \int n(\mathbf{r})\, d^3r = \sum_i \frac{1}{e^{\beta(\epsilon_i-\mu)} - 1}. \tag{1.67}$$

These equations (Eqs 1.64–1.67) are a closed set, which must be solved self-consistently. At zero temperature all of the particles are in the condensate, and

$$n(\mathbf{r}) = N|\psi_0(\mathbf{r})|^2, \tag{1.68}$$

where $\psi_0(\mathbf{r})$ is the ground state wave function. These coupled equations must be solved self-consistently to find the wavefunctions, density $n(\mathbf{r})$ and the effective potential $V_{\text{eff}}(\mathbf{r})$. They are known as the **Gross–Pitaevskii** equations.

Solving this coupled set of nonlinear equations must be done numerically. Nevertheless the solution again shows a form of BEC. If the lowest energy state in the potential well has energy ϵ_0, then at some critical temperature T_c the occupation of this one quantum state, N_0, suddenly increases from a small number (of order 1) to a large number (of order N). From the rules of statistical mechanics we cannot, strictly speaking, call this a thermodynamic phase transition, since the total number of particles is finite, and there is no **thermodynamic limit**. Despite this, the total number of atoms in the trap can be large ($10^4 - 10^6$) and so in practice the critical temperature T_c is quite sharp and well defined.

Bose–Einstein condensation in trapped ultra-cold gases was first observed in 1995. The 2001 Nobel prize for physics was awarded to Cornell, Ketterle, and Weiman for this achievement. This discovery followed decades of work by many groups in which the technology of trapping and cooling atoms with magnetic and laser traps was developed and refined. In 1995 three different groups of researchers achieved BEC, using the different alkali atoms ^{87}Rb, ^{23}Na, and ^{7}Li. The different traps used in the experiments combined magnetic trapping methods (as described previously) with laser trapping and cooling methods (which we will not describe here). The temperatures at which BEC were observed were of order $0.5 - 2\ \mu$K, depending on the alkali atom used and the atomic density achieved in the trap. The dramatic results of one such

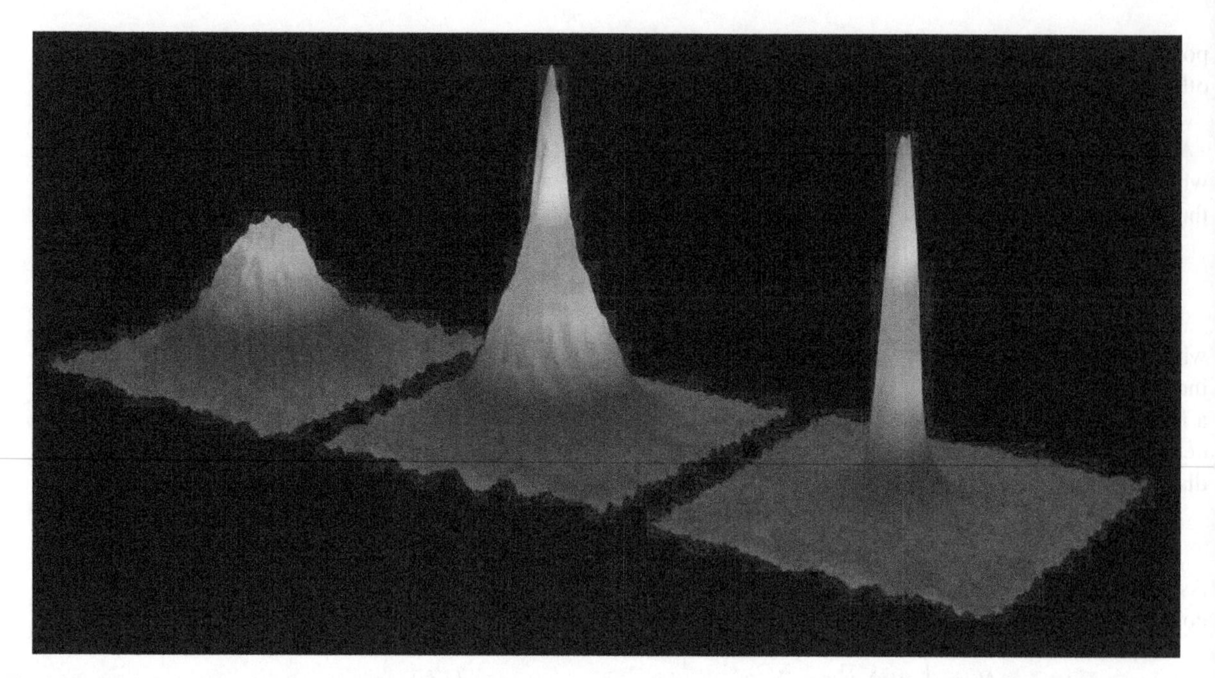

Fig. 1.10 Velocity distribution of atoms in a BEC. On the left we see the broad Maxwell–Boltzmann type distribution in the normal gas above T_c. In the center we see the gas just below T_c, with some fraction of the atoms beginning to appear in the sharp condensate peak at zero velocity. On the right we see the velocity distribution at temperature well below T_c where almost all atoms are condensed into the zero-velocity peak. The intrinsic width of the peak is governed by the harmonic oscillator-like ground state wave function of the atoms in the trap. Reprinted figure with permission from Ketterle (2002). Copyright (2002) by the American Physical Society.

experiment are shown in Fig. 1.10. The figure shows the velocity distribution of the atoms in the gas, first at temperatures above T_c (left), just below T_c (centre), and then well below T_c (right). The sharp peak at the center can be identified with the atoms in the condensate, N_0, while the remaining atoms which are not in the condensate $N - N_0$ have a broad distribution of velocities typical of a normal gas. The picture is actually obtained by suddenly turning off the trap magnetic field and then allowing the atoms to move freely for a certain interval of time. The distance the atoms have moved in this time interval can be measured when the atoms are illuminated with laser light at the frequency of a strong optical absorbtion line of the atom. The spatial distribution of the light absorbed by the atoms then shows their positions, and hence their velocities at the moment the trap was turned off.

Interestingly the central condensate peak seen in Fig. 1.10 is not infinitely narrow. This is simply a consequence of the trapping potential, $V_{\text{trap}}(\mathbf{r})+V_{\text{eff}}(\mathbf{r})$ in Eq. 1.65. The condensate occurs in the ground state wave function, $\psi_0(\mathbf{r})$, and this single state has a finite width momentum distribution as can be seen in the momentum representation of the wave function,

$$\psi_0(\mathbf{r}) = \frac{1}{(2\pi)^3} \int A_{\mathbf{k}} e^{i\mathbf{k}\cdot\mathbf{r}} d^3k. \tag{1.69}$$

The finite width of the momentum distribution in the ground state can be simply estimated from the uncertainty principle, as $\Delta p \sim \hbar/\Delta x$, where Δx is the effective width of the ground state wave function in the trap potential minimum of Fig. 1.9. In contrast the width of the velocity distribution of the noncondensed atoms in the trap, the broad background in Fig. 1.10, can be estimated from the usual Maxwell–Boltzmann velocity distribution of a gas. A crude estimate is $\frac{1}{2}m\langle v_x^2\rangle = \frac{1}{2}k_BT$, from the equipartition of energy theorem. Hence for the broad normal component $\Delta v \sim (k_BT/m)^{1/2}$, compared with $\Delta v \sim \hbar/(m\Delta x)$ for the sharper condensate peak.

Following the achievement of BEC in atoms gases, many different experiments have been performed. The system is ideal for many experiments, since all of the physical parameters of the experiment can be controlled and it is possible to manipulate the BEC in many different ways. It has also been possible to do similar experiments on atoms with Fermi–Dirac statistics, but of course these do not have an analog of BEC. It turns out that the interatomic interaction, Eq. 1.62, is very important in the properties of the atomic BEC, and so strictly one is studying a system of **weakly interacting** bosons, and not an ideal Bose gas. These weak two-body interactions are very significant. In particular, it can be shown that an ideal BEC is **not a superfluid**. When there are no interactions the critical velocity for superfluid flow (discussed in the next chapter) is zero. It is only when the interactions are finite that it becomes possible to sustain a true superfluid state, for example, to have a liquid that can flow with zero viscosity or which can sustain persistent currents which are unaffected by external perturbations. In the atomic BEC, which have been obtained experimentally the small, but finite, residual interactions between the particles mean that these are indeed effectively true superfluids. Persistent currents and even superfluid vortices have been observed.

Finally, experiments have also explored the consequences of **macroscopic quantum coherence** in BEC, showing quantum superpositions and interference between systems with macroscopic (10^5 or 10^6) numbers of particles. Such macroscipic superposition states can be used as an actual physical realization of the Schrödinger cat problem in quantum measurement theory! Just as the Schödinger cat is placed in a quantum superposition of "dead" and "live" cat quantum states, in exactly the same way the BEC can be placed in a superposition of two quantum states each of which differ from each other in a macroscopic number of particle coordinates.

Further reading

For a short article on the life of S.N. Bose and the discovery of Bose–Einstein statistics from New Sceintist see: "The man who chopped up light" by Home and Griffin (1994).

The fundamental princples of themodynamics in classical and quantum systems which we have used in this chapter are described in many books, for example: Mandl (1988), or Huang (1987). These books also discuss in detail the ideal bose gas, and Bose–Einstein condensation.

In writing the section on BEC in atomic gases I have made considerable use of the detailed reviews by: Leggett (2001) and Dalfovo *et al.* (1999). The Nobel prize lecture given by Ketterle (2002) also describes the events leading up to the 1995 discovery of BEC. The review by Phillips (1998) describes the principles of laser atom traps, which operate on somewhat different principles from the magnetic traps described above in Section 1.4.

Two books with comprehensive discussions about all aspects of BEC are Pethick and Smith (2001) and Pitaevskii and Stringari (2003).

Exercises

(1.1) Extend the counting argument given above for the number of ways to place N_s particles in M_s available quantum states for the case of fermion particles. Show that if only one particle can be placed in each quantum state, then Eq. 1.1 becomes

$$W_s = \frac{M_s!}{N_s!(M_s - N_s)!}.$$

(1.2) For the case of the Fermi–Dirac state counting given above in Exercise 1.1, maximize the total entropy $S = k_B \ln W$, and hence show that for fermions

$$N_s = M_s \frac{1}{e^{\beta(\epsilon_s - \mu)} + 1},$$

the Fermi–Dirac distribution.

(1.3) The Bose–Einstein distribution can be derived easily using for formalism of the **grand canonical ensemble** as follows.

(a) Write the energies the many-particle quantum states in the form

$$E = \sum_{\mathbf{k}} n_{\mathbf{k}} \epsilon_{\mathbf{k}}$$

where $n_{\mathbf{k}} = 0, 1, \ldots$. Hence show that the grand partition function for the bose gas can be written as a product over all the **k** states

$$\mathcal{Z} = \prod_{\mathbf{k}} \mathcal{Z}_{\mathbf{k}}$$

and show that

$$\mathcal{Z}_{\mathbf{k}} = \frac{1}{1 - e^{-\beta(\epsilon_{\mathbf{k}} - \mu)}}.$$

(b) Show that the average particle number in state **k** is

$$\langle n_{\mathbf{k}} \rangle = -k_B T \frac{\partial \ln \mathcal{Z}_{\mathbf{k}}}{\partial \epsilon_{\mathbf{k}}}$$

and hence confirm that this equals the Bose–Einstein distribution.

(1.4) It is possible to realize a two-dimensional Bose gas by trapping helium atoms on the surface of another material, such as graphite.

(a) Show that for a two-dimensional gas of area, A, the number of quantum states is

$$\frac{A}{(2\pi)^2} d^2k$$

per unit volume in k-space.

(b) Show that the corresponding density of states $g(\epsilon)$ is constant,

$$g(\epsilon) = \frac{m}{2\pi\hbar^2}.$$

(1.5) For the two-dimensional gas from Exercise 1.4, show that the analog of Eq. 1.29 is

$$n = \frac{mk_B T}{2\pi\hbar^2} \int_0^\infty \frac{ze^{-x}}{1 - ze^{-x}} dx.$$

Do the integral exactly (using an easy substitution) to obtain

$$n = -\frac{mk_B T}{2\pi\hbar^2} \ln(1 - z).$$

Hence write down an explicit expression for μ as a function of n and T, and show that μ never becomes zero. This proves that the two-dimensional ideal Bose gas never becomes a BEC!

(1.6) Find the entropy per particle in the three-dimensional ideal Bose gas in the temperature range $0 \le T \le T_c$, using

$$s(T) = \int_0^T \frac{C_V}{T'} dT'$$

(obtained from $ds = du/T$ and $du = C_V\, dT$). From your result, show that the total entropy of N particles obeys

$$S(T) = N_0 s(0) + N_n s(T_c),$$

where $N = N_0 + N_n$. This result shows that we can consider the state below T_c as a **statistical mixture** (like a liquid–gas saturated vapor) of N_0 particles in the "condensate" with entropy per particle $s(0)$, and N_n "normal" fluid particles with entropy $s(T_c)$. This implies that we can view BEC as a first-order phase transition.

(1.7) Using the wave functions of the different $S = 2$ states for $M_S = 2, 1, 0, -1, -2$ given in Eq. 1.55 show that the first order in B_z the magnetic field changes the energy levels by

$$\Delta E = +\frac{1}{2}\mu_B M_S B_z.$$

For the $S = 1$ energy levels in Eq. 1.56 show that

$$\Delta E = -\frac{1}{2}\mu_B M_S B_z.$$

Show that this is consistent with the energy level scheme sketched in Fig. 1.4.

(1.8) (a) Approximating the trapping potential in a magnetic trap by a three-dimensional harmonic oscillator potential

$$V_{\text{trap}}(\mathbf{r}) = \frac{1}{2}m\omega^2(x^2 + y^2 + z^2)$$

in Eq. 1.65 and ignoring V_{eff}, show that the single particle quantum states have energies

$$\epsilon_{n_x n_y n_z} = \hbar\omega\left(n_x + n_y + n_z + \tfrac{3}{2}\right).$$

(b) Find the total number of quantum states with energies less than ϵ, and thus show that the density of states is

$$g(\epsilon) = \frac{\epsilon^2}{2(\hbar\omega)^3}.$$

when ϵ is large.

(c) Write down the analog of Eq. 1.29 for this density of states, and show that the BEC temperature for N atoms in the trap is of order $T_c \sim N^{1/3}\hbar\omega/k_{\text{B}}$ when N is large.

Superfluid helium-4

2

2.1 Introduction

There are only two **superfluids** which can be studied in the laboratory. These are the two isotopes of helium: ^{4}He and ^{3}He. Unlike all other substances they are unique because they remain in the liquid state even down to absolute zero in temperature. The combination of the light nuclear masses and the relatively weak van der Waals interactions between the particles means that at low pressure they do not freeze into a crystalline solid. All other elements and compounds eventually freeze into solid phases at low temperatures.

Despite the identical electronic properties of the ^{4}He and ^{3}He atoms, they have completely different properties at low temperatures. This is not due to the difference in nuclear mass, but arises because ^{4}He is a spin zero boson, while ^{3}He is a spin $1/2$ fermion (due to the odd number of spin 1/2 constituents in the nucleus). As we shall see below, although they both form superfluid phases, the origin and physical properties of these superfluids is completely different. Superfluidity in ^{4}He occurs below 2.17 K, and was first discovered and studied in the 1930s.[1]On the other hand, the superfluidity of ^{3}He only occurs below about 2 mK, three orders of magnitude lower in temperature than for ^{4}He. Another surprising difference is that ^{3}He has more than one distinct superfluid phase. Two of the superfluid transitions in ^{3}He were first observed by Osheroff, Richardson, and Lee in 1972, and they received the 1996 Nobel prize in physics for this work.

In this chapter we shall focus on the properties of ^{4}He. We shall postpone discussion of the properties of ^{3}He until later, since it is more closely related to exotic superconductors than ^{4}He. In this chapter we first discuss the concept of a **quantum fluid**, and discuss why helium is unique in its ability to remain liquid down to absolute zero in temperature. Next we shall introduce the main physical properties of superfluid ^{4}He. Then we discuss the concept of the **macroscopic wavefunction** or **off diagonal long range order**, which provides the main link to the previous chapter on Bose-Einstein condensation (BEC). Finally, we discuss Landau's **quasiparticle** theory of ^{4}He which shows the very important consequencs of the strong particle–particle interactions. It it these interactions which make the theory of superfluid ^{4}He much more difficult than that of BEC in atomic gases.

For completeness we should also note that superfluids other than helium are at least theoretically possible. Inside neutron stars one has a dense "fluid" of neutrons. These are neutral spin half fermions, and so could have superfluid phases rather like ^{3}He. Another possibility might be for superfluid states

[1]Kapitsa eventually firmly established the existence of superfluidity in 1938, receiving the Nobel prize for this work over 40 years later in 1978, which is perhaps one of the longest ever delays in the award of a Nobel prize! Surprisingly, Kapitsa shared the 1978 Nobel prize with Penzias and Wilson who discovered the cosmic microwave backround. Sharing the prize like this was unusual, since there is no obvious connection between their work and Kapitsa's. Perhaps the connection is simply that the microwave backround of the universe is at 2.7 K, which happens to be close to the superfluid temperature of helium at 2.17 K!

of molecular or atomic hydrogen. At normal atmospheric pressures hydrogen condenses into a solid phase and not a superfluid. Applying pressure a number of different solid phases are found, but to date no superfluid phases have been found. But it is predicted that at extremely high pressures (perhaps like those found at the center of the planet Jupiter) a variety of possible new liquid phases could occur.[2] These possibilities are very interesting theoretically, and may have implications for astronomy and nuclear physics, but we shall not consider them further in this book.

[2]Superconducting phases of metallic hydrogen have also been predicted to occur. If they exist, these will also require extremely high pressures that have so far been impossible to obtain in the laboratory. High pressure diamond anvil cells have compressed hydrogen to the point where it becomes metallic, but the crystal structures found have still not achieved the ultimate theoretically predicted limit of a monatomic face centered cubic crystal which could be a room temperature superconductor!

2.2 Classical and quantum fluids

We begin with an overview of the concept of a **quantum fluid**. This is a substance which remains fluid (i.e. gas or liquid) at such low temperatures that the effects of quantum mechanics play a dominant role. This contrasts strongly with normal **classical fluids** in which quantum mechanics is essentially irrelevant, and where the physical properties are purely determined by the laws of classical statistical mechanics.

In order to see why quantum mechanics is more or less irrelevant for most fluids in nature, let us consider a typical single-component gas of particles. In a system of rare-gas atoms such as helium or neon, the atoms all have mass, m, and interact with each other predominantly via a pairwise interatomic interaction potential, $V(r)$. For rare gases the interatomic potential contains a short ranged repulsion and a weak but long ranged van der Waals attraction (also called the dispersion force), proportional to $1/r^6$ at large separations. The repulsion is best approximated by an exponential function of r, giving an overall potential of the form,

$$V(r) = ae^{-br} - \frac{C}{r^6}, \tag{2.1}$$

where a, b, and C are constants which can be calculated or determined empirically. Alternatively, the potential near the attractive minimum can be reasonably well represented by the simpler (but less accurate) Lennard–Jones type potential,

$$V(r) = \epsilon_0 \left(\frac{d^{12}}{r^{12}} - 2\frac{d^6}{r^6} \right). \tag{2.2}$$

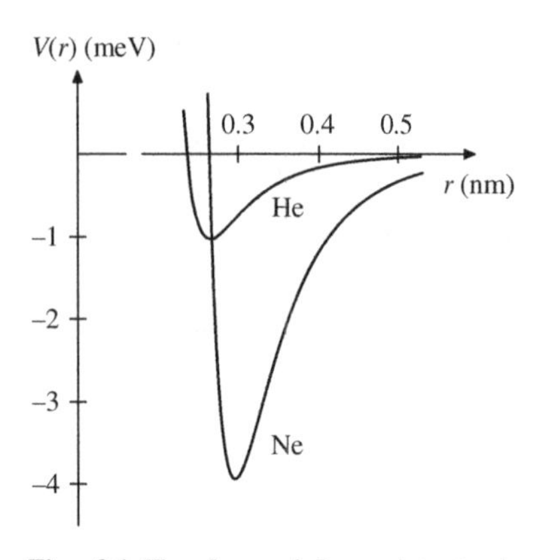

Fig. 2.1 The Lennard–Jones interatomic potential of helium compared with neon. One can see that both atoms are about the same size, given the similar positions of the potential minima, but the binding potential well is about four times deeper for neon than for helium.

Here the the parameters d and ϵ_0 define the position and depth for the attractive potential well. For helium the potential minimum is at $\epsilon_0 = 1.03$ meV at an interatomic separation of $d = 0.265$ nm. For neon the attractive well is about four times stronger, at $\epsilon_0 = 3.94$ meV. The corresponding separation at $d = 0.296$ nm is only slightly larger than for helium. Thus the helium and neon atoms can be thought of as approximately the same size (with neon slightly larger), but the neon atoms bind together much more strongly, as sketched in Fig. 2.1. The weaker attraction for helium, combined with the lighter atomic mass ($4u$ compared with 20) makes helium a quantum liquid, while neon is essentially purely classical. The other rare gases, such as argon and xenon, become progressively more strongly binding as the mass becomes bigger, and so they become even further into the classical regime.

Supposing for now that we can treat the gas as a classical gas, that is, using standard classical statistical mechanics, we would write the classical

Hamiltonian

$$\mathcal{H}(\mathbf{p}_1,\ldots,\mathbf{p}_N,\mathbf{r}_1\ldots,\mathbf{r}_N)=\sum_{i=1,N}\frac{\mathbf{p}_i^2}{2m}+\frac{1}{2}\sum_{i\neq j}V(\mathbf{r}_i-\mathbf{r}_j), \tag{2.3}$$

where any given configuration of the gas is defined by specifying all of the N particle momenta $\mathbf{p}_1,\ldots,\mathbf{p}_N$ and all of the particle coordinates $\mathbf{r}_1,\ldots,\mathbf{r}_N$. Classically at temperature T the probability of each possible configuration is given by the Boltzmann probability density

$$P(\mathbf{p}_1,\ldots,\mathbf{p}_N,\mathbf{r}_1\ldots,\mathbf{r}_N)=\frac{1}{Z_N}e^{-\beta\mathcal{H}}, \tag{2.4}$$

where again $\beta = 1/k_BT$. Here, Z_N, is the classical partition function for N particles,

$$Z_N=\frac{1}{N!}\int d^3p_1,\ldots,d^3p_Nd^3r_1,\ldots,d^3r_Ne^{-\beta\mathcal{H}}. \tag{2.5}$$

Here the factor of $1/N!$ takes into account the fact that the particles are indistinguishable. If it is omitted the famous Gibbs paradox occurs.[3] The $1/N!$ implies that any two configurations of particles $\mathbf{p}_1,\ldots,\mathbf{p}_N,\mathbf{r}_1,\ldots,\mathbf{r}_N$ which only differ from each other by a permutation of the particles are taken to be identical. Note that in classical statistical physics indistinguishability of particles is completely taken into account by the $1/N!$ factor, and there is no need to distinguish further whether the identical particles are bosons or fermions.

The Boltzmann probability given above leads directly to the famous Maxwell–Boltzmann distribution of velocities for the atoms. This arises because for classical systems the partition function, Eq. 2.5, can be factorized exactly into two terms, one which is a kinetic energy contribution and another which depends on the potential energy alone,

$$\begin{aligned}Z_N&=\left(\prod_i\int e^{-p_i^2/2mk_BT}d^3p_i\right)Q_N\\&=(2\pi mk_BT)^{3N/2}Q_N\end{aligned} \tag{2.6}$$

where

$$Q_N=\frac{1}{N!}\int d^3r_1\ldots d^3r_N\exp\left(-\frac{1}{2}\beta\sum_{i\neq j}V(\mathbf{r}_i-\mathbf{r}_j)\right). \tag{2.7}$$

The fact that the momentum part of Eq. 2.6 is a product of independent factors for each particle implies that the momenta of the particles are statistically independent of one another. The probability that any given particle has momentum in the region d^3p of momentum space is therefore

$$P(\mathbf{p})d^3p=\frac{1}{(2\pi mk_BT)^{3/2}}e^{-p^2/2mk_BT}d^3p. \tag{2.8}$$

The fraction of particles with momentum $\mathbf{p}$ of mangitude between p and $p+dp$ is

$$P_{\mathrm{MB}}(p)dp=\frac{4\pi p^2}{(2\pi mk_BT)^{3/2}}e^{-p^2/2mk_BT}dp, \tag{2.9}$$

the Maxwell–Boltzmann momentum distribution. Note that this result is true for any classical system, whether gas, liquid, or solid. It is valid for an arbitrarily strong particle–particle interaction $V(r)$ and it is not just a result which is valid for an ideal-gas.

[3] The Gibbs paradox is the only consequence of particle indistingushability in classical statistical physics. A classical monatomic gas is separated into two halves by a partition wall. Upon removing the partition there appears to be an entropy increase of $k_B \ln 2$ per particle, since each particle can now be found randomly on either side of the partition wall. This is incorrect, since no irreversible thermodynamic process took place. Gibbs postulated that a factor $1/N!$ was necessary in the partition function in order to prevent this excess entropy. Over 50 years later it was found that the $1/N!$ is indeed present when taking the classical limit of the Bose–Einstein or Fermi–Dirac quantum ideal gases.

Now we are in a position to estimate whether or not quantum mechanics will be important for the liquid. From the Maxwell–Boltzmann distribution we see that a typical particle will have momentum of order

$$p = (2mk_\mathrm{B}T)^{1/2}, \tag{2.10}$$

corresponding approximately to the maximum of $P_\mathrm{MB}(p)$ in Eq. 2.9. Quantum mechanically the particles will therefore have typical de Broglie wavelengths, $\lambda = h/p$, of order the **thermal de Broglie wavelength**, defined conventionally by

$$\lambda_\mathrm{dB} = \left(\frac{2\pi\hbar^2}{mk_\mathrm{B}T}\right)^{1/2}. \tag{2.11}$$

We can expect that quantum effects will be significant when this wavelength is comparable with other typical length scales in the liquid, while quantum effects can be neglected when λ_dB is negligible.

We can easily calculate λ_dB for any given substance. For example, the rare-gas neon, Ne, condenses into liquid at about 27 K and freezes at about 24 K at standard pressure. The atomic mass is 20, and so $\lambda_\mathrm{dB} \approx 0.07$ nm. But, as we have seen, the interatomic potential for neon has its minimum at about $d \approx 0.3$ nm, and so the typical interatomic distances are significantly greater than the thermal de Broglie wavelength. Quantum effects are therefore relatively unimportant throughout the whole of the gas and liquid phases. This is typical of almost all of the elements in the periodic table.

On the other hand for helium we find a very different situation. Helium has an interatomic potential well much weaker than neon ($\epsilon_0 = 1.03$ meV compared with 3.9 meV for Ne). Consequently helium becomes liquid at only 4 K. Combined with the smaller mass for helium this results in $\lambda_\mathrm{dB} \approx 0.4$ nm for liquid ^{4}He. This is actually greater than the typical interatomic distances, given by for example, $d \approx 0.27$ nm. So we can expect that quantum mechanical effects are always important for liquid ^{4}He. For the lighter isotope, ^{3}He, there is an even larger thermal de Broglie wavelength, λ_dB, and so again quantum effects are also dominant. For this reason the two isotopes of helium are the only substances which are completely dominated by quantum effects in their liquid states.

We can see how dramatically the quantum properties of ^{4}He influence its phase diagram, by comparing helium with neon, as shown in Fig. 2.2. For almost all substances the temperature–pressure phase diagram is similar to the one shown for neon, on the right-hand-side. There are gas, liquid, and solid phases, which meet at a triple point, (T_3, P_3). The gas and liquid phases are separated by a line which ends in the gas–liquid critical point, (T_c, P_c) (shown as a dot in Fig. 2.2). For neon $T_3 = 24.57$ K and $T_c = 44.4$ K. Therefore, as we have seen, quantum effects are small throughout the liquid part of the phase diagram. In contrast, as shown in the left-hand-side of Fig. 2.2 the phase diagram of ^{4}He is quite different. There still exists a gas–liquid critical point, at 5.18 K. But the liquid–gas phase transition line never intersects with the liquid–solid phase boundary, and so there is no triple point. Instead, for pressures below about 2.5 MPa there is no solid phase at any temperature, and helium remains liquid even down to absolute zero of temperature. Helium-4 is also very unique in having two different kinds of liquid phases, denoted by I and II in Fig. 2.2. The liquid I is a normal liquid, while helium II is superfluid. The simple estimate

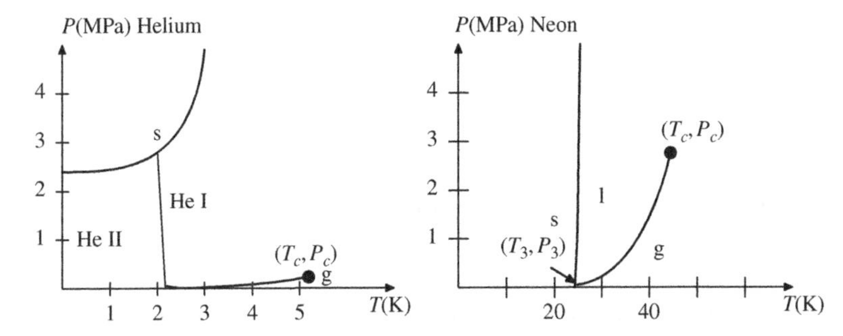

Fig. 2.2 The phase diagram of ^{4}He and neon compared. Left: helium remains liquid even down to absolute zero in temperature for pressures below about 2.5 MPa. There are two different liquid phases: the normal liquid, He I, and the superfluid phase, He II. The liquid–gas critical point is shown as a solid circle. For contrast, the phase diagram of neon (right side) has a phase diagram which is typical of most other substances. The liquid and gas phases exist at high temperature and at low temperatures only the solid phase occurs. Neon has a triple point, (T_3, P_3), while helium does not.

of λ_{dB} above shows that quantum effects are important in both of the liquid phases.

The phase diagram for the other isotope of helium, ^{3}He, is similar to the one for ^{4}He in Fig. 2.2. The main difference being that the superfluid phase does not occur until about 2 mK. As we shall see below, this difference is because ^{3}He is a fermion while ^{4}He is a boson.

But why is it that liquid helium does not crystallize, even at absolute zero? The point is that quantum fluids have **zero point motion** and hence they have nonzero kinetic energy, however low the temperature. In a solid phase each atom is localized at some particular site in the crystal. It must have some uncertainty in position, Δx, which is less than the crystal lattice spacing, a. By the uncertainty principle, it has some uncertainty in momentum, Δp, and hence some finite kinetic energy. To make a rough estimate we can assume that each atom in the crystal vibrates around its equilibrium position as an independent quantum harmonic oscillator (this is the Einstein oscillator phonon model). The zero point energy is

$$E_0 = \tfrac{3}{2}\hbar\omega_0, \tag{2.12}$$

per atom, where ω_0 is the vibrational frequency of the atom displaced from its equilibrium crystal lattice site. Using the Lennard–Jones pair potential model and assuming an fcc crystal lattice we can make the estimate

$$\omega_0 = \sqrt{\frac{4k}{m}}, \tag{2.13}$$

where

$$k = \frac{1}{2}\frac{d^2V(r)}{dr^2} = \frac{36\epsilon_0}{r_0^2} \tag{2.14}$$

is the spring constant of the interatomic pair potential (Fig. 2.1) at the equilibrium distance, r_0. Using the Lennard–Jones parameters for helium, given above, leads to a zero point energy of about $E_0 \approx 7$ meV. This would be equivalent to a thermal motion corresponding to about 70 K, and is far too great to allow the liquid phase to condense into a solid. The solid phase only occurs when external pressure is applied, as can be seen in Fig. 2.2. On the other hand, making the same estimate for the parameters for neon gives a smaller zero point energy of about $E_0 \approx 4$ meV, which is comparable to the thermal motion at the melting point of 24 K.

2.3 The macroscopic wave function

As shown in Fig. 2.2, liquid ^{4}He has two distinct liquid phases, He I and II. He I is a **normal liquid** phase, characterized by fairly standard liquid state properties. But He II is a **superfluid**, characterized by fluid flow with zero viscosity, infinite thermal conductivity, and other unusual properties.

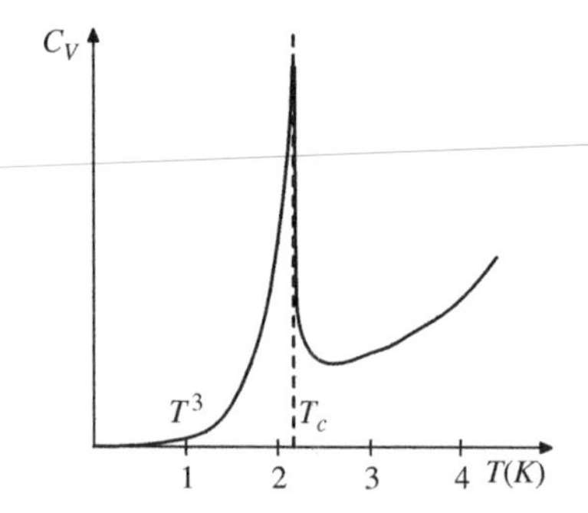

Fig. 2.3 Specific heat of ^{4}He. At the critical temperature T_c there is a singularity shaped like the Greek symbol λ. This λ transition belongs to the three-dimensional XY model universality class.

At the boundary between the He I and II phases one observes the characteristic singularity in specific heat, as shown in Fig. 2.3. The shape resembles the Greek letter lambda, λ, and because of this the transition is often referred to as the **lambda point**. Clearly, from Fig. 2.3, there are a number of differences compared with the specific heat of a BEC as shown in Fig. 1.7. First, at low temperatures the specific heat is of the form,

$$C_V \sim T^3 \tag{2.15}$$

in He II, and not $T^{3/2}$ as is it in the BEC case of Fig. 1.7.

More dramatically, the nature of the specific heat at T_c is quite different in ^{4}He compared with BEC. In the BEC there is a simple change of slope, or cusp, in C_V at the critical point, as shown in Fig. 1.7. But in ^{4}He there is a much sharper feature. The specific heat near T_c is very close to being logarithmic, $C_V \sim \ln|T - T_c|$, but is actually a very weak power law behavior. It has a characteristic form

$$C_V = \begin{cases} C(T) + A_+|T - T_c|^{-\alpha} & (T > T_c) \\ C(T) + A_-|T - T_c|^{-\alpha} & (T < T_c), \end{cases} \tag{2.16}$$

where $C(T)$ is a smooth (non-singular) function of T near T_c. The parameter α is the critical exponent, which is measured to have a value close to -0.009.[4] In fact this power law behavior, and even the actual measured values of the constants α, A_+, and A_-, are in essentially perfect agreement with theoretical predictions for a universality class of thermodynamic phase transitions, called the three-dimensional XY-model class. The theory of such phase transitions is based on the hypothesis of scaling, and the critical exponents are calculated using the methods of renormalization group analysis. This topic goes well beyond the scope of this book (see, for example, the books on thermodynamics listed at the end of Chapter 1). But the main point is that many different physical systems end up having identical sets of critical exponents, such as α in Eq. 2.16. The so called XY model is characterized by systems that have a form of order which can be described by a two-dimensional unit vector,

[4] Microgravity experiments for measuring critical exponents, such as the parameter α, are proposed for flights of the space shuttle. These would allow high precision measurements and ensure that any effects of gravity are negligible. The problem occurs because there is a tiny change in pressure due to gravity between the top and bottom of a liquid helium sample. In turn this means that T_c is slightly different between the top and bottom, which leads to a tiny smearing of the singularity in specific heat.

$$\mathbf{n}(\mathbf{r}) = (n_x, n_y) = (\cos\theta, \sin\theta), \tag{2.17}$$

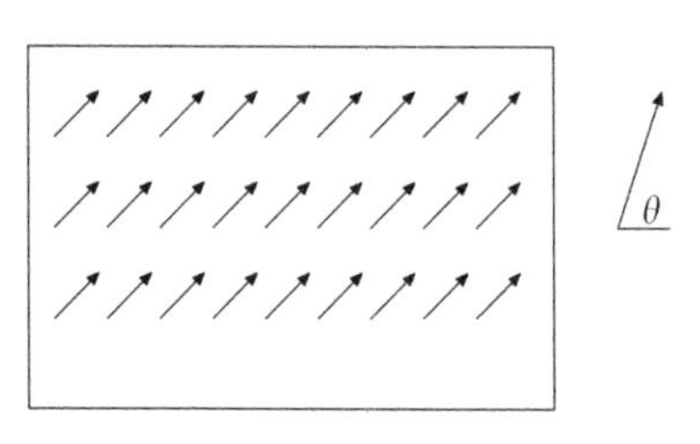

Fig. 2.4 Ordering in the XY model. Each point in space has a unit vector $\hat{\mathbf{n}}$. Above T_c these are random on long length scales, while below T_c they develop long ranged order with a common (arbitrary) direction. The direction, θ of the vector $\hat{\mathbf{n}} = (\cos\theta, \sin\theta)$ corresponds to the phase of the macroscopic wave function $\psi(\mathbf{r}) = |\psi(\mathbf{r})|e^{i\theta}$.

for an angle θ at every point in space, $\mathbf{r}$. The helium λ-transition separates two thermodynamic phases, the normal liquid (He I) in which this XY vector $\mathbf{n}(\mathbf{r})$ is spatially random, and one (He II) in which there is an ordering of $\mathbf{n}(\mathbf{r})$, like the ordering in a magnet. This idea is illustrated in Fig. 2.4.

What is the physical interpretation of this unit vector, $\mathbf{n}(\mathbf{r})$, or the angle θ? We can motivate the existence of a phase angle θ by postulating the existence of a **macroscopic wave function**. This is similar to the condensate wave function $\psi_0(\mathbf{r})$, in the Gross–Pitaevskii approach to BEC introduced in the previous chapter, for example, Eq. 1.65. But the ideas are much more subtle here, since we must take into account the strong interparticle interactions $V(\mathbf{r})$. For helium these interactions are strong, and cannot be neglected or approximated with a simple mean-field approach.

Nevertheless, let us imagine that we can characterize the state of the helium by a local wave function $\psi_0(\mathbf{r})$ at every point in space. The wave function corresponds to a **condensate**, or macroscopic number of particles. We can choose to normalize the wave function so that the density of particles in the condensate is

$$n_0 = |\psi_0(\mathbf{r})|^2. \tag{2.18}$$

With this definition, integrating over the volume of the sample we see that the macroscopic wave function is normalized so that

$$N_0 = n_0 V = \int |\psi_0(\mathbf{r})|^2 d^3r, \tag{2.19}$$

where N_0 is the total number of particles in the condensate. Above T_c this will be essentially zero, while below T_c it will be some finite fraction of the total (macroscopic) particle number, N.

In general the wave function is complex, and its phase, $\theta(\mathbf{r})$ corresponds to the angle in the XY-model of Fig. 2.4. Therefore we can write the complex wave function as

$$\psi_0(\mathbf{r}) = \sqrt{n_0(\mathbf{r})}e^{i\theta(\mathbf{r})}, \tag{2.20}$$

where $\sqrt{n_0(\mathbf{r})} = |\psi_0(\mathbf{r})|$ is its modulus and $\theta(\mathbf{r})$ is its phase. With this definition, we can see that if $|\psi_0(\mathbf{r})| = 0$ it is not possible to define the phase θ. But if $|\psi_0(\mathbf{r})|$ is nonzero, then we can view the wave function phase θ as a natural physical parameter of the system. We can postulate that the phase transition, T_c, is the temperature where this wave function phase first becomes ordered throughout the fluid. In terms of the theory of phase transitions we would say that the macroscopic wave function $\psi_0(\mathbf{r})$ is the **order parameter** of the He II phase. Its proper microscopic interpretation will be developed more fully in the following sections.

2.4 Superfluid properties of He II

Although the above picture of the macroscopic wave function given above is highly simplified, it also contains the essential truth about **superflow** in liquid ^{4}He. Recall that in the BEC we always had condensation into the zero momentum quantum state. In this case it is the single particle wave function $e^{i\mathbf{k}.\mathbf{r}}/\sqrt{V}$ for $\mathbf{k} = 0$ which becomes occupied with a macroscopic number of particles. This wave function is simply a constant, $1/\sqrt{V}$, or it can be defined equally well with any arbitrary constant phase θ, as $e^{i\theta}/\sqrt{V}$. But in either case the wave function phase θ is constant in space. But what if we consider a system where somehow the condensate is in some more general state, where the phase $\theta(\mathbf{r})$ is not constant? In this case there is a superflow.

Superflow arises whenever the condensate phase, $\theta(\mathbf{r})$ varies in space. If we assume that we have a condensate wave function, $\psi_0(\mathbf{r})$, we can apply the usual formula from quantum mechanics to obtain the current density for particle flow,

$$\mathbf{j}_0 = \frac{\hbar}{2mi}\left[\psi_0^*(\mathbf{r})\nabla\psi_0(\mathbf{r}) - \psi_0(\mathbf{r})\nabla\psi_0^*(\mathbf{r})\right]. \tag{2.21}$$

This current density, $\mathbf{j}_0$, is the number of particles flowing per unit area per second in the condensate (note that it is not an electrical current density, since no charge flow occurs because the particles are neutral). Using $\psi_0(\mathbf{r}) = \sqrt{n_0}e^{i\theta}$ and the product rule for differentiation we have

$$\nabla\psi(\mathbf{r}) = e^{i\theta}\nabla\sqrt{n_0} + i\sqrt{n_0}e^{i\theta}\nabla\theta,$$

and its complex conjugate

$$\nabla\psi^*(\mathbf{r}) = e^{-i\theta}\nabla\sqrt{n_0} - i\sqrt{n_0}e^{-i\theta}\nabla\theta.$$

Substituting these into the current formula Eq. 2.21 we find

$$\mathbf{j}_0 = \frac{\hbar}{m}n_0\nabla\theta. \tag{2.22}$$

Since the condensate density is n_0 and the net current of particles equals a density times a velocity, we see that this can be interpreted as showing that the condensate flows with particle velocity $\mathbf{v}_s$ where,

$$\mathbf{v}_s = \frac{\hbar}{m}\nabla\theta. \tag{2.23}$$

This equation is the fundamental defining relation for the **superfluid velocity**, $\mathbf{v}_s$. The condensate particles have density n_0, and flow with an average velocity $\mathbf{v}_s$, hence giving the condensate contribution to the current flow,[5]

$$\mathbf{j}_0 = n_0\mathbf{v}_s. \tag{2.24}$$

Unlike fluid flow in normal liquids, **superflow** is a movement of particles without dissipation. The existence of such supercurrents was demonstrated in a number of key experiments in the 1930s. Kapitsa showed that liquid He II could flow through narrow capillaries without any apparent resistance due to fluid viscosity, as shown in Fig. 2.5(a). In any normal fluid the flow velocity in such a capillary would depend on the viscosity of the fluid, η, the pressure difference, $\Delta P = P_2 - P_1$ and the length and diameter of the tube. From dimensional analysis one can easily see that the typical flow velocity of a viscous fluid in a cylindrical tube of length L and radius R is of order, v, where

$$\frac{\Delta P}{L} \sim \eta\frac{v}{R^2}, \tag{2.25}$$

(viscosity has units of Force × Time / Distance^2). But Kapitsa found that for He II the pressure difference was always $\Delta P = 0$ whatever the flow velocity.[6] Therefore superflow is fluid flow with zero viscosity, $\eta = 0$.

On the other hand some other experiments seemed to imply a finite viscosity. A circular disk suspended in fluid can be made to undergo torsional oscillations about its axis. The oscillation frequency depends on the torsional stiffness of the support, and on moment of inertia of the disk. But in a viscous fluid such a disk will drag with it a thin layer of fluid along with its motion, effectively increasing the inertia of the disk. By making a stack of closely spaced disks,

[5] This formula gives only the direct contribution of the condensate to the total current, $\mathbf{j}_0$. As we shall see below, the total current includes both this condensate part and other contributions. The total particle current can be written $\mathbf{j}_s = n_s\mathbf{v}_s$, and $n_s \neq n_0$ as will be seen below.

[6] Actually the viscosity is only zero provided that the flow velocity is less than the **critical velocity**, v_c. If the flow velocity exceeds v_c, dissipation occurs and so the viscosity becomes nonzero.

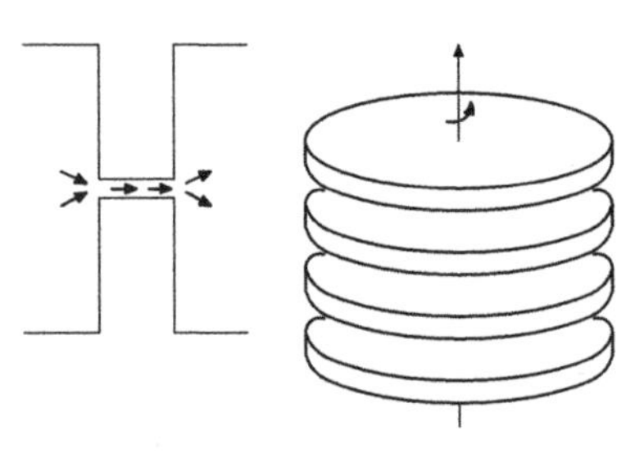

Fig. 2.5 Superfluid properties of He II. (a) Flow through narrow capillaries without resistance due to viscosity. (b) Inertia of a torsional oscillation of a stack of disks.

all the fluid between the disks should be dragged along and contribute to the rotational moment of inertia. By immersing such a stack of disks in He II, Andronikashvilli was able to show that a fraction of the liquid helium did contribute to the inertia, but a fraction did not.

This experiment provided motivation for the **two-fluid model** of liquid helium. If the total particle density of the fluid is n, then we can divide this into two components

$$n = n_s + n_n. \tag{2.26}$$

The superfluid component, n_s, flows with zero viscosity. This component does not contribute to the interia of the rotating disks, since it is not dragged along by the disk motion. On the other hand, the normal component, n_n, acts like a conventional viscous fluid. It contributes to the moment of inertia of the disk stack. The **superfluid density** is defined as the mass density of the superfluid part of the fluid

$$\rho_s = mn_s. \tag{2.27}$$

The temperature dependence of the two-fluid components, n_s and n_n is sketched in Fig. 2.6. At very low temperatures, near $T = 0$, it is found empirically that almost all of the fluid is in the condensate, so $n_s \sim n$ and $n_n \sim 0$. In this temperature region experiments show that,

$$n_s(T) \approx n - AT^4, \tag{2.28}$$

where A is a constant. Therefore at absolute zero temperature all of the particles will take part in the superflow, and this number gradually decreases as the temperature is increased.

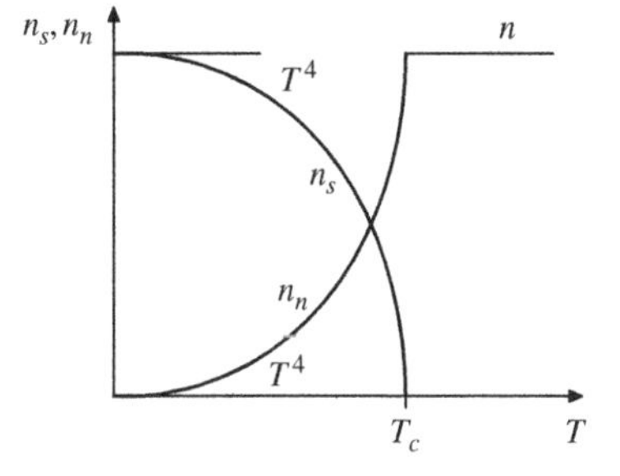

Fig. 2.6 Temperature dependence of the superfluid and normal components of liquid He II, as measured, for example, from in the torsional oscillation disk stack experiment.

On the other hand, close to the critical temperature T_c nearly all of the fluid is normal, so $n_s \approx 0$ and $n_n \approx n$. It is found experimentally that near T_c the superfluid component, n_s, vanishes like a power law

$$n_s \sim \begin{cases} B(T_c - T)^{\upsilon} & T < T_c, \\ 0 & T > T_c. \end{cases} \tag{2.29}$$

The exponent υ is another critical exponent, like the α in specific heat, and has a value of about 0.67. Again this measured value is in perfect agreement with theoretical predictions based on the three-dimensional *XY* model.

The superfluid experiments show that the two fluid components can move relative to each other without any friction. In the capillary flow experiment, the normal component feels friction from the walls and remains at rest, while the superfluid component flows freely down the capillary. In the rotating stack of disks experiment, the normal component is dragged along with the disks, while the superfluid component remains at rest. It is therefore possible to define separate velocities for the two fluid components, leading to a novel kind of hydrodynamics. In this **two-fluid hydrodynamics** there are two types of current flow, $\mathbf{j}_s$ and $\mathbf{j}_n$, correspond to the superfluid and normal fluid particle current densities. The total current density $\mathbf{j}$ obeys,

$$\begin{aligned} \mathbf{j} &= \mathbf{j}_s + \mathbf{j}_n, \\ \mathbf{j}_s &= n_s \mathbf{v}_s, \\ \mathbf{j}_n &= n_n \mathbf{v}_n, \end{aligned} \tag{2.30}$$

and $\mathbf{v}_s$ and $\mathbf{v}_n$ are the velocities of the two fluid components.

This two-fluid hydrodynamics has some interesting consequences. The normal fluid component carries with it entropy, because of the usual random particle motions in the normal fluid. But the condensate is essentially a single many-body quantum state, and therefore carries no entropy. The flow of heat in a super fluid is thus given entirely by the normal component. The heat current density is therefore,

$$\mathbf{Q} = Ts\mathbf{v}_n, \tag{2.31}$$

where $\mathbf{Q}$ is the heat current, and s is the entropy per unit volume.

The fact that only the normal fluid component carries entropy leads to an unusual thermo-mechanical effect. If we connect two volumes of superfluid by a very thin capillary (a superleak) then the superfluid can flow but the normal fluid cannot (because it effectively has finite viscosity). The normal fluid does not carry entropy, and so no heat can flow through the leak (since $\mathbf{v}_n = 0$ in Eq. 2.31). Therefore there is a particle flow but no heat flow. There is no mechanism for the temperatures on either side to come to equilibrium, and so T_1 remains different from T_2 indefinitely. On the other hand, detailed balance requires that the particle flow leads to equilibrium in chemical potential, $\mu_1 = \mu_2$. Using the thermodynamic relations

$$G = \mu N, \tag{2.32}$$

$$dG = -SdT + VdP, \tag{2.33}$$

where $G = U - TS + PV$ is the Gibbs Free energy, one can easily show that the temperature difference $T_2 - T_1$ is accompanied by a pressure difference

$$P_2 - P_1 = s(T_2 - T_1), \tag{2.34}$$

where again $s = S/V$ is the entropy per unit volume in the normal fluid. This is illustrated in Fig. 2.7. The figure also illustrates the spectacular **fountain effect**, in which a temperature difference is used to power a fountain of liquid helium (Fig. 2.7). A tube is blocked by a plug of dense gauze material, so that only superflow can pass. This tube is heated, leading to a temperature difference. The temperature difference drives a net superflow through the tube, which fountains out above the free surface.

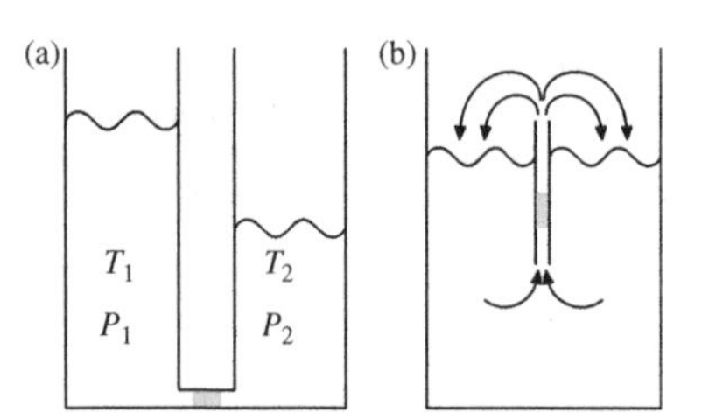

Fig. 2.7 (a) Thermo-mechanical effects in superfluid helium: two superfluid containers connected by a narrow superleak maintain a constant pressure and temperature difference. (b) The fountain effect: superfluid flows through a small heated superleak tube coming out in a fountain above the surface. A temperature difference has directly driven a pressure difference and a net superflow.

Another surprising superflow property is the ability for a beaker of superfluid helium spontaneously to empty itself! This is because of a microscopic **wetting layer** of liquid helium on the surface of the beaker. These occur in many fluids and are typically a few microns (10^{-6}m) thick. For example, looking at wine in a glass one can often see a thin visible fluid layer coating the sides of the glass just above the meniscus. As the alcohol and water evaporate a thin deposit of solid material is left on the glass. In the case of a superfluid, however, there is no viscosity to prevent the superfluid flowing up through the wetting layer and out over the rim of the beaker. Eventually the beaker could empty itself completely through this mechanism.

2.5 Flow quantization and vortices

The existence of the macroscopic wave function in He II leads to a **quantization** of the superflow. We had the definition,

$$\mathbf{v}_s = \frac{\hbar}{m}\nabla\theta, \tag{2.35}$$

for the superfluid velocity. This implies that superflow is a potential flow, in the language of classical fluid mechanics. Taking the curl, we see that

$$\nabla \times \mathbf{v}_s = 0, \tag{2.36}$$

so the flow is **irrotational**.

Now consider a flow along a closed tube, as shown in Fig. 2.8. Consider any closed path which extends around the whole tube length, as shown. We can define the flow **circulation** by the integral

$$\kappa = \oint \mathbf{v}_s \cdot d\mathbf{r} \tag{2.37}$$

taken around the closed path. The irrotational property, Eq. 2.36, implies that the value of this integral is independent of which path we use for the integral. Any path which wraps around the whole tube exactly once will give the same value for κ. Inserting Eq. 2.35 we find

$$\kappa = \frac{\hbar}{m}\oint \nabla\theta \cdot d\mathbf{r} = \frac{\hbar}{m}\Delta\theta. \tag{2.38}$$

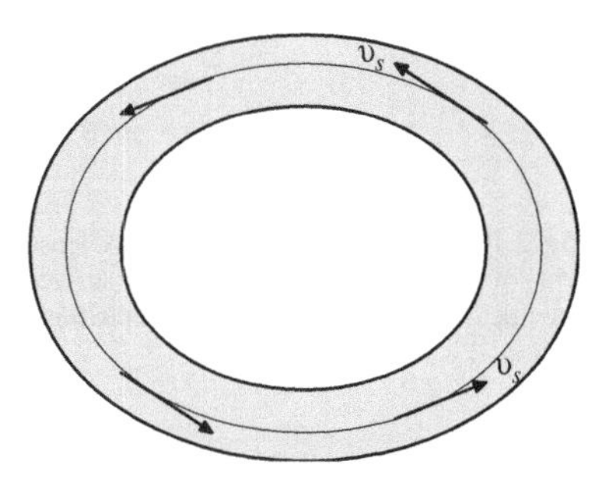

Fig. 2.8 Flow quantization in a superfluid. The flow is everywhere irrotational (zero curl), but a net circulation occurs around a closed flow path. The circulation is quantized in units of h/m.

Here $\Delta\theta$ is the change in the phase angle theta after going around the tube. But if the macroscopic wave function, $\psi_0(\mathbf{r}) = \sqrt{n_0}e^{i\theta(\mathbf{r})}$ is to be defined uniquely we must have

$$\psi(\mathbf{r}) = \psi(\mathbf{r})e^{i\Delta\theta}, \tag{2.39}$$

which implies

$$\Delta\theta = 2\pi n, \tag{2.40}$$

where n is an integer. The circulation of the flow is therefore quantized, in units of the quantum of circulation, h/m,

$$\kappa = \frac{h}{m}n. \tag{2.41}$$

The quantum number n corresponds to the topological **winding number** for the phase θ around the closed loop. It is the number of times that θ winds through 2π on going around the closed path. This idea is illustrated in Fig. 2.9. We start at a point in the fluid where $\theta = 0$, and then move continuously around the ring. After we have gone once around, we find that the angle θ has changed by a multiple of 2π (once in Fig. 2.9). The dashed line in Fig. 2.9 has become wound around the central line exactly once. This winding cannot be removed by any local (and smooth) changes to the phase $\theta(\mathbf{r})$, and so it is a topological property of the macroscopic wave function.

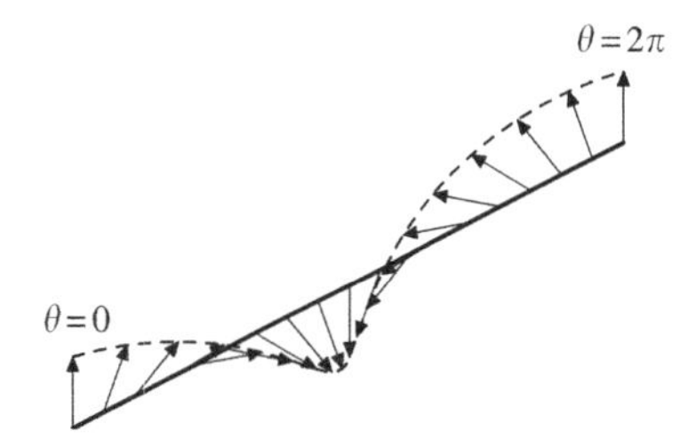

Fig. 2.9 Winding number of the superfluid phase θ. In this example, the phase θ starts at zero, and continuously increases until it equals 2π. The dashed line winds around the central line exactly once. Hence this example shows a winding number of 1. For superflow around a closed loop we identify the two points at the ends of the line, and hence the phase change must be an exact multiple of 2π, as in the case shown. A phase slip occurs if this winding suddenly changes from one value to another.

This flow quantization has been confirmed in a number of experiments. The most direct method is to actually rotate the circular tube. Alternatively one can use a system of helium trapped between two concentric rotating cylinders. If one starts with the fluid at rest, then initially $\kappa = 0$ and so $n = 0$. Rotating the cylinders or tube will accelerate the normal component until it has the same angular velocity as the apparatus, but leave the superfluid component at rest. If the system is now cooled slowly, then particles from the normal fluid will gradually be added to the condensate. Their angular momentum will be transferred to the condensate, which will begin to rotate. But the circulation κ does not increase smoothly, but in sudden jumps of h/m. These jumps are called *phase-slip* events. The winding of the phase θ suddenly increases by one unit of 2π (Fig. 2.9). Such phase slips correspond to changes in the quantum

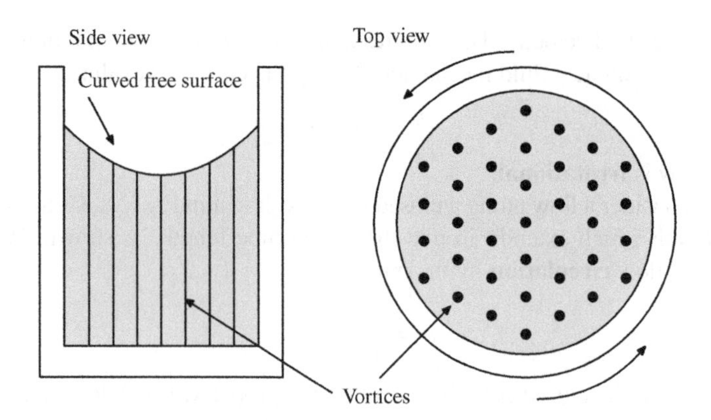

Fig. 2.10 Flow in a rotating cylinder. A dense array of vortices develops, and the free surface becomes curved due to the usual centrifugal effect.

number n, and so are analogous to the transitions between different quantum states in an atom.

We can also set up a **persistent current** in such a superflow. Starting with a rotating normal fluid flow above T_c, and then cooling, we are left with a circulating superflow. Since there is no viscosity, this flow can, in principle, continue indefinitely.

If we rotate a whole cylindrical container of liquid helium, rather than a tube, then the condition $\nabla \times \mathbf{v}_s = 0$ might suggest that it should be impossible to have any form of circulation. In that case we would expect that the free surface of the superfluid would remain perfectly flat, rather than becoming curved due to centrifugal forces. In fact the free surface of a rotating ^{4}He superfluid does indeed become curved, exactly as expected for a normal rotating classical fluid, as sketched in Fig. 2.10. So how can this be possible? The irrotational flow condition $\nabla \times \mathbf{v}_s = 0$ would imply that

$$\oint \mathbf{v}_s \cdot d\mathbf{r} = 0 \tag{2.42}$$

around any closed path, and hence there should be no macroscopic rotation of the fluid. The answer to this paradox is that the fluid must contain **vortices**.

A vortex is a circulating flow which can satisfy the condition $\nabla \times \mathbf{v}_s = 0$ almost everywhere, and still allow a net rotation. In cylindrical polar coordinates

$$\nabla \times \mathbf{v}_s = \frac{1}{r}\begin{vmatrix} \mathbf{e}_r & r\mathbf{e}_\phi & \mathbf{e}_z \\ \frac{\partial}{\partial r} & \frac{\partial}{\partial \phi} & \frac{\partial}{\partial z} \\ v_r & rv_\phi & v_z \end{vmatrix}, \tag{2.43}$$

where $\mathbf{e}_r$, $\mathbf{e}_\phi$, and $\mathbf{e}_z$ are unit vectors in the r, ϕ, and z directions at a point $\mathbf{r} = (r, \phi, z)$. Here v_r, v_ϕ, and v_z are the components of $\mathbf{v}_s$ in these directions, $\mathbf{v}_s = v_r\mathbf{e}_r + v_\phi\mathbf{e}_\phi + v_z\mathbf{e}_z$. A circulating flow with cylindrical symmetry will obey $\nabla \times \mathbf{v}_s = 0$ if,

$$\frac{1}{r}\frac{\partial}{\partial r}(rv_\phi) = 0. \tag{2.44}$$

Therefore the flow velocity around the vortex must be of the form

$$\mathbf{v}_s = \frac{\kappa}{2\pi r}\mathbf{e}_\phi, \tag{2.45}$$

where κ is the net circulation. The flow quantization property ensures that $\kappa = n(h/m)$ where h/m is the quantum of circulation and n is an integer. In

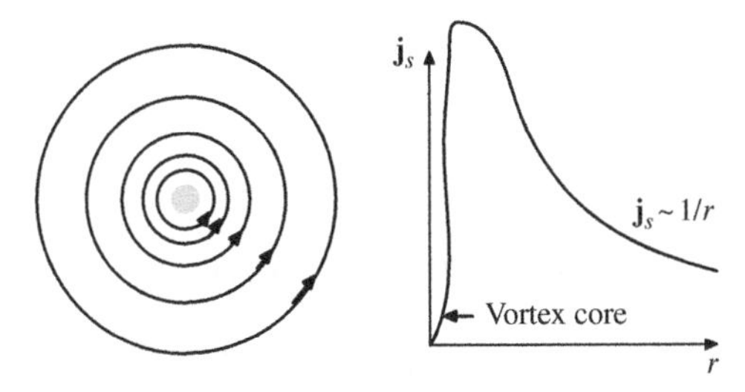

Fig. 2.11 Flow in a single superfluid vortex.

practice only $n = \pm 1$ vortices are observed, since these have the lowest energy. The flow around a vortex is illustrated in Fig. 2.11.

Equation 2.45 satisfies $\nabla \times \mathbf{v}_s = 0$ except at $r = 0$. This is the **vortex core**. In practice helium vortices have a small core region, of order about 1 Å (10^{-10} m) in radius. In this vortex core the macroscopic wave function $\psi_0(\mathbf{r})$ is zero, and so the phase θ is not defined. The fact that the condensate wave function becomes zero also avoids any infinite singularity in the supercurrent at the centre of the vortex, since $\mathbf{j}_s = n_s \mathbf{v}_s$ will also be zero in the core region. The radial variation of the flow velocity is sketched in Fig. 2.11. The circulating current decreases as $1/r$ outside of the small core region.

The existence of vortices explains how a rotating cylinder of superfluid maintains a macroscopic circulation of flow, and hence a centrifugally curved free surface. The rotating cylinder of super-fluid helium must contain a large number of vortices. Each individual vortex contributes h/m to the overall circulation of the fluid. If the cylinder has radius R and angular velocity ω then the total circulation around the edge of the cylinder will be

$$\kappa = \oint \mathbf{v}_s \cdot \mathbf{dr} = (2\pi R)(\omega R), \tag{2.46}$$

taking an integration contour around the whole cylinder, radius R. Since this circulation is quantized,

$$\kappa = \frac{h}{m} N_v, \tag{2.47}$$

where N_v is the total number of vortices in the rotating fluid. Combining these equations we find that there are

$$\frac{N_v}{\pi R^2} = \frac{2m\omega}{h} \tag{2.48}$$

vortices per unit area in the rotating superfluid.

At small rotation rates vortices in helium tend to form quite well ordered triangular arrays, as sketched in Fig. 2.10. But it is also possible to generate tangles of vortices by sudden changes in rotation speed, or other perturbations. When the vortices become tangled, then the flow has a strong random element, varying rapidly from point to point in space. This is a novel form of **turbulence**.

Interestingly, vortices have also been observed in atomic BEC, of the type discussed in chapter 1. It is possible to impart angular momentum to the atoms in the trap by rotating the magnetic fields which provide the trap potential $V_{\text{trap}}(\mathbf{r})$. For example the trap can be made into an elliptical shape, and the ellipse major axis can be made to rotate at a given rate. As the trapped gas cools into the BEC state, then vortices form to accommodate the rotation. One can even see that these vortices also form into quite well ordered triangular arrays, as shown in

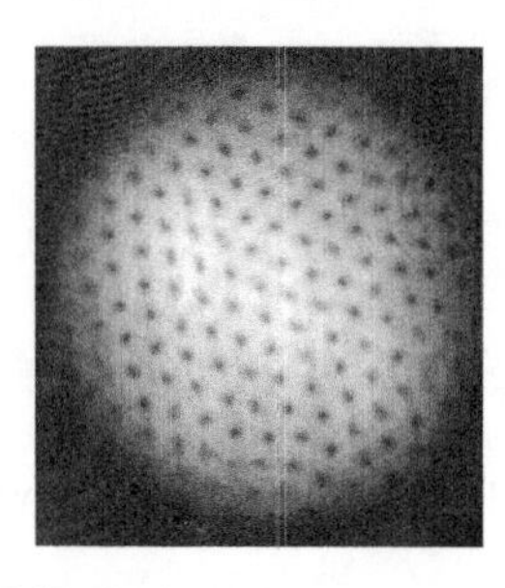

Fig. 2.12 Vortices in a rotating BEC. Reprinted figure with permission from Raman, Abo-Shaeer, Vogels, Xu and Ketterle (2001). Copyright (2001) by the American Physical Society.

the remarkable pictures shown in Fig. 2.12 (Abo-Shaeer *et al.* 2001; Ketterle 2003).

2.6 The momentum distribution

The description of the properties of superfluid helium given above was mostly phenomenological, but relied on the concept of a macroscopic wave function. The existence of a single wave function into which all of the particles become condensed seems quite natural following the discussion of the ideal Bose gas in Chapter 1. But unfortunately things are not quite so simple, because helium is a **strongly interacting** quantum liquid.

Quantum mechanically we describe a system of N interacting particles by its many-particle Hamiltonian. Assuming pairwise interactions between the particles this is,

$$\hat{H} = \sum_{i=1,N} -\frac{\hbar^2}{2m}\nabla_i^2 + \frac{1}{2}\sum_{i\neq j} V(\mathbf{r}_i - \mathbf{r}_j). \tag{2.49}$$

This quantum Hamiltonian is simply derived from the classical Hamiltonian, Eq. 2.3 with the usual replacement $\mathbf{p} \to -i\hbar\nabla$. This Hamiltonian is an operator which acts on N-particle wave functions of the following form

$$\Psi(\mathbf{r}_1, \mathbf{r}_2, \ldots, \mathbf{r}_N).$$

The fact that we are working with spin zero boson particles is expressed by the fact that the wave function must be **symmetric** under exchange (permutation) of the particles. Thus if we exchange the coordinates $\mathbf{r}_i$ and $\mathbf{r}_j$ then,

$$\Psi(\ldots, \mathbf{r}_i, \ldots, \mathbf{r}_j, \ldots) = \Psi(\ldots, \mathbf{r}_j, \ldots, \mathbf{r}_i, \ldots). \tag{2.50}$$

The possible wave functions of the system are the eigenstates of the Hamiltonian,

$$\hat{H}\Psi_n(\mathbf{r}_1, \ldots, \mathbf{r}_N) = E_n^{(N)}\Psi_n(\mathbf{r}_1, \ldots, \mathbf{r}_N) \tag{2.51}$$

where $E_n^{(N)}$ for $n = 0, 1, 2, \ldots,$ are the energy levels of the N particles.

If the system is at zero temperature, we should find the N particle ground state, $\Psi_0(\mathbf{r}_1, \ldots, \mathbf{r}_N)$. At finite temperatures we should, in principle, find all the energy levels $E_n^{(N)}$, then the usual rules of statistical physics tell us that the quantum state n has the Boltzmann probability

$$P_n = \frac{1}{Z_N} e^{-E_n^{(N)}/k_B T}, \tag{2.52}$$

where the partition function is

$$Z_N = \sum_n e^{-E_n^{(N)}/k_B T}. \tag{2.53}$$

In principle all thermodynamic quantities can be found from the partition function. For example, the Helmholtz free energy is given by $F = -k_B T \ln Z_N$. Usually it is often more convenient to work with systems which do not have a definite particle number, N, and so we work in the **grand canonical ensemble**, where we calculate the grand canonical partition function

$$\mathcal{Z} = \sum_{n,N} e^{-\beta(E_n^{(N)} - \mu N)}. \tag{2.54}$$

and the corresponding grand canonical potential

$$\Omega(T,\mu) = -k_B T \ln \mathcal{Z}, \tag{2.55}$$

where μ is the chemical potential. All thermodynamic quantities can be calculated from the grand partition function, for example, the average number of particles is

$$\langle N \rangle = k_B T \frac{\partial \ln \mathcal{Z}}{\partial \mu}, \tag{2.56}$$

and the average total energy is

$$U = \langle \hat{H} \rangle = \mu \langle N \rangle - \frac{\partial \ln \mathcal{Z}}{\partial \beta}, \tag{2.57}$$

where, as usual, $\beta = 1/k_B T$.

This mathematical framework provides, in principle, a systematic and direct way to calculate all observable properties of interacting many-particle systems. If the particle–particle interactions are zero, it is easy to solve exactly. If the interactions $V(\mathbf{r})$ are weak, then some form of perturbation theory can be used. This is especially useful in the **weakly-interacting Bose gas**. As we have seen in Chapter 1, BEC in atom traps are effectively weakly interacting systems, because the density of atoms is very low. In this case perturbation theories are essentially exact.

But, unfortunately, helium is a strongly interacting quantum fluid. The van der Waals potential between helium atoms, $V(\mathbf{r})$ in Fig. 2.1, is weak in the attractive region, but is large and positive in the small r region. Effectively the helium atoms are like hard-spheres packed very densely. In this case perturbation theory can give some useful insights, but cannot be expected to be more than qualitative in its predictions. Fortunately, nowadays it is possible to more or less exactly solve complex many-body systems such as Eq. 2.51 using modern numerical techniques. The Quantum Monte Carlo method (QMC) makes it possible to more or less directly calculate observables for many-boson systems with large numbers of particles (eg several hundred). By systematically comparing the results of calculations with increasing numbers of particles, one can make accurate predictions for the $N \to \infty$ thermodynamic limit.

Quantum Monte Carlo calculations have directly shown that superfluid helium does indeed have a **condensate** of atoms in the ground state (Ceperley 1995). But the meaning of this statement is slightly different from the case of the ideal Bose gas. In the ideal Bose gas all of the particles are in the zero momentum state at $T = 0$. For helium at zero temperature a macroscopic number of particles occupy the zero momentum state, but not all of them. To make these statements more precise it is helpful to define the **one particle density matrix**,

$$\rho_1(\mathbf{r}_1 - \mathbf{r}_1') = N \int \Psi_0^*(\mathbf{r}_1, \mathbf{r}_2, \ldots, \mathbf{r}_N)\Psi_0^*(\mathbf{r}_1', \mathbf{r}_2, \ldots, \mathbf{r}_N) d^3r_2, \ldots, d^3r_N. \tag{2.58}$$

This is a correlation function between the many-particle wave function at particle coordinates $\mathbf{r}_1, \mathbf{r}_2, \ldots, \mathbf{r}_N$ and $\mathbf{r}_1', \mathbf{r}_2, \ldots, \mathbf{r}_N$. By integrating over all but one coordinate, $\mathbf{r}_2, \ldots, \mathbf{r}_N$, we are averaging over all configurations of the other particles except the first one. Of course, there is nothing special about the first particle coordinate, $\mathbf{r}_1$. Using the symmetry under exchange of boson wave functions, we would find an identical result whichever set of the $N-1$ particle coordinates we integrate over.

Some general properties of the density matrix can be seen easily. The fact that it should only depend on the coordinate difference $\mathbf{r}_1 - \mathbf{r}_1'$, as written, is a consequence of overall translational symmetry, that is, in an infinite sized system we can add a constant vector, $\mathbf{R}$, to all coordinates simultaneously without changing any physical observables. Furthermore, normalization of the many-particle wave function implies that if we set $\mathbf{r}_1 = \mathbf{r}_1'$ and integrate over $\mathbf{r}_1$ we must obtain

$$\int \rho_1(0) d^3 r_1 = N, \tag{2.59}$$

implying that

$$\rho_1(0) = \frac{N}{V} = n, \tag{2.60}$$

the particle number density.

Now, consider the density matrix in the case of a noninteracting Bose gas. Then, at $T = 0$ all of the particles occupy a single one-particle state, say $\psi_0(\mathbf{r})$. The corresponding many-particle wave function is

$$\Psi_0(\mathbf{r}_1, \mathbf{r}_2, \ldots, \mathbf{r}_N) = \psi_0(\mathbf{r}_1)\psi_0(\mathbf{r}_2)\psi_0(\mathbf{r}_3), \ldots, \psi_0(\mathbf{r}_N). \tag{2.61}$$

One can easily see that this if even under exchange of any pair of particles $\mathbf{r}_i$ and $\mathbf{r}_j$, and therefore has the correct boson symmetry. Evaluating the one-body density matrix for this trial ground state we obtain

$$\rho_1(\mathbf{r}_1 - \mathbf{r}_1') = N\psi_0^*(\mathbf{r}_1)\psi_0(\mathbf{r}_1') \int |\psi_0(\mathbf{r}_2)|^2, \ldots, |\psi_0(\mathbf{r}_N)|^2 d^3 r_2, \ldots, d^3 r_N,$$

$$= N\psi_0^*(\mathbf{r}_1)\psi_0(\mathbf{r}_1'). \tag{2.62}$$

The state $\psi_0(\mathbf{r})$ will normally be a zero momentum plane wave state, which is just a constant

$$\psi_0(\mathbf{r}) = \frac{1}{\sqrt{V}} e^{i\mathbf{k}\cdot\mathbf{r}} = \frac{1}{\sqrt{V}} \tag{2.63}$$

when condensation is in the usual $\mathbf{k} = 0$ state. Therefore the density matrix of the noninteracting Bose gas is a constant and equal to the particle density,

$$\rho_1(\mathbf{r}_1 - \mathbf{r}_1') = \frac{N}{V} = n, \tag{2.64}$$

at zero temperature.

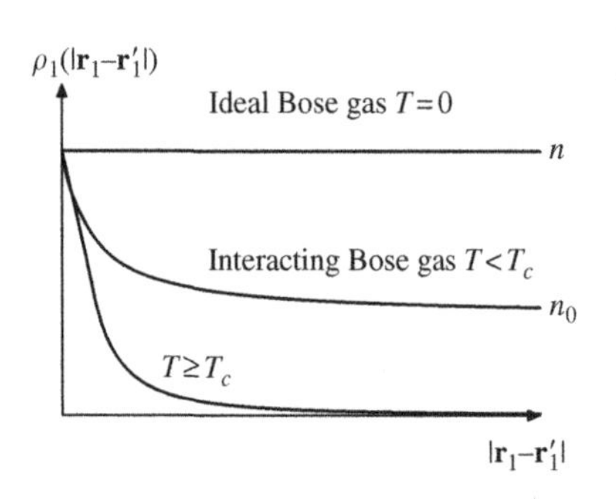

Fig. 2.13 One particle density matrix $\rho_1(\mathbf{r}_1 - \mathbf{r}_1')$. For the ideal Bose gas at $T = 0$ is is just a constant, n. For the interacting Bose gas at $T < T_c$ it is n at $\mathbf{r}_1 = \mathbf{r}_1'$, and approaches a smaller constant n_0 for large $|\mathbf{r}_1 - \mathbf{r}_1'|$. At temperatures above T_c the condensate density is zero, $n_0 = 0$, signifying a normal liquid state.

But in liquid ^{4}He we cannot assume that the particle interactions are negligible. The density matrix of the interacting helium gas has been calculated using QMC techniques. The results show that at zero temperature the density matrix is not constant for all distances, unlike Eq. 2.64. But it does approach a constant value at large separations, as shown in Fig. 2.13. This constant value provides the proper definition for the **condensate density**, n_0, in an interacting Bose gas,

$$n_0 = \lim_{|\mathbf{r}_1 - \mathbf{r}_1'| \to \infty} \rho_1(\mathbf{r}_1 - \mathbf{r}_1'), \tag{2.65}$$

This is now a rigorous definition, unlike our introduction of this quantity earlier from the concept of the macroscopic wave function $\psi_0(\mathbf{r})$. In fact the density matrix also allows us to define a suitable effective single particle macroscopic wave function. If we **define** $\psi_0(\mathbf{r})$, by extending Eq. 2.62 to a general interacting particle case as

$$\rho_1(\mathbf{r}_1 - \mathbf{r}_1') \sim \psi_0^*(\mathbf{r}_1)\psi_0(\mathbf{r}_1') \tag{2.66}$$

in the limit where $|\mathbf{r}_1 - \mathbf{r}_1'|$ is large, then this macroscopic wave function will have the normalization

$$|\psi_0(\mathbf{r})|^2 = n_0, \tag{2.67}$$

which we assumed earlier for the condensate wave function.

The same picture applies at all temperatures below T_c. For any temperatures below T_c the density matrix approaches a constant value, $n_0(T)$ at large values of $|\mathbf{r}_1 - \mathbf{r}_1'|$, defining the condensate density as a function of temperature $n_0(T)$. But at temperatures near to T_c the condensate density approaches zero. For $T \geq T_c$ the condensate density vanishes, and so the density matrix goes to zero in the limit $|\mathbf{r}_1 - \mathbf{r}_1'| \to \infty$,

$$\lim_{|\mathbf{r}_1 - \mathbf{r}_1'| \to \infty} \rho_1(\mathbf{r}_1 - \mathbf{r}_1') = \begin{cases} n_0(T) & T < T_c, \\ 0 & T \geq T_c. \end{cases} \tag{2.68}$$

This behavior is shown in Fig. 2.13.[7]

[7] The temperature dependence of the density matrix is explored in detail for the case of the ideal Bose gas in Exercise 2.6.

Experimental confirmation of this picture is provided by measurements of the **momentum distribution**. Recall that at the beginning of this chapter we discussed the Maxwell–Bolzmann distribution of particles in a classical gas. In Eq. 2.8 we defined the quantity $P(\mathbf{p})$ corresponding to the probability that any given particle has momentum in the region d^3p of momentum space. The same quantity can be defined for a quantum fluid. At zero temperature it is an expectation value,

$$P(\mathbf{p})d^3p = \frac{Vd^3p}{(2\pi\hbar)^3}\langle\Psi|\hat{n}_{\mathbf{k}}|\Psi\rangle, \tag{2.69}$$

where $|\Psi\rangle = \Psi(\mathbf{r}_1, \ldots, \mathbf{r}_N)$ is the N-body wave function, and $\hat{n}_{\mathbf{k}}$ is the **number operator** for quantum state $\mathbf{k}$ (defined more carefully in Chapter 5). As usual, here $\mathbf{p} = \hbar\mathbf{k}$ and $Vd^3p/(2\pi\hbar)^3$ is the number of quantum states in the region d^3p of momentum. The expectation value can be found from the one-body density matrix. It is simply a Fourier transform,

$$n_{\mathbf{k}} = \langle\Psi|\hat{n}_{\mathbf{k}}|\Psi\rangle = \int \rho_1(\mathbf{r})e^{i\mathbf{k}\cdot\mathbf{r}}d^3r. \tag{2.70}$$

As we can see in Fig. 2.13 $\rho_1(\mathbf{r})$ is a smooth function of $\mathbf{r}$. At temperatures above T_c it tends to zero for large $|\mathbf{r}|$, while below T_c it tends to a constant value n_0 as $|\mathbf{r}| \to \infty$. Therefore, it can be written as,

$$\rho_1(\mathbf{r}) = n_0 + \Delta\rho_1(\mathbf{r}), \tag{2.71}$$

where $\Delta\rho_1(\mathbf{r})$ always goes to zero at large $|\mathbf{r}|$. Carrying out the Fourier transform, we find

$$\begin{aligned} n_{\mathbf{k}} &= \int n_0 e^{i\mathbf{k}\cdot\mathbf{r}}d^3r + \int \Delta\rho_1(\mathbf{r})e^{i\mathbf{k}.\mathbf{r}}d^3r, \\ &= n_0 V\delta_{\mathbf{k},0} + f(\mathbf{k}). \end{aligned} \tag{2.72}$$

Here $\delta_{\mathbf{k},0} = 1$ if $\mathbf{k} = 0$ and it is zero if $\mathbf{k} \neq 0$. The second term, $f(\mathbf{k})$ is the Fourier transform of $\Delta\rho_1(\mathbf{r})$. Since we know that $\Delta\rho_1(\mathbf{r})$ goes to zero at large $\mathbf{r}$, we can deduce that $f(\mathbf{k})$ must be a smooth function of $\mathbf{k}$ near to the point $\mathbf{k} = 0$.

The corresponding momentum distribution of the particles can be written,

$$P(\mathbf{p}) = n_0 V \delta(\mathbf{p}) + \frac{V}{(2\pi\hbar)^3} f(\mathbf{p}/\hbar). \tag{2.73}$$

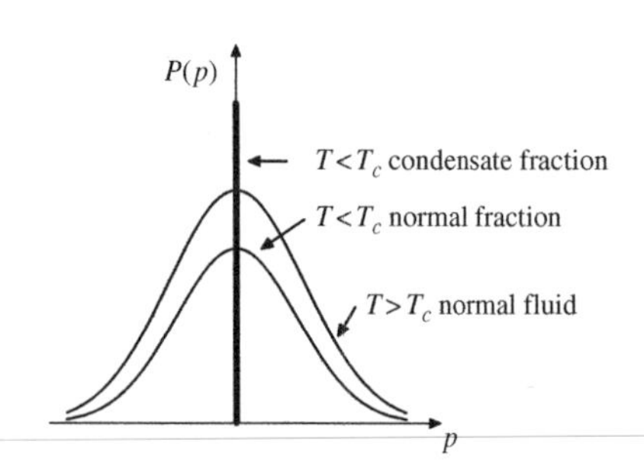

Fig. 2.14 Momentum density, $P(\mathbf{p})$, in the interacting Bose gas. At temperatures above T_c the momentum distribution resembles the classical Maxwell–Boltzmann distribution. However, at temperatures below T_c, part of the distribution becomes concentrated in a delta function peak at zero momentum. The weight of this delta function defines the condensate density, n_0.

The two contributions are: first a contribution from the condensate, which is a Dirac delta function at zero momentum. This corresponds to a total of $N_0 = n_0 V$ particles in the $\mathbf{k} = 0$ state. And second, there is a contribution corresponding to the remaining $N - N_0$ particles. This second contribution is a smooth function of momentum, $\mathbf{p}$. These are shown in Fig. 2.14. Qualitatively the smooth part is not unlike the Maxwell–Boltzmann momentum distribution, with an approximately Gaussian spread of different particle momenta. But its physical origin is quite different, it is not due to thermal motion, but to zero point kinetic energy of the particles in the quantum liquid of helium. Therefore it is present even at absolute zero of temperature. Upon raising the temperature the condensate part gradually decreases, until at T_c the condensate eventually disappears with $n_0 = 0$. At the same time the smooth contribution gradually evolves until it eventually approaches the Maxwell–Boltzmann form of Eq. 2.8.

This picture has been confirmed experimentally, since the momentum distribution $P(\mathbf{p})$ can be measured by neutron scattering experiments. These not only show good qualitative agreement with the above picture, but even agree quantitatively with theoretical predictions from Quantum Monte Carlo simulations. For liquid helium at low temperatures, both experiment and theory agree that, even at $T = 0$, the condensate part is in fact only about 10% of the total fluid,

$$n_0 \approx 0.1n. \tag{2.74}$$

This is quite different from a noninteracting (or weakly interacting) Bose gas where we would expect $n_0 \approx n$, as indicated in Fig. 2.13. The fact that it is such a small fraction for helium shows how important the strong interactions are in determining the many-particle quantum ground state.

2.7 Quasiparticle excitations

One should be careful not to confuse the **condensate density**, n_0, described above and the **superfliud density**, n_s defined in the previous section. Obviously they are closely related, and are both equal to n in a noninteracting Bose gas. However, their definitions are quite different physically. The condensate density n_0 is defined from the one-particle density matrix, or equivalently from the delta function peak in the momentum distribution. It is the number of particles present in the zero momentum state. On the other hand the superfluid density is defined by the two-fluid model of superflow. The definition is the equation for particle current $\mathbf{j}_s = n_s \mathbf{v}_s = n_s \hbar(\nabla\theta)/m$. Thus while the condensate density is a property of the ground state, the superfluid density is a property of the superflow. In fact, as we shall see, at zero temperature **all** of the particles participate in superflow. Therefore at $T = 0$ we have

$$n_s = n \qquad n_0 \approx 0.1n \tag{2.75}$$

for superluid ^{4}He. No contradiction is present, even though this is surprising at first sight.

Why do we know that all the particles participate in superflow? The argument uses the principle of **Galilean invariance**. In other words we view the system from the point of view of a uniformly moving observer. Since relativistic effects will be negligible, the transformation from stationary (laboratory rest frame) coordinates to moving frame coordinates is the **Galilean transformation**,

$$\mathbf{r}' = \mathbf{r} - \mathbf{v}t. \tag{2.76}$$

Consider a classical particle from the point of view of an observer in such moving frame of reference. The momentum of a particle of mass m is shifted by

$$\mathbf{p}' = \mathbf{p} - m\mathbf{v}. \tag{2.77}$$

Its kinetic energy, $E = p^2/2m$, is also shifted by

$$E' = E - \mathbf{p}\cdot\mathbf{v} + \tfrac{1}{2}mv^2. \tag{2.78}$$

Quantum mechanically, the Schrödinger equation also obeys this Galilean invariance, when we make the usual replacement $\mathbf{p} \to -i\hbar\nabla$.

Using this idea we can construct a trial many-body wave function for a uniformly moving superfluid. If $\Psi_0(\mathbf{r}_1, \ldots, \mathbf{r}_N)$ is the ground state wave function for a condensate at rest, then we can construct a trial wave function of a uniformly moving superfluid as follows. If the superfluid as a whole is moving with velocity $\mathbf{v}$, then each individual particle must have an additional momentum $m\mathbf{v}$ due to the motion. Defining $\hbar\mathbf{q} = m\mathbf{v}$, then the wave function must have an extra factor $e^{i\mathbf{q}\cdot\mathbf{r}}$ corresponding to this momentum. The corresponding many-particle wave function must be,

$$\Psi(\mathbf{r}_1, \ldots, \mathbf{r}_N) = e^{i\mathbf{q}.(\mathbf{r}_1+\mathbf{r}_2\cdots+\mathbf{r}_N)}\Psi_0(\mathbf{r}_1, \ldots, \mathbf{r}_N). \tag{2.79}$$

If the wave vector $\mathbf{q}$ is very small (of order $1/L$ where L is the macroscopic length of the sample), then this trial function will be almost exactly an eigenstate of the Hamiltonian. Clearly it also has the correct exchange symmetry required for a boson wave function. We can use this trial wave function to calculate the total momentum of the moving fluid

$$\langle\hat{\mathbf{P}}\rangle = N\hbar\mathbf{q}, \tag{2.80}$$

where $\hat{\mathbf{P}} = \hat{\mathbf{p}}_1 + \hat{\mathbf{p}}_2 + \cdots$ is the total momentum operator. Also, the total energy of this trial state is

$$\langle\hat{H}\rangle \approx E_0 + N\frac{\hbar q^2}{2m} = E_0 + \frac{1}{2M}\langle\hat{\mathbf{P}}\rangle^2, \tag{2.81}$$

since the superfluid has zero net momentum in the moving reference frame. Here E_0 is the ground state energy, and $M = Nm$ is the total mass. Clearly these are the classical Hamiltonian equations of motion for a macroscopic object of mass M and momentum $\langle\hat{\mathbf{P}}\rangle$, moving with velocity $\mathbf{v}_s$, where

$$\mathbf{v}_s = \frac{1}{M}\langle\hat{\mathbf{P}}\rangle = \frac{\hbar\mathbf{q}}{m}. \tag{2.82}$$

The implication is that at zero temperature the whole mass of the fluid, M, contributes to the superflow, and not just the fraction in the zero momentum condensate.

This phenomenon is sometimes referred to as the **rigidity** of the ground state wave function. The small perturbation in momentum $\hbar\mathbf{q}$ gives every particle in the fluid a small boost, and the whole quantum state moves rigidly

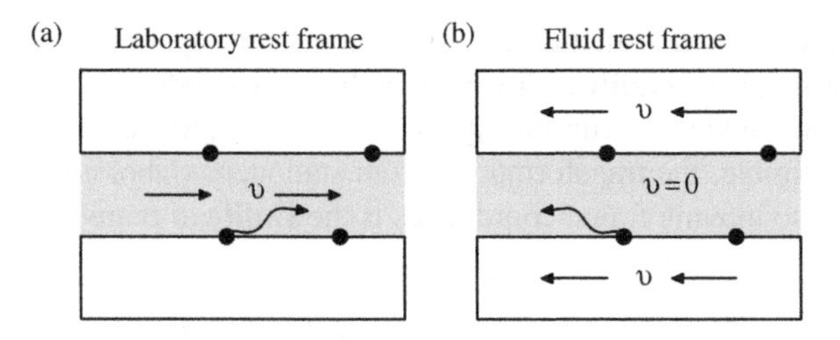

Fig. 2.15 Friction in a normal fluid flowing in a tube with rough walls. (a) In the frame of reference of the wall, and (b) in the frame of reference of the fluid motion. Friction is possible if a quasiparticle can become excited in the fluid by interaction with defects on the surface, as shown.

with a constant velocity. An observer in a Galilean frame of reference moving with same velocity as the fluid would see exactly the same ground state wave function as a stationary observer would see in a fluid at rest.[8] Clearly **every particle** contributes to the superflow at zero temperature, and hence $n_s = n$ at $T = 0$.

[8] In fact Galilean invariance also applies to a normal fluid state. To complete the argument for the existence of superfluidity, we need to also show that the fluid does not experience friction due to interactions with the walls. This is the property which truly distinguishes a moving superfluid from a moving normal fluid. The latter experiences energy dissipation and friction, while the superfluid does not.

The Russian theorist Lev Landau made a brilliant theoretical deduction from the above idea. The question is why should the superfluid flow without friction, while a normal fluid would experience viscosity and hence dissipation of energy? He imagined a fluid flowing in a narrow tube. For a normal fluid, the friction and viscosity arise because the flowing fluid particles experience random scattering events from the atomically rough walls of the tube, as shown in Fig.2.15. This is the mechanism which transfers momentum from the fluid to the walls, providing the viscous friction force on the fluid. But why should this not also happen for a superfluid? Consider again the situation from a frame of reference moving with the fluid. In this case it is the walls that are moving. The rough walls are time dependent perturbations, which can be dealt with using Fermi's Golden rule for time dependent perturbation theory. Treating the potential energy due to the defect as a potential $V(\mathbf{r})$ in the laboratory frame, this is equivalent to a time dependent potential $V(\mathbf{r}' + \mathbf{v}t)$ in the rest frame of the fluid. A single quantum particle with initial momentum $\mathbf{p}_i$ and energy ϵ_i can be scattered elastically to the final state $\mathbf{p}_f$ and energy ϵ_f only if

$$\epsilon_f = \epsilon_i - \mathbf{v} \cdot (\mathbf{p}_i - \mathbf{p}_f). \tag{2.83}$$

Applying this to the case of a particle initially in the condensate at $\mathbf{p} = 0$, we see that we can create an **elementary excitation** of momentum $\mathbf{p}$ and energy $\epsilon(\mathbf{p})$ only if

$$\epsilon(\mathbf{p}) = \mathbf{v} \cdot \mathbf{p}. \tag{2.84}$$

In a normal fluid the energy of an excited particle of momentum $\mathbf{p}$ is $\epsilon(\mathbf{p}) = p^2/2m$. In this case it is always possible to find some values of $\mathbf{p}$ which satisfy Eq. 2.84. The condition,

$$\frac{p^2}{2m} = \mathbf{v} \cdot \mathbf{p}$$

is obeyed on a cone of momentum vectors $\mathbf{p}$, obeying $|\mathbf{p}| = 2mv\cos\phi$, with angle ϕ to the direction of $\mathbf{v}$. Therefore the rough walls can always impart momentum to the fluid, leading to viscous friction. This would be true in either a classical fluid, or in a strongly interacting normal quantum fluid, like He I just above the λ-point. It would also always be true in a perfectly noninteracting ideal Bose gas. For this reason the ideal Bose gas is not believed to be a true superfluid, even at zero temperature.

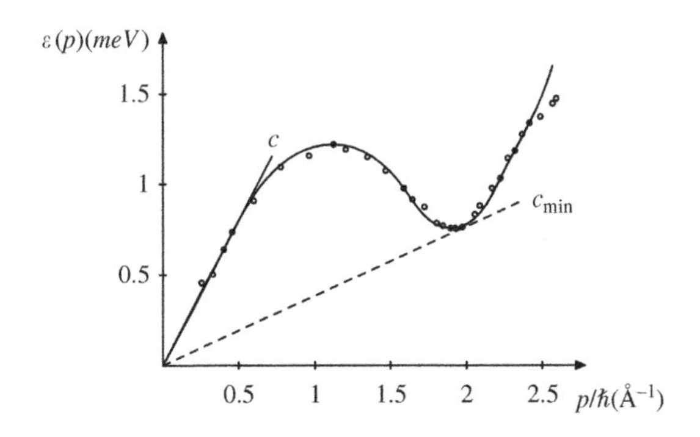

Fig. 2.16 Energy of a **quasiparticle** in superfluid helium. Landau's argument shows that there is no dissipation due to quasiparticle excitations for superfluids moving with velocities less than, minimum slope, $c_{\min}$. In practice the critical velocity of the superfluid flow is considerably less than this limit, since other excitations, notably vortices, can be generated and these lead to dissipation and viscosity.

True superfluid behavior only occurs if it is impossible to satisfy Eq. 2.84, and therefore no scattering can occur. From this argument Landau postulated that the energy spectrum of particle excitations, $\epsilon(\mathbf{p})$, must be very different in a superfluid compared with normal fluid. He proposed that in He II the single particle excitation energy, $\epsilon(\mathbf{p})$, must have the shape shown in Fig. 2.16. This unusual shape has since been confirmed by many experiments, most directly by neutron scattering. In fact the points on the graph were obtained in experimental measurements, using neutron scattering.

In Fig. 2.16 there are three main regions of the graph. At small momenta, $\mathbf{p}$, the energy is approximately linear in $|\mathbf{p}|$,

$$\epsilon(\mathbf{p}) = c|\mathbf{p}|. \tag{2.85}$$

Such linear behavior is typical of phonons in solids, and so this part of the excitation spectrum is called the phonon-part of the spectrum. At very large momentum the spectrum approaches a conventional normal liquid

$$\epsilon(\mathbf{p}) = \frac{p^2}{2m^*}. \tag{2.86}$$

This is because very high momentum particles will become ballistic and move more or less independently of the other fluid particles. The fact that the particles interact strongly leads to the effective mass m^* in place of the bare mass of a ^{4}He atom. But probably the most surprising part of the spectrum in Fig. 2.16 is the minimum. This part is called the **roton** part of the spectrum,

$$\epsilon(\mathbf{p}) = \Delta + \frac{(p - p_0)^2}{2\mu}. \tag{2.87}$$

The physical interpretations of the very different motions in the phonon and roton regions are summarized in Fig. 2.17. At low momentum a single helium atom couples strongly to the many-particle condensate. As it moves, the condensate moves rigidly with it, leading to a motion almost like a solid-body, and hence a phonon-like energy spectrum. On the other hand, at very high momentum the atom can move relatively independently of the rest of the fluid particles. In contrast, at the intermediate momenta of the roton minimum the moving particle couples strongly to its neighbors. As the atom moves the neighbors must move out of the way. The neighbors must move to the side and end up behind the moving particle, so they actually move in a circular **backflow**. The net effect is a forward motion of one particle, accompanied by a ring of particles rotating backwards. Feynman has likened the motion to the motion of

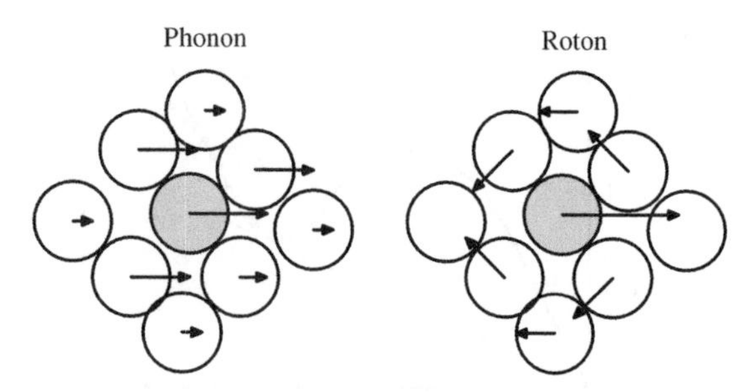

Fig. 2.17 Physical interpretation of the phonon and roton parts of the quasiparticle spectrum. The phonon motion corresponds to de Broglie wavelengths, $p = h/\lambda$ greater than a single atomic size, and leads to coupled motion of groups of atoms moving together, rather like a phonon in a solid. The roton corresponds to de Broglie wavelengths of order the interparticle separation. It corresponds to a central particle moving forward, while the close packed neighbors must move out of the way in a circular motion. Feynman notes that this combination of linear and circular motion has some nice similarity to a moving smoke-ring.

a smoke-ring! The smoke ring itself moves forward, while the smoke particles themselves move in circular motion constantly rotating backwards around the rim of the smoke ring (Feynman 1972).

The excitations described by $\epsilon(\mathbf{p})$ are clearly not single particles, but complex many-body motions. Nevertheless, they behave to some extent like effective particles, or **quasiparticles**. They are the **elementary excitations** of the system, like the phonons in a solid or the spin waves in a magnet. By assuming the excitation spectrum had the shape given above, Landau was able to calculate accurately many of the empirically determined properties of liquid He II. A very simple calculation gives the specific heat at low temperatures,

$$C_V \sim T^3,$$

in agreement with experiments (the calculation is essentially the same at the Debye model for the specific heat due to phonons in a solid). Also, the fact that the low temperature superfluid density varies as T^4,

$$n_s(T) = n - AT^4$$

can also be deduced from the existence of the linear phonon part of the spectrum.

Most importantly, Landau showed that the quasiparticle spectrum of Fig. 2.17 explains the absence of viscosity in a moving superfluid fluid, hence proving the most fundamental superfluid property. The point is that with this functional form for $\epsilon(\mathbf{p})$ the scattering equation, Eq. 2.84, cannot be satisfied for any values of momentum, $\mathbf{p}$, if the superfluid velocity, $\mathbf{v}_s$ is small. From the plot Fig. 2.16 it is clear that

$$\epsilon(\mathbf{p}) > c_{\text{min}}|\mathbf{p}|, \tag{2.88}$$

where c_{min} is the slope of the minimum slope tangent shown in Fig. 2.16. Consequently, a low energy quasiparticle starting with near zero initial momentum $\mathbf{p}$ cannot be scattered into any possible final states when

$$|\mathbf{v}_s| < c_{\text{min}}. \tag{2.89}$$

Therefore the superfluid flows without dissipation due to scattering of quasiparticles, provided that the flow velocity is less than the ideal critical velocity, c_{min}. In practice the actual critical velocity is really very much less than this theoretical limit. There are other excitations, such as vortices, which can also be created by the flow of the superfliud past rough walls (just as vortices are created by, for example, the air flowing over aeroplane wings). Once the flow velocity is great enough to nucleate vortices, then they immediately lead to a finite effective friction and hence a finite viscosity. In fact, very fast flows will become **turbulent** due to vortex nucleation, and the fluid flow becomes turbulent, in analogy to turbulence in classical fluid dynamics.

Finally, another interesting consequence of this model, is the existence of two types of sound modes, called **first** and **second sound**. First sound, is an oscillation of the total density n. The sound velocity is the equal to the slope of the phonon quasiparticle spectrum, $c \approx 238\ \mathrm{ms}^{-1}$, from Fig. 2.16. On the other hand, second sound is an oscillation in which the superfluid and normal fluid densities, n_s and n_n oscillate out of phase, and so the total density, n, remains constant. We can think of this as an "entropy wave," since the normal fluid component has entropy, while the superfluid does not. Landau predicted that at low temperatures the velocity of second sound is $c_s = c/\sqrt{3}$, in excellent agreement with experiments. It is possible to develop a full set of hydrodynamic equations to describe first and second sound, but we shall not do this here, since it would take us too far away from the main subject areas of this book.

2.8 Summary

In this chapter has covered a lot of ground, and so let us briefly summarize the main points. Firstly we developed the idea of a quantum fluid, and showed why helium is so unique in its ability to remain liquid even down to zero temperature. Secondly we introduced some of the main physical phenomena associated with He II, such as the λ transition in specific heat, superflow properties and flow quantization. At the same time we developed some of the underlying theoretical concepts, especially the idea of the macroscopic wave function and its phase θ. This idea was developed more fully from the concept of the single particle density matrix and the corresponding momentum distribution. Finally we discussed the origin of superfluidity, and Landau's quasiparticle excitation model.

In the discussion in this chapter we have tried to avoid the more formal concepts of many-body perturbation theory. We have even avoided introducing second quantized field operators, $\hat{a}^{+}$ and $\hat{a}$. These become essential in any more rigorous discussion of the many-particle quantum states which are needed to understand superfluidity. The physical ideas introduced in this chapter can be understood without the need for the full machinery of field theory. Nevertheless, in later chapters we shall return to some of the topics discussed here, and then we shall have to make use of these quantum field theory methods.

Further reading

Properties of normal fluids are covered in the volume *Soft Condensed Matter*, Jones (2002), in the Oxford Master Series in Condensed Matter Physics. The general ideas of an order parameter for a phase transition, such as the macroscopic wave function discussed here, are developed in Chakin and Lubensky (1995).

A text with a very good and detailed experimental overview of properties of superfluid helium is Tilley and Tilley (1990). A quite accessible theoretical description and experimental overview is provided by Feynman (1972).

Experimental measurements of momentum distributions in liquids, including He II, are discussed in Silver and Sokol (1989). This is a specialist conference proceedings, but contains several clear introductory articles. The book also includes introductory discussions of quantum Monte Carlo techniques and results for calculations on liquid helium. More detailed descriptions of the quantum Monte Carlo method and results for liquid helium are given by Ceperley (1995).

The observations of regular triangular vortex lattices in BEC was originally made by Abo-Shaeer *et al.* (2001). Some very beautiful pictures can be found on the Ketterle group web site (Ketterle 2003).

Exercises

(2.1) For a classical fluid we derived the Maxwell–Boltzmann momentum distribution, $P_{\mathrm{MB}}(p)$, given by Eq. 2.9. Confirm that this is a correctly normalized probability distribution

$$\int_0^\infty P_{\mathrm{MB}}(p)dp = 1,$$

and that the mean kinetic energy is

$$\int_0^\infty \frac{p^2}{2m} P_{\mathrm{MB}}(p)dp = \frac{3}{2}k_{\mathrm{B}}T.$$

You will need the standard integrals

$$\int_0^\infty x^2 e^{-x^2} dx = \frac{\sqrt{\pi}}{4}$$

and

$$\int_0^\infty x^4 e^{-x^2} dx = \frac{3\sqrt{\pi}}{8}.$$

(2.2) Using Eq. 2.45 for the superflow velocity $\mathbf{v}_s$ in a vortex, show that the flow is both irrotational and incompressible:

$$\nabla \times \mathbf{v}_s = 0 \qquad \nabla \cdot \mathbf{v}_s = 0.$$

Hint: use the formula for curl in cylindrical polar cordinates given in Eq. 2.43 and the corresponding expression for divergence,

$$\nabla \cdot \mathbf{a} = \frac{1}{r}\frac{\partial}{\partial r}(ra_r) + \frac{1}{r}\frac{\partial}{\partial \phi}a_\phi + \frac{\partial}{\partial z}a_z.$$

(2.3) The particle flow around a vortex, Eq. 2.45, is in precise analogy to the magnetic field in a wire carrying electrical current. Show that if we make the equivalence $\mathbf{v}_s \leftrightarrow \mathbf{B}$, then the analog of κ is given by Ampère's law

$$\mu_0 I = \oint \mathbf{B} \cdot d\mathbf{r}.$$

Show that this analogy can be extended so that the kinetic energy per unit volume of superfluid

$$dE = \tfrac{1}{2}\rho_s v_s^2 d^3r$$

(where $\rho_s = mn_s$ is the mass density of the superfluid) is equivalent to the electromagnetic energy density of the wire

$$dE = \frac{1}{2\mu_0}B^2 d^3r.$$

Hence find exact equivalences between the physical parameters of superflow in a vortex, $\{\kappa, \rho_s, \mathbf{v}_s\}$, and the corresponding parameters for a wire $\{I, \mu_0, \mathbf{B}\}$.

(2.4) Given that the force per unit length between two parallel current carrying wires a distance R apart is,

$$F = \frac{\mu_0 I_1 I_2}{2\pi R},$$

use the electromagnetic analogy from problem 2.3 to find the force per unit length between to parallel superfluid vortices with circulations κ_1 and κ_2. Hence show that the interaction energy of two parallel vortices is

$$\Delta E = \rho_s \frac{\kappa_1 \kappa_2}{2\pi} \ln R/R_0$$

per unit length, where R_0 is a constant.

(2.5) Derive the formulas for the average particle number and energy in the cannonical ensemble, given in Eqs. 2.56 and 2.57.

(2.6) We can calculate the momentum distribution $P(\mathbf{p})$ in the ideal Bose gas at temperature T by replacing the average particle number $\bar{n}_{\mathbf{k}}$ in Eq. 2.69 by the Bose–Einstein distribution

$$n_{\mathbf{k}} = \frac{1}{e^{\beta(\epsilon_{\mathbf{k}}-\mu)} - 1}.$$

(a) Given that the single particle density matrix is

$$\rho_1(\mathbf{r}) = \frac{1}{(2\pi)^3}\int n_{\mathbf{k}} e^{-i\mathbf{k}\cdot\mathbf{r}} d^3k,$$

show that the spherical symmetry of space implies that

$$\rho_1(r) = \frac{4\pi}{(2\pi)^3}\int_0^\infty n_{\mathbf{k}} \frac{\sin(kr)}{kr} k^2 dk.$$

(b) The large r behavior of $\rho_1(r)$ is clearly given by the integrand above at small k. Assuming that $T > T_c$ expand the Bose–Einstein distribution for small k,

$$n_{\mathbf{k}} = \frac{1}{a + bk^2 + \cdots},$$

and find the constants a and b. Then, using the integral

$$\int_0^\infty \frac{x}{c^2 + x^2} \sin x dx = \frac{\pi}{2} e^{-c}$$

(those who are good with contour integration, might derive this themselves!), show that for large r,

$$\rho_1(r) \sim \frac{\text{const}}{r} e^{-r/d}.$$

(c) The result from part (b) shows that above T_c the density matrix goes to zero exponentially with distance. Show that length scale involved is

$$d = \left(\frac{\hbar^2}{2m\mu}\right)^{1/2}$$

near to T_c. Hence show that as the temperature approaches T_c from above, $T \to T_c^+$, this length diverges at T_c. A diverging length scale of this kind is usually associated with **critical phenomena** of phase transitions.

(d) Repeat the above argument of parts (b) and (c) for the case $T < T_c$. First show that now

$$\rho_1(r) = n_0 + \frac{1}{(2\pi)^3}\int_0^\infty \bar{n}_k \frac{\sin(kr)}{kr} 4\pi k^2 dk.$$

Then, given the integral

$$\int_0^\infty \frac{\sin x}{x} dx = \frac{\pi}{2},$$

show that

$$\rho_1(r) = n_0 + \frac{\text{const}}{r}$$

for large r. Hence, unlike part (c), we can conclude that there is no characteristic length scale involved in BEC below T_c. Such behavior is usually typical of systems at a **critical point**.

Superconductivity

3

3.1 Introduction

This chapter describes some of the most fundamental experimental facts about superconductors, together with the simplest theoretical model: the **London equation**. We shall see how this equation leads directly to the expulsion of magnetic fields from superconductors, the Meissner–Ochsenfeld effect, which is usually considered to be the fundamental property which defines superconductivity.

The chapter starts with a brief review of the Drude theory of conduction in normal metals. We shall also show how it is possible to use the Drude theory to make the London equation plausible. We shall also explore some of the consequences of the London equation, in particular the existence of vortices in superconductors and the differences between type I and II superconductors.

3.2 Conduction in metals

The idea that metals are good electrical conductors because the electrons move freely between the atoms was first developed by Drude in 1900, only 5 years after the original discovery of the electron.

Although Drude's original model did not include quantum mechanics, his formula for the conductivity of metals remains correct even in the modern quantum theory of metals. To briefly recap the key ideas in the theory of metals, we recall that the wave functions of the electrons in crystalline solids obey **Bloch's theorem**,[1]

$$\psi_{n\mathbf{k}}(\mathbf{r}) = u_{n\mathbf{k}}(\mathbf{r})e^{i\mathbf{k}\cdot\mathbf{r}}, \tag{3.1}$$

where $u_{n\mathbf{k}}(\mathbf{r})$ is a function which is periodic, $\hbar\mathbf{k}$ is the crystal momentum, and $\mathbf{k}$ takes values in the first Brillouin zone of the reciprocal lattice. The energies of these Bloch wave states give the **energy bands**, $\epsilon_{n\mathbf{k}}$, where n counts the different electron bands. Electrons are fermions, and so at temperature T a state with energy ϵ is occupied according to the **Fermi–Dirac** distribution

$$f(\epsilon) = \frac{1}{e^{\beta(\epsilon-\mu)}+1}. \tag{3.2}$$

The chemical potential, μ, is determined by the requirement that the total density of electrons per unit volume is

$$\frac{N}{V} = \frac{2}{(2\pi)^3}\sum_n \int \frac{1}{e^{\beta(\epsilon_{n\mathbf{k}}-\mu)}+1} d^3k, \tag{3.3}$$

[1] See, for example, the text *Band theory and electronic properties of solids* by J. Singleton (2001), or other textbooks on Solid State Physics, such as Kittel (1996), or Ashcroft and Mermin (1976).

where the factor of 2 is because of the two spin states of the $s = 1/2$ electron. Here the integral over **k** includes all of the first Brillouin zone of the reciprocal lattice and, in principle, the sum over the band index n counts all of the occupied electron bands.

In all of the metals that we are interested in here the temperature is such that this Fermi gas is in a highly degenerate state, in which $k_B T \ll \mu$. In this case $f(\epsilon_{n\mathbf{k}})$ is nearly 1 in the region "inside" the Fermi surface, and is 0 outside. The Fermi surface can be defined by the condition $\epsilon_{n\mathbf{k}} = \epsilon_F$, where $\epsilon_F = \mu$ is the Fermi energy. In practice, for simplicity, in this book we shall usually assume that there is only one conduction band at the Fermi surface, and so we shall ignore the band index n from now on. In this case the density of **conduction electrons**, n, is given by

$$n = \frac{2}{(2\pi)^3} \int \frac{1}{e^{\beta(\epsilon_{\mathbf{k}}-\mu)} + 1} d^3k, \tag{3.4}$$

where $\epsilon_{\mathbf{k}}$ is the energy of the single band which crosses the Fermi surface. In cases where the single band approximation is not sufficient, it is quite easy to add back a sum over bands to the theory whenever necessary.

Metallic conduction is dominated by the thin shell of quantum states with energies $\epsilon_F - k_B T < \epsilon < \epsilon_F + k_B T$, since these are the only states which can be thermally excited at temperature T. We can think of this as a low density gas of "electrons" excited into empty states above ϵ_F and of "holes" in the occupied states below ϵ_F. In this **Fermi gas** description of metals the electrical conductivity, σ, is given by the Drude theory as,

$$\sigma = \frac{ne^2\tau}{m}, \tag{3.5}$$

where m is the effective mass of the conduction electrons,[2] $-e$ is the electron charge and τ is the average lifetime for free motion of the electrons between collisions with impurities or other electrons.

[2]Note that the band mass of the Bloch electrons, m, need not be the same as the bare mass of an electron in vacuum, m_e. The effective mass is typically 2–3 times greater. In the most extreme case, the **heavy fermion materials** m can be as large as $50-100m_e$!

The conductivity is defined by the **constitutive equation**

$$\mathbf{j} = \sigma\boldsymbol{\mathcal{E}}. \tag{3.6}$$

Here **j** is the electrical current density which flows in response to the external electric field, $\mathcal{E}$. The resistivity ρ obeys

$$\boldsymbol{\mathcal{E}} = \rho\mathbf{j}, \tag{3.7}$$

and so ρ is simply the reciprocal of the conductivity, $\rho = 1/\sigma$. Using the Drude formula we see that

$$\rho = \frac{m}{ne^2}\tau^{-1}, \tag{3.8}$$

and so the resistivity is proportional to the **scattering rate**, τ^{-1} of the conduction electrons. In the SI system the resistivity has units of Ωm, or is more often quoted in Ωcm.

Equation 3.5 shows that the electrical conductivity depends on temperature mainly via the different scattering processes which enter into the mean lifetime τ. In a typical metal there will be three main scattering processes, scattering by impurities, by electron–electron interactions and by electron–phonon collisions. These are independent processes, and so we should add the scattering rates to obtain the total effective scattering rate

$$\tau^{-1} = \tau_{\text{imp}}^{-1} + \tau_{\text{el-el}}^{-1} + \tau_{\text{el-ph}}^{-1}, \quad (3.9)$$

where τ_{imp}^{-1} is the rate of scattering by impurities, $\tau_{\text{el-el}}^{-1}$ the electron–electron scattering rate, and $\tau_{\text{el-ph}}^{-1}$ the electron–phonon scattering rate. Using Eq. 3.8 we see that the total resistivity is just a sum of independent contributions from each of these different scattering processes,

$$\rho = \frac{m}{ne^2}\left(\tau_{\text{imp}}^{-1} + \tau_{\text{el-el}}^{-1} + \tau_{\text{el-ph}}^{-1}\right). \quad (3.10)$$

Each of these lifetimes is a characteristic function of temperature. The impurity scattering rate, τ_{imp}^{-1}, will be essentially independent of temperature, at least for the case of nonmagnetic impurities. The electron–electron scattering rate, $\tau_{\text{el-el}}^{-1}$, is proportional to T^2, where T is the temperature. While at low temperatures (well below the phonon Debye temperature) the electron–phonon scattering rate, $\tau_{\text{el-ph}}^{-1}$, is proportional to T^5. Therefore we would expect that the resistivity of a metal is of the form

$$\rho = \rho_0 + aT^2 + \cdots \quad (3.11)$$

at very low temperatures. The zero temperature resistivity, the residual resistivity, ρ_0, depends only on the concentration of impurities.

For most metals the resistivity does indeed behave in this way at low temperatures. However, for a superconductor something dramatically different happens. Upon cooling the resistivity first follows the simple smooth behavior, Eq. 3.11, but then suddenly vanishes entirely, as sketched in Fig. 3.1. The temperature where the resistivity vanishes is called the critical temperature, T_c. Below this temperature the resistivity is not just small, but is, as far as can be measured, exactly zero.

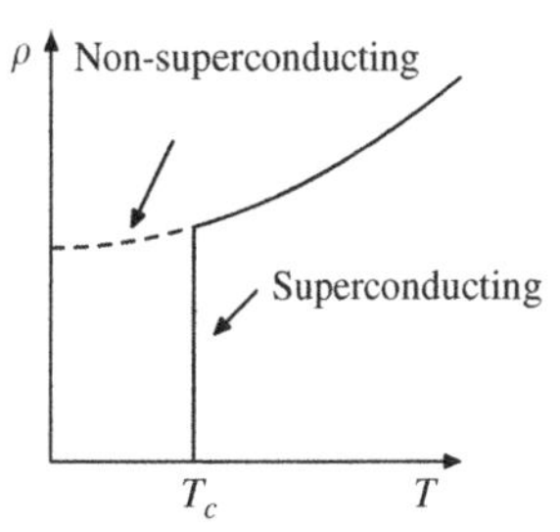

Fig. 3.1 Resistivity of a typical metal as a function of temperature. If it is a non-superconducting metal (such as copper or gold) the resistivity approaches a finite value at zero temperature, while for a superconductor (such as lead, or mercury) all signs of resistance disappear suddenly below a certain temperature, T_c.

This phenomenon was a complete surprise when it was first observed by H. Kammerling Onnes in 1911. He had wanted to test the validity of the Drude theory by measuring the resistivity at the lowest temperatures possible. The first measurements on samples of platinum and gold were quite consistent with the Drude model. But then he then turned his attention to mercury, because of its especially high purity. Based on Eq. 3.11 one could expect a very small, perhaps even zero, residual resisitivity in exceptionally pure substances. But what Kammerling Onnes actually observed was completly unexpected, and not consistent with Eq. 3.11. Surprisingly he discovered that all signs of resistance appeared to vanish suddenly below about 4 K. This was quite unexpected from the Drude model, and was, in fact, the discovery of a new state of matter: superconductivity.

3.3 Superconducting materials

A number of the elements in the periodic table become superconducting at low temperatures, as summarized in Table 3.1. Of the elements, Niobium (Nb) has

Table 3.1 Some selected superconducting elements and compounds

Substance	T_c (K)	
Al	1.2	
Hg	4.1	First superconductor, discovered 1911
Nb	9.3	Highest T_c of an element at normal pressure
Pb	7.2	
Sn	3.7	
Ti	0.39	
Tl	2.4	
V	5.3	
W	0.01	
Zn	0.88	
Zr	0.65	
Fe	2	High pressure
H	300	Predicted, under high pressure
O	30	High pressure, maximum T_c of any element
S	10	High pressure
Nb_3Ge	23	A15 structure, highest known T_c before 1986
$Ba_{1-x}Pb_xBiO_3$	12	First perovskite oxide structure
$La_{2-x}Sr_xCuO_4$	35	First high T_c superconductor
$YBa_2Cu_3O_{7-\delta}$	92	First superconductor above 77 K
$HgBa_2Ca_2Cu_3O_{8+\delta}$	135–165	Highest T_c ever recorded
$K_3\,C_{60}$	30	Fullerene molecules
YNi_2B_2C	17	Borocarbide superconductor
MgB_2	38	Discovery announced in January 2001
Sr_2RuO_4	1.5	Possible p-wave superconductor
UPt_3	0.5	"Heavy fermion" exotic superconductor
$(TMTSF)_2ClO_4$	1.2	Organic molecular superconductor
ET-BEDT	12	Organic molecular superconductor

the highest critical temperature T_c of 9.2 K at atmospheric pressure. Interestingly, while some common metals such as aluminium (1.2 K), tin (3.7 K), and lead (7.2 K) become superconducting, other equally good, or better, metals (such as copper, silver, or gold) show no evidence for superconductivity at all. It is still a matter of debate whether or not they would eventually become superconducting if made highly pure and cooled to sufficiently low temperatures. As recently as 1998 it was discovered that platinum becomes superconducting, but only when it is prepared into small nano-particles at temperatures of a few milliKelvin.

Another recent discovery is that quite a few more elments also become superconducting when they are subjected to extremely high pressures. Samples must be pressurized between two anvil shaped diamonds. Using this technique it is possible to obtain such high pressures that substances which are normally insulators become metallic, and some of these novel metals become superconducting. Sulfur and oxygen both become superconducting at surprisingly high temperatures. Even iron becomes superconducting under pressure. At normal pressures iron is, of course, magnetic, and the magnetism prevents superconductivity from occuring. However, at high pressures a nonmagnetic phase can be found, and this becomes superconducting. For many years the "holy grail" for this sort of high pressure work has been to look for superconductivity in metallic hydrogen. It has been predicted that metallic hydrogen could become superconducting at as high a temperature as 300 K, which would be the first room temperature superconductor! To date, high pressure phases of metallic hydrogen have indeed been produced, but, so far at least, superconductivity has not been found.

Superconductivity appears to be fairly common in nature, and there are perhaps several hundred known superconducting materials. Before 1986 the highest known T_c values were in the A-15 type materials, including Nb_3Ge with $T_c = 23$ K. This, and the closely related compound Nb_3Sn ($T_c = 18$ K) are widely used in the superconducting magnet industry.

In 1986 Bednorz and Müller discovered that the material $La_{2-x}Ba_xCuO_4$ becomes superconducting with a T_c which is maximum at 38 K for $x \approx 0.15$. Within a matter of months the related compound $YBa_2Cu_3O_7$ was discovered to have $T_c = 92$ K, ushering in the era of "high temperature superconductivity."[3] This breakthrough was especially important in terms of possible commercial applications of superconductivity, since these superconductors are the first which can operate in liquid nitrogen (boiling point 77 K) rather than requiring liquid helium (4 K). Other high temperature superconductors have been discovered in chemically related systems. Currently $HgBa_2Ca_2Cu_3O_{8+\delta}$ has the highest confirmed value of T_c at 135 K at room pressure, shown in Fig. 3.2, rising to 165 K when the material is subjected to high pressures. The reason why these particular materials are so unique is still not completely understood, as we shall see in later chapters of this book.

[3]Bednorz and Müller received the 1987 Nobel prize for physics, within a year of publication of their results. At the first major condensed matter physics conference after these discoveries, the 1987 American Physical Society March Meeting held in New York city, there was a special evening session devoted to the discoveries. The meeting hall was packed with hundreds of delegates sitting in the gangways, others had to watch the proceedings on TV screens in the hallways. The number of speakers was so great that the session lasted all through the night until the following morning, when the hall was needed for next official session of the conference! The following day's New York Times newspaper headline reported the meeting as the "Woodstock of Physics."

As well as high temperature superconductors, there are also many other interesting superconducting materials. Some of these have exotic properties which are still not understood and are under very active investigation. These include other oxide-based superconducting materials, organic superconductors, C_{60} based fullerene superconductors, and "heavy fermion" superconductors (typically compounds containing the elements U or Ce) which are dominated by strong electron–electron interaction effects. Other superconductors have surprising properties, such as coexitence of magnetism and superconductivity, or evidence of exotic "unconventional" superconducting phases. We shall discuss some of these strange materials in Chapter 7.

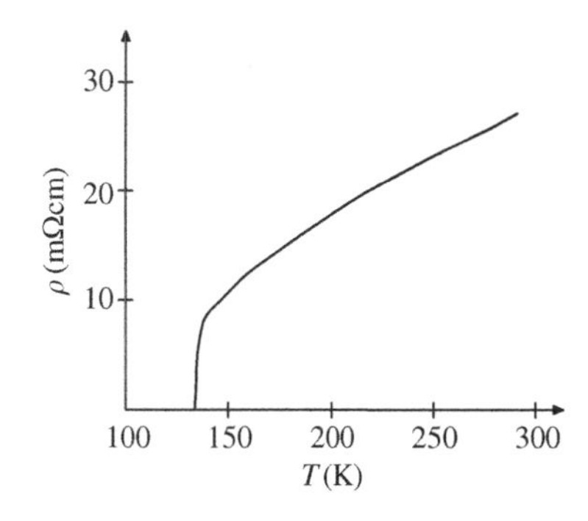

Fig. 3.2 Resistivity of $HgBa_2Ca_2Cu_3O_{8+\delta}$ as a function of temperature (adapted from data of Chu 1993). Zero resistance is obtained at about 135 K, the highest known T_c in any material at normal pressure. In this material T_c approaches a maximum of about 165 K under high pressure. Note the rounding of the resistivity curve just above T_c, which is due to superconducting fluctuation effects. Also, well above T_c the resistivity does not follow the expected Fermi liquid behavior.

3.4 Zero-resistivity

As we have seen, in superconductors the resistivity, ρ, becomes zero, and so the conductivity σ appears to become infinite below T_c. To be consistent with the constitutive relation, Eq. 3.6, we must always have zero electric field,

$$\boldsymbol{\mathcal{E}} = 0,$$

at all points inside a superconductor. In this way the current, $\mathbf{j}$, can be finite. So we have current flow without electric field.

Notice that the change from finite to zero resistivity at the superconducting critical temperature T_c is very sudden, as shown in Fig. 3.1. This represents a thermodynamic phase transition from one state to another. As for other phase transitions, such as from liquid to gas, the properties of the phases on either side of the transition can be completely different. The change from one to the other occurs sharply at a fixed temperature rather than being a smooth cross-over from one type of behavior to another. Here the two different phases are referred to as the "normal state" and the "superconducting state." In the normal state

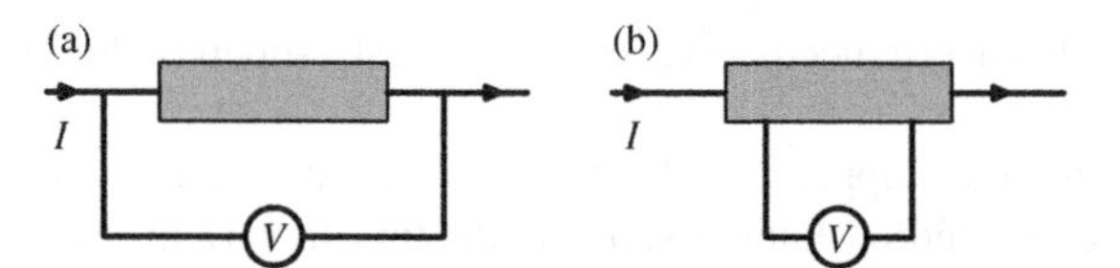

Fig. 3.3 Measurement of resistivity by (a) the two terminal method, (b) the four terminal method. The second method, (b), is much more accurate since no current flows through the leads measuring the voltage drop across the resistor, and so the resistances of the leads and contacts are irrelevant.

the resistivity and other properties behaves similarly to a normal metal, while in the superconducting state many physical properties, including resistivity, are quite different.

In some cases, notably the high temperature superconductors, looking closely at the $\rho(T)$ curve near to T_c shows a small range of temperatures where the resistance starts to decrease before becoming truly zero. This is visible in Fig. 3.2 as a slight bend just above T_c. This bend is due to thermodynamic critical fluctuations associated with the phase transition. The precise thermodynamic phase transition temperature T_c can be defined as the temperature where the resistivity first becomes exactly zero.[4]

[4]One could perhaps imagine the existence of materials where the resistivity approached zero smoothly without a thermodynamic phase transition. For example in a completely pure metal with no impurities one might expect the $\rho \to 0$ as temperature approaches absolute zero. Such a system would not be classified as a superconductor in the standard terminology, even though it might have infinite conductivity. The word superconductor is used only to mean a material with a definite phase transition and critical temperature T_c. A true superconductor must also exhibit the Meissner–Ochsenfeld effect.

The key characteristic of the superconducting state is that the resistivity is **exactly** zero,

$$\rho = 0, \tag{3.12}$$

or the conductivity, σ, is infinite. But how do we know that the resistivity is exactly zero? After all, zero is rather difficult to distinguish from some very very small, but finite, number.

Consider how one might actually measure the resistivity of a superconductor. The simplest measurement would be a basic "two terminal" geometry shown in Fig. 3.3(a). The sample resistance, R, is related to the resisitvity

$$R = \rho \frac{L}{A} \tag{3.13}$$

and L is the sample length and A is its cross sectional area. But the problem with the two-terminal geometry shown in Fig. 3.3(a) is that even if the sample resistance is zero the overall resistance is finite, because the sample resistance is in series with resistances from the connecting leads and from the electrical contacts between the sample and the leads. A much better experimental technique is the four terminal measurement of Fig. 3.3(b). There are four leads connected to the sample. Two of them are used to provide a current, I, through the sample. The second pair of leads are then used to measure a voltage, V. Since no current flows in the second pair of leads the contact resistances will not matter. The resistance of the part of the sample between the second pair of contacts will be $R = V/I$ by Ohm's law, at least in the idealized geometry shown. In any case if the sample is superconducting we should definitely observe $V = 0$ when I is finite implying that $\rho = 0$. (Of course the current I must not be too large. All superconductors have a critical current, I_c, above which the superconductivity is destroyed and the resistance becomes finite again).

The most convincing evidence that superconductors really have $\rho = 0$ is the observation of **persistent currents**. If we have a closed loop of superconducting wire, such as the ring shown in Fig. 3.4 then it is possible to set up a current, I, circulating in the loop. Because there is no dissipation of energy

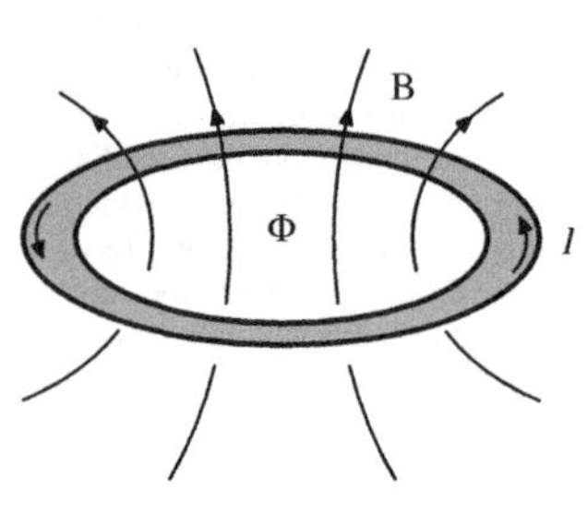

Fig. 3.4 Persistent current around a superconducting ring. The current maintains a constant magnetic flux, Φ, through the superconducting ring.

due to finite resistance, the energy stored in the magnetic field of the ring will remain constant and the current never decays.

To see how this persistent current can be set up, consider the flux of magnetic field through the center of the superconducting ring. The flux is defined by the surface integral

$$\Phi = \int \mathbf{B} \cdot \mathbf{dS} \tag{3.14}$$

where $\mathbf{dS}$ is a vector perpendicular to the plane of the ring. Its length dS, is an infinitesimal element of the area enclosed by the ring. But, by using the Maxwell equation

$$\nabla \times \boldsymbol{\mathcal{E}} = -\frac{\partial \mathbf{B}}{\partial t} \tag{3.15}$$

and Stokes's theorem

$$\int (\nabla \times \boldsymbol{\mathcal{E}}) dS = \oint \boldsymbol{\mathcal{E}} \cdot \mathbf{dr} \tag{3.16}$$

we can see that

$$-\frac{d\Phi}{dt} = \oint \boldsymbol{\mathcal{E}} \cdot \mathbf{dr} \tag{3.17}$$

where the line integral is taken around the closed path around the inside of the ring. This path can be taken to be just inside the superconductor, and so $\boldsymbol{\mathcal{E}} = 0$ everywhere along the path. Therefore

$$\frac{d\Phi}{dt} = 0 \tag{3.18}$$

and hence the magnetic flux through the ring stays constant as a function of time.

We can use this property to set up a persistent current in a superconducting ring. In fact it is quite analogous to the way we saw in Chapter 2 how to set up a persistent superfluid flow in ^{4}He. The difference is that now we use a magnetic field rather than rotation of the ring. First we start with the superconductor at a temperature above T_c, so that it is in its normal state. Then apply an external magnetic field, $\mathbf{B}_{\text{ext}}$. This passes easily through the superconductor since the system is normal. Now cool the system to below T_c. The flux in the ring is given by $\Phi = \int \mathbf{B}_{\text{ext}} \cdot \mathbf{dS}$. But we know from Eq. 3.18 that this remains constant, no matter what. It is constant even if we turn off the source of external magnetic field, so that now $\mathbf{B}_{\text{ext}} = 0$. The only way the superconductor can keep Φ constant is to generate its own magnetic field $\mathbf{B}$ through the center of the ring, which it must achieve by having a circulating current, I, around the ring. The value of I will be exactly the one required to induce a magnetic flux equal to Φ inside the ring. Further, because Φ is constant the current I must also be constant. We therefore have a set up circulating persistent current in our superconducting ring.

Furthermore if there were any electrical resistance at all in the ring there would be energy dissipation and hence the current I would decay gradually over time. But experiments have been done in which persistent currents were observed to remain constant over a period of years. Therefore the resistance must really be exactly equal to zero to all intents and purposes!

3.5 The Meissner–Ochsenfeld effect

Nowadays, the fact that the resistivity is zero, $\rho = 0$, is not taken as the true definition of superconductivity. The fundamental proof that superconductivity occurs in a given material is the demonstration of the Meissner–Ochsenfeld effect.

This effect is the fact that a superconductor **expels** a weak external magnetic field. First, consider the situation illustrated in Fig. 3.5 in which a small spherical sample of material is held at temperature T and placed in a small external magnetic field, $\mathbf{B}_{\text{ext}}$. Suppose initially we have the sample in its normal state, $T > T_c$, and the external field is zero, as illustrated in the top part of the diagram in Fig. 3.5. Imagine that we first cool to a temperature below T_c (left diagram) while keeping the field zero. Then later as we gradually turn on the external field the field inside the sample must remain zero (bottom diagram). This is because, by the Maxwell equation Eq. 3.15 combined with $\boldsymbol{\mathcal{E}} = 0$ we must have

$$\frac{\partial \mathbf{B}}{\partial t} = 0 \tag{3.19}$$

at all points inside the superconductor. Thus by applying the external field to the sample after it is already superconducting we must arrive at the state shown in the bottom diagram in Fig. 3.5 where the magnetic field $\mathbf{B} = 0$ is zero everywhere inside the sample.

But now consider doing things in the other order. Suppose we take the sample above T_c and first turn on the external field, $\mathbf{B}_{\text{ext}}$. In this case the magnetic field will easily penetrate into the sample, $\mathbf{B} = \mathbf{B}_{\text{ext}}$, as shown in the right hand picture in Fig. 3.5. What happens then we now cool the sample? The Meissner–Ochsenfeld effect is the observation that upon cooling the system to below T_c the magnetic field is **expelled**. So that by cooling we move from the situation depicted on right to the one shown at the bottom of Fig. 3.5. This fact cannot be deduced from the simple fact of zero resistivity ($\rho = 0$) and so this is a new and separate physical phenomenon assosciated with superconductors.

There are several reasons why the existence of the Meissner–Ochsenfeld affecting a sample is taken as definitive proof of superconductivity. At a practical level it is perhaps clearer to experimentally demonstrate the flux expulsion than zero resistivity, because, for example, it is not necessary to attach any

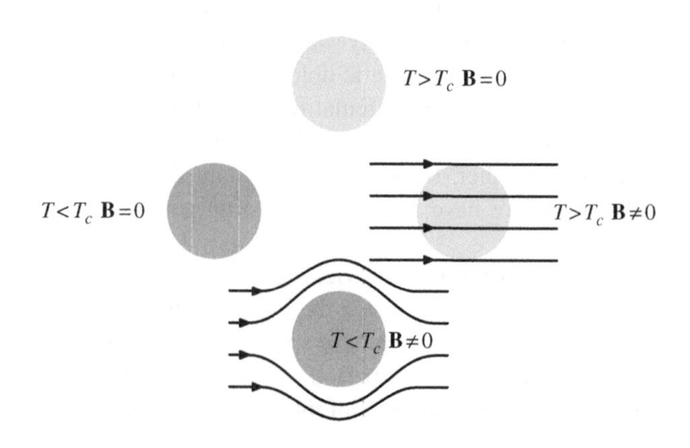

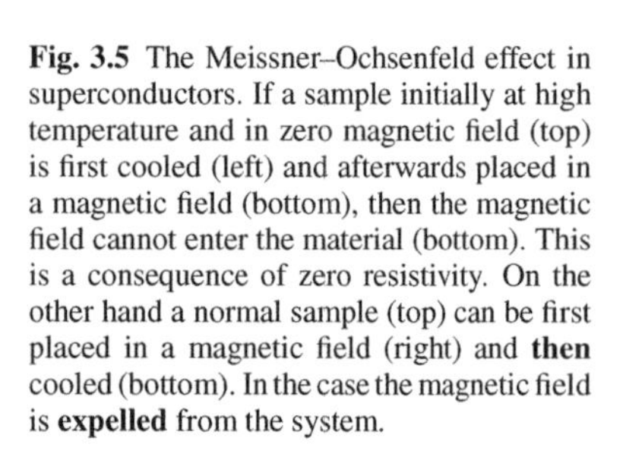
Fig. 3.5 The Meissner–Ochsenfeld effect in superconductors. If a sample initially at high temperature and in zero magnetic field (top) is first cooled (left) and afterwards placed in a magnetic field (bottom), then the magnetic field cannot enter the material (bottom). This is a consequence of zero resistivity. On the other hand a normal sample (top) can be first placed in a magnetic field (right) and **then** cooled (bottom). In the case the magnetic field is **expelled** from the system.

electrical leads to the sample. A more fundamental reason is that the Meissner–Ochsenfeld effect is a property of **thermal equilibrium**, while resistivity is a nonequilibrium transport effect. In fact one can see in Fig. 3.5 that we reach the same final state of the system (bottom picture in Fig. 3.5) whether we first cool and then apply the field, or the other way around. Therefore the final state of the system does not depend on the history of the sample, which is a necessary condition for thermal equilibrium. It is perhaps possible to imagine exotic systems for which the resistivity vanishes, but for which there is no Meissner–Ochsenfeld effect. In fact some quantum Hall effect states possess this property. But, for the purposes of this book, however, we shall always define a **superconductor** as a system which exhibits the Meissner–Ochsenfeld effect.

3.6 Perfect diamagnetism

In order to maintain $\mathbf{B} = 0$ inside the sample whatever (small) external fields are imposed as required by the Meissner–Ochsenfeld effect there obviously must be screening currents flowing around the edges of the sample. These produce a magnetic field which is equal and opposite to the applied external field, leaving zero field in total.

The simplest way to describe these screening currents is to use Maxwell's equations in a magnetic medium (see Blundell (2001) or other texts on magnetic materials). The total current is separated into the externally applied currents (e.g. in the coils producing the external field), $\mathbf{j}_{\mathrm{ext}}$, and the internal screening currents, $\mathbf{j}_{\mathrm{int}}$,

$$\mathbf{j} = \mathbf{j}_{\mathrm{ext}} + \mathbf{j}_{\mathrm{int}}. \tag{3.20}$$

The screening currents produce a magnetization in the sample, $\mathbf{M}$ per unit volume, defined by

$$\nabla \times \mathbf{M} = \mathbf{j}_{\mathrm{int}}. \tag{3.21}$$

As in the theory of magnetic media (Blundell 2001) we also define a magnetic field $\mathbf{H}$ in terms of the external currents only

$$\nabla \times \mathbf{H} = \mathbf{j}_{\mathrm{ext}}. \tag{3.22}$$

The three vectors $\mathbf{M}$ and $\mathbf{H}$ and $\mathbf{B}$ are related by[5]

$$\mathbf{B} = \mu_0(\mathbf{H} + \mathbf{M}). \tag{3.23}$$

Maxwell's equations also tell us that

$$\nabla \cdot \mathbf{B} = 0. \tag{3.24}$$

[5]Properly the name "magnetic field" is applied to $\mathbf{H}$ in a magnetic medium. Then the field $\mathbf{B}$ is called the magnetic induction or the magnetic flux density. Many people find this terminology confusing. Following Blundell (2001), in this book we shall simply call them the "H-field" and "B-field," respectively, whenever there is a need to distinguish between them.

The magnetic medium Maxwell's equations above are supplemented by boundary conditions at the sample surface. From Eq. 3.24 it follows that the component of $\mathbf{B}$ perpendicular to the surface must remain constant; while from the condition Eq. 3.22 one can prove that components of $\mathbf{H}$ parallel to the surface remain constant. The two boundary conditions are therefore,

$$\Delta \mathbf{B}_{\perp} = 0, \tag{3.25}$$

$$\Delta \mathbf{H}_{\parallel} = 0. \tag{3.26}$$

Note that we are using SI units here. In SI units **B** is in Tesla, while **M** and **H** are in units of Amperes per metre, Am^{-1}. The magnetic constant is $\mu_0 = 4\pi \times 10^{-7}$. One should take note that many books and research papers on superconductivity still use the older cgs units. In cgs units **B** and **H** are in gauss and oersteds, respectively. 1 gauss $= 10^{-4}$ T, 1 oersted $= 10^3/4\pi$ Am^{-1} and in cgs units

$$\mathbf{B} = \mathbf{H} + 4\pi\mathbf{M}$$

and

$$\nabla \times \mathbf{H} = 4\pi\mathbf{j}.$$

In these units the susceptibility of a superconductor is $\chi = -1/(4\pi)$ rather than the SI value of -1.

Note that there is no μ_0 or ϵ_0 in the cgs system of units. Instead, the speed of light, $c = 1/\sqrt{\epsilon_0\mu_0}$, often appears explicitly. For example, the Lorentz force on a charge q particle, moving with velocity **v** in a magnetic field **B** is

$$\mathbf{F} = \frac{1}{c} q\mathbf{v} \times \mathbf{B}$$

in cgs units, compared with the SI unit equivalent

$$\mathbf{F} = q\mathbf{v} \times \mathbf{B}.$$

Also note that the unit of electrical charge is the Coulomb (C) in SI units, but it is the statcoulomb in cgs units, where 1 statcoulomb $= 3.336 \times 10^{-10}$ C.

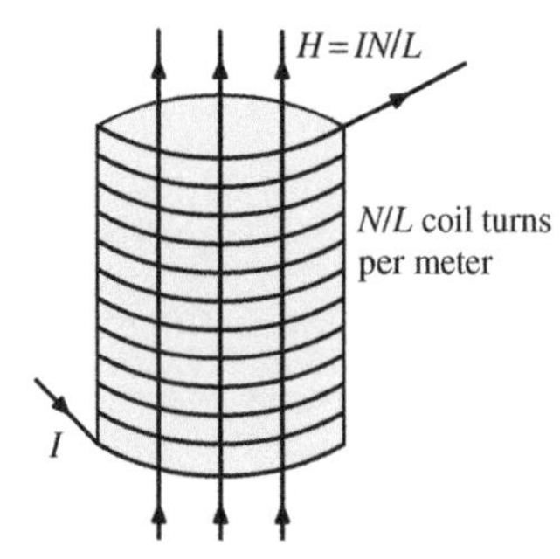

Fig. 3.6 Measurement of **M** as a function of **H** for a sample with solenoidal geometry. A long solenoid coil of N/L turns per metre leads to a uniform field $H = IN/L$ Amperes per metre inside the solenoid. The sample has magnetization, M, inside the solenoid, and the magnetic flux density is $B = \mu_0(H + M)$. Increasing the current in the coils from I to $I + dI$, by dI leads to an inductive e.m.f. $\mathcal{E} = -d\Phi/dt$ where $\Phi = NBA$ is the total magnetic flux threading the N current turns of area A. This inductive e.m.f. can be measured directly, since it is simply related to the differential self-inductance of the coil, $\mathcal{L}$, via $\mathcal{E} = -\mathcal{L}dI/dt$. Therefore, by measuring the self-inductance $\mathcal{L}$ one can deduce the B-field and hence M as a function of I or H.

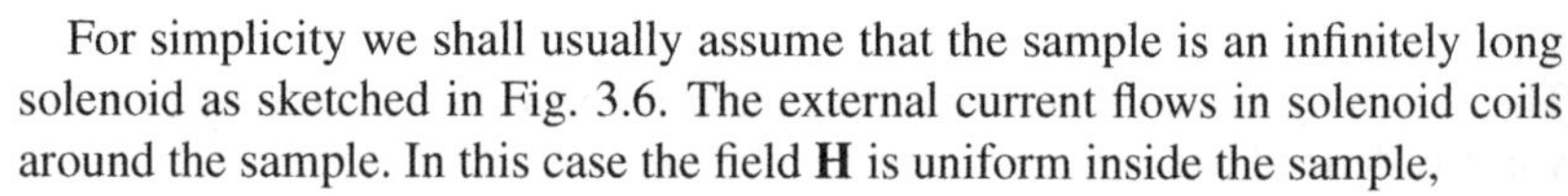

For simplicity we shall usually assume that the sample is an infinitely long solenoid as sketched in Fig. 3.6. The external current flows in solenoid coils around the sample. In this case the field **H** is uniform inside the sample,

$$\mathbf{H} = I\frac{N}{L}\mathbf{e}_z \tag{3.27}$$

where I is the current flowing through the solenoid coil and there are N coil turns in length L. $\mathbf{e}_z$ is a unit vector along the solenoid axis.

Imposing the Meissner condition $\mathbf{B} = 0$ in Eq. 3.23 immediately leads to the magnetization

$$\mathbf{M} = -\mathbf{H}. \tag{3.28}$$

The magnetic susceptibility is defined by

$$\chi = \left.\frac{dM}{dH}\right|_{H=0} \tag{3.29}$$

and so we find that for superconductors

$$\chi = -1 \tag{3.30}$$

(or $-1/4\pi$ in cgs units!).

Solids with a negative value of χ are called diamagnets (in contrast positive χ is a paramagnet). Diamagnets screen out part of the external magnetic field, and so they become magnetized oppositely to the external field. In superconductors the external field is completely screened out. Therefore we can say that superconductors are **perfect diamagnets**.

The best way to detect superconductivity in some unknown sample is therefore to measure its susceptibility. If the sample is fully superconducting then χ as a function of T will be something like the sketch given in Fig. 3.7. Thus by measuring χ one will find $\chi = -1$ in a superconductor, evidence for perfect diamagnetism, or the Meissner effect. This is usually considered much more reliable evidence for superconductivity in a sample than zero resistance alone would be.[6]

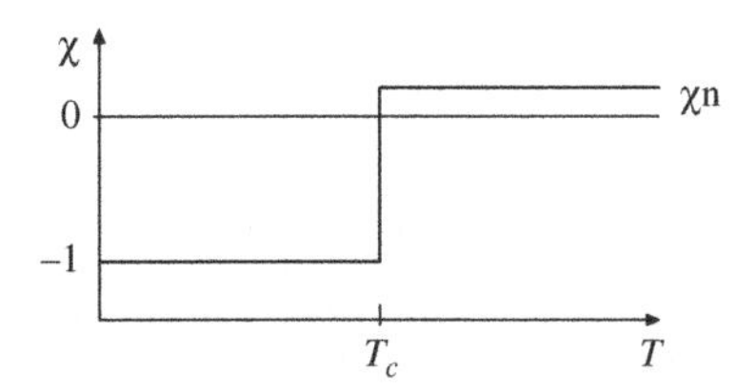

Fig. 3.7 Magnetic susceptibility, χ, of a superconductor as a function of temperature. Above T_c it is a constant normal state value, χ_n, which is usually small and positive (paramagetic). Below T_c the susceptibility is large and negative, $\chi = -1$, signifying perfect diamagnetism.

[6]Note that for non-cylindrical sample shapes one must take into account the appropriate demagnetizing factors, Blundel (2001), in measuring susceptibility.

3.7 Type I and type II superconductivity

This susceptibility χ is defined in the limit of very weak external fields, **H**. As the field becomes stronger it turns out that either one of two possible things can happen.

The first case, called a **type I superconductor**, is that the **B** field remains zero inside the superconductor until suddenly the superconductivity is destroyed. The field where this happens is called the **critical field**, H_c. The way the magnetization M changes with H in a type I superconductor is shown in Fig. 3.8. As shown, the magnetization obeys $M = -H$ for all fields less than H_c, and then becomes zero (or very close to zero) for fields above H_c.

Many superconductors, however, behave differently. In a **type II supercon ductor** there are two different critical fields, denoted H_{c1}, the **lower critical field**, and H_{c2} the **upper critical field**. For small values of applied field H the Meissner–Ochsenfeld effect again leads to $M = -H$ and there is no magnetic flux density inside the sample, $B = 0$. However, in a type II superconductor once the field exceeds H_{c1}, magnetic flux does start to enter the superconductor and hence $B \neq 0$, and M is closer to zero than the full Meissner–Ochsenfeld value of $-H$. Upon increasing the field H further the magnetic flux density gradually increases, until finally at H_{c2} the superconductivity is destroyed and $M = 0$. This behavior is sketched on the right-hand-side of Fig. 3.8.

As a function of the temperature the critical fields vary, and they all approach zero at the critical temperature T_c. The typical phase diagrams of type I and type II superconductors as a function of H and T are shown in Fig. 3.9.

The physical explanation of the thermodynamic phase between H_{c1} and H_{c2} was given by Abrikosov. He showed that the magnetic field can enter the superconductor in the form of **vortices**, as shown in Fig. 3.10. Each vortex consists of a region of circulating supercurrent around a small central core which has essentially become normal metal. The magnetic field is able to pass through the

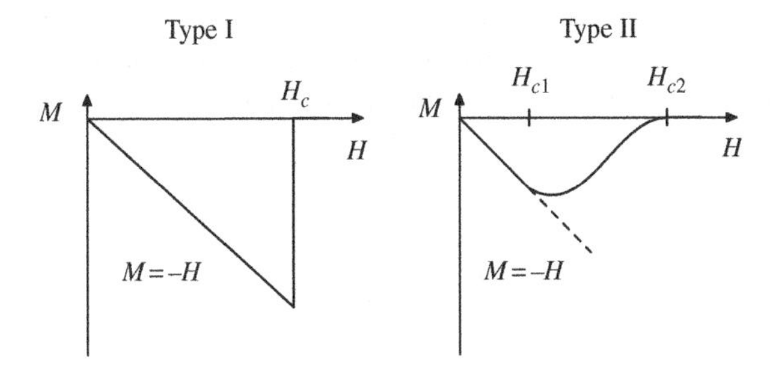

Fig. 3.8 The magnetization M as a function of H in type I and type II superconductors. For type I perfect Meissner diamagnetism is continued until H_c, beyond which superconductivity is destroyed. For type II materials perfect diamagnetism occurs only below H_{c1}. Between H_{c1} and H_{c2} Abrikosov vortices enter the material, which is still superconducting.

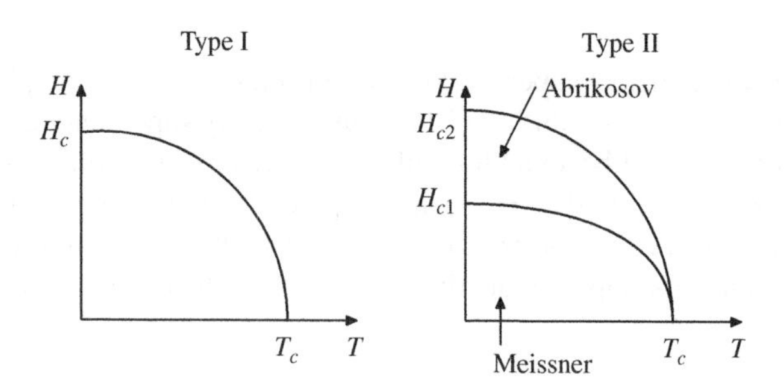

Fig. 3.9 The $H-T$ phase diagram of type I and type II superconductors. In type II superconductors the phase below H_{c1} is normally denoted the Meissner state, while the phase between H_{c1} and H_{c2} is the Abrikosov or mixed state.

sample inside the vortex cores, and the circulating currents serve to screen out the magnetic field from the rest of the superconductor outside the vortex.

It turns out that each vortex carries a fixed unit of magnetic flux, $\Phi_0 = h/2e$ (see below), and hence, if there are a total of N_v vortices in a sample of total area, A, then the average magnetic flux density, B, is

$$B = \frac{N_v}{A}\frac{h}{2e}. \tag{3.31}$$

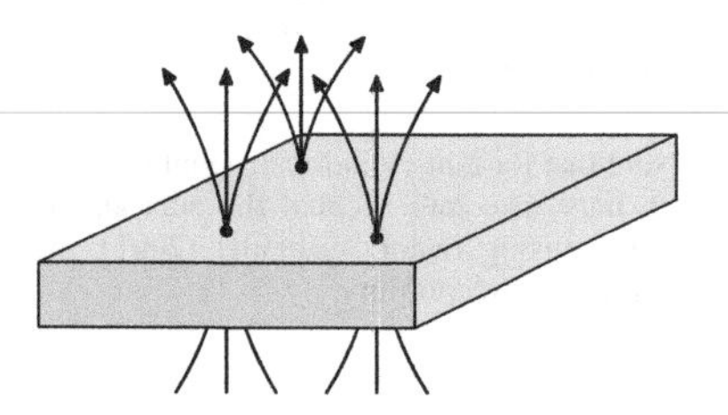

Fig. 3.10 Vortices in a type II superconductor. The magnetic field can pass through the superconductor, provided it is channelled through a small "vortex core." The vortex core is normal metal. This allows the bulk of the material to remain superconducting, while also allowing a finite average magnetic flux density B to pass through.

It is instructive to compare this result for the number of vortices per unit area,

$$\frac{N_v}{A} = \frac{2eB}{h} \tag{3.32}$$

with the similar expression derived earlier for the density of vortices in rotating superfluid ^{4}He, Eq. 2.48. There is in fact a direct mathematical analogy between the effect of a uniform rotation at angular frequency ω in a neutral superfluid, and the effect of a magnetic field, B, in a superconductor.

3.8 The London equation

The first theory which could account for the existence of the Meissner–Ochsenfeld effect was developed by two brothers, F. London and H. London, in 1935. Their theory was originally motivated by the two-fluid model of superfluid ^{4}He. They assumed that some fraction of the conduction electrons in the solid become superfluid while the rest remain normal. They then assumed that the superconducting electrons could move without dissipation, while the normal electrons would continue to act as if they had a finite resistivity. Of course the superfluid electrons always "short circuit" the normal ones and make the overall resistivity equal to zero. As in the theory of superfluid ^{4}He discussed in Chapter 2, we denote the number density of superfluid electrons by n_s and the density of normal electrons by $n_n = n - n_s$, where n is the total density of electrons per unit volume.

Although this model is simple several of its main predictions are indeed correct. Most importantly it leads to the **London equation** which relates the electrical current density inside a superconductor, $\mathbf{j}$, to the magnetic vector potential, $\mathbf{A}$, by

$$\mathbf{j} = -\frac{n_s e^2}{m_e}\mathbf{A}. \tag{3.33}$$

This is one of the most important equations describing superconductors. Nearly 20 years after it was originally introduced by the London brothers it was eventually derived from the full microscopic quantum theory of superconductivity by Bardeen Cooper and Schrieffer.

Let us start to make the London equation Eq. 3.33 plausible by reexamining the Drude model of conductivity. This time consider the Drude theory for finite frequency electric fields. Using the complex number representation of the a.c. currents and fields, the d.c. formula becomes modified to:

$$\mathbf{j}e^{-i\omega t} = \sigma(\omega)\boldsymbol{\mathcal{E}}e^{-i\omega t} \tag{3.34}$$

where the conductivity is also complex. Its real part corresponds to currents which are in phase with the applied electrical field (resistive), while the imaginary part corresponds to out of phase currents (inductive and capacitive).

Generalizing the Drude theory to the case of finite frequency, the conductivity turns out to be

$$\sigma(\omega) = \frac{ne^2\tau}{m}\frac{1}{1 - i\omega\tau}, \tag{3.35}$$

Ashcroft and Mermin (1976). This is essentially like the response of a damped harmonic oscillator with a resonant frequency at $\omega = 0$. Taking the real part we get

$$\mathrm{Re}[\sigma(\omega)] = \frac{ne^2}{m}\frac{\tau}{1 + \omega^2\tau^2}, \tag{3.36}$$

a Lorentzian function of frequency. Note that the width of the Lorentzian is $1/\tau$ and its maximum height is τ. Integrating over frequency, we see that the area under this Lorentzian curve is a constant

$$\int_{-\infty}^{+\infty} \mathrm{Re}[\sigma(\omega)]d\omega = \frac{\pi ne^2}{m} \tag{3.37}$$

independent of the lifetime τ.

Now it is interesting to consider what would be the corresponding Drude model $\sigma(\omega)$ in the case of a perfect conductor, where there is no scattering of the electrons. We can we can obtain this by taking the limit $\tau^{-1} \to 0$ in the Drude model. Taking this limit Eq. 3.35 gives:

$$\sigma(\omega) = \frac{ne^2}{m}\frac{1}{\tau^{-1} - i\omega} \to -\frac{ne^2}{i\omega m} \tag{3.38}$$

at any finite frequency, ω. There is no dissipation since the current is always out of phase with the applied electric field and $\sigma(\omega)$ is always imaginary. There is a purely inductive response to an applied electric field. The real part of the conductivity $\mathrm{Re}[\sigma(\omega)]$ is therefore zero at any finite frequency, ω in this $\tau^{-1} \to 0$ limit. But the sum rule, Eq. 3.37, must be obeyed whatever the value of τ. Therefore the real part of the conductivity, $\mathrm{Re}[\sigma(\omega)]$ must be a function which is zero almost everywhere but which has a finite integral. This must be, of course, a Dirac delta function,

$$\mathrm{Re}[\sigma(\omega)] = \frac{\pi ne^2}{m}\delta(\omega). \tag{3.39}$$

One can see that this is correct by considering the $\tau^{-1} \to 0$ limit of the Lorentzian peak in $\mathrm{Re}[\sigma(\omega)]$ in Eq. 3.36. The width of the peak is of order τ^{-1} and goes to zero, but the maximum height increases keeping a constant total area because of the sum rule. The τ^{-1} goes to zero limit is thus a Dirac delta function located at $\omega = 0$.

Inspired by the two fluid model of superfluid ^{4}He, the London brothers assumed that we can divide the total electron density, n, into a normal part, n_n and a superfluid part, n_s,

$$n = n_s + n_n. \tag{3.40}$$

They assumed that the "normal" electrons would still have a typical metallic damping time τ, but the superfluid electrons would move without dissipation, corresponding to $\tau = \infty$. They assumed that this superfluid component will give rise to a Dirac delta function peak in the conductivity located at $\omega = 0$ and a purely imaginary response elsewhere,

$$\sigma(\omega) = \frac{\pi n_s e^2}{m_e}\delta(\omega) - \frac{n_s e^2}{i\omega m_e}. \tag{3.41}$$

Note that we effectively **define** n_s by the weight in this delta function peak, and (by convention) we use the bare electron mass in vacuum, m_e, rather than the effective band mass, m, in this definition.

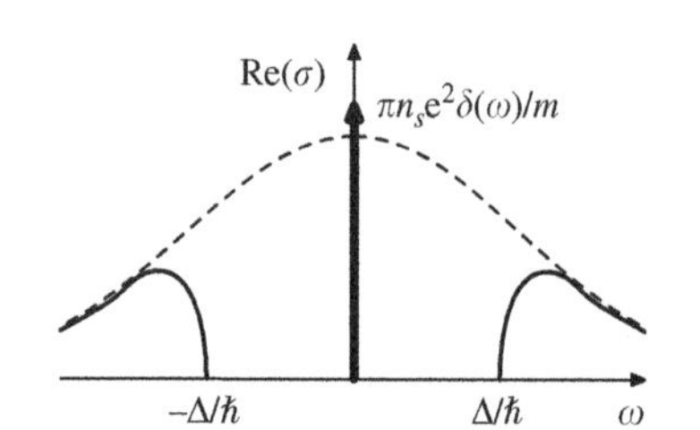

Fig. 3.11 The finite frequency conductivity of a normal metal (dashed line) and a superconductor (solid line). In the superconducting case an energy gap leads to zero conductivity for frequencies below $\Delta/\hbar$. The remaining spectral weight becomes concentrated in a Dirac delta function at $\omega = 0$.

In fact the experimentally measured finite frequency conductivity $\mathrm{Re}\,\sigma(\omega)$ in superconductors does indeed have a delta function located at zero frequency. But other aspects of the two fluid model conductivity assumed by London and London are not correct. In particular the "normal" fluid component is not simply like the conductivity of a normal metal. In fact the complete $\mathrm{Re}[\sigma(\omega)]$ of a superconductor looks something like the sketch in Fig. 3.11. There is a delta function peak located at $\omega = 0$, and the amplitude of the peak defines n_s, the superfluid density or condensate density. At higher frequencies the real part of the conductivity is zero, $\mathrm{Re}[\sigma(\omega)] = 0$, corresponding to dissipationless current flow. However, above a certain frequency, corresponding to $\hbar\omega = 2\Delta$ (where 2Δ is the "energy gap") the conductivity again becomes finite. The presence of an energy gap was observed shortly before the Bardeen Cooper and Schrieffer (BCS) theory was completed, and the energy gap was a central feature of the theory, as we shall see later.

Derivation of the London equation

If we restrict our attention to frequencies below the energy gap, then the conductivity is exactly given by Eq. 3.41. In this regime we can derive the London equation relating the supercurrent $\mathbf{j}$ to the magnetic field $\mathbf{B}$.

Taking the curl of both sides of the equation $\mathbf{j} = \sigma\boldsymbol{\mathcal{E}}$ we find

$$\begin{aligned}(\nabla \times \mathbf{j})e^{-i\omega t} &= \sigma(\omega)(\nabla \times \boldsymbol{\mathcal{E}})e^{-i\omega t}, \\ &= -\sigma(\omega)\frac{d(\mathbf{B}e^{-i\omega t})}{dt}, \\ &= i\omega\sigma(\omega)\mathbf{B}e^{-i\omega t}, \\ &= -\frac{n_s e^2}{m_e}\mathbf{B}e^{-i\omega t}, \end{aligned} \tag{3.42}$$

where in the final step we used Eq. 3.38 for the finite frequency conductivity of the superconductor.

We now take the $\omega = 0$ limit of the above equations. The last line in Eq. 3.42 effectively relates a d.c. current, $\mathbf{j}$ to a static external magnetic field $\mathbf{B}$ by,

$$\nabla \times \mathbf{j} = -\frac{n_s e^2}{m_e}\mathbf{B}. \tag{3.43}$$

This equation completely determines $\mathbf{j}$ and $\mathbf{B}$ because they are also related by the static Maxwell equation:

$$\nabla \times \mathbf{B} = \mu_0 \mathbf{j}. \tag{3.44}$$

Combining these two equations gives

$$\nabla \times (\nabla \times \mathbf{B}) = -\mu_0 \frac{n_s e^2}{m_e}\mathbf{B} \tag{3.45}$$

or

$$\nabla \times (\nabla \times \mathbf{B}) = -\frac{1}{\lambda^2}\mathbf{B}, \tag{3.46}$$

where λ has dimensions of length, and is the **penetration depth** of the superconductor,

$$\lambda = \left(\frac{m_e}{\mu_0 n_s e^2}\right)^{1/2}. \tag{3.47}$$

It is the distance inside the surface over which an external magnetic field is screened out to zero, given that $B = 0$ in the bulk.

Finally, the London equation can also be rewritten in terms of the magnetic vector potential $\mathbf{A}$ defined by

$$\mathbf{B} = \nabla \times \mathbf{A}, \tag{3.48}$$

giving

$$\mathbf{j} = -\frac{n_s e^2}{m_e}\mathbf{A}, \tag{3.49}$$

$$= -\frac{1}{\mu_o \lambda^2}\mathbf{A}. \tag{3.50}$$

Note that this only works provided that we choose the correct **gauge** for the vector potential, $\mathbf{A}$. Recall that $\mathbf{A}$ is not uniquely defined from Eq. 3.48 since $\mathbf{A} + \nabla\chi(\mathbf{r})$ leads to exactly the same $\mathbf{B}$ for any scalar function, $\chi(\mathbf{r})$. But conservation of charge implies that the current and charge density, ρ, obey the continuity equation

$$\frac{\partial \rho}{\partial t} + \nabla \cdot \mathbf{j} = 0. \tag{3.51}$$

In a static, d.c., situation the first term is zero, and so $\nabla \cdot \mathbf{j} = 0$. Comparing with the London equation in the form, Eq. 3.49 we see that this is satisfied provided that the gauge is chosen so that $\nabla \cdot \mathbf{A} = 0$. This is called the **London gauge**.

For superconductors this form of the London equation effectively replaces the normal metal $\mathbf{j} = \sigma\boldsymbol{\mathcal{E}}$ constitutive relation by something which is useful when σ is infinite. It is interesting to speculate about whether or not it would be possible to find other states of matter which are perfect conductors with $\sigma = \infty$, but which do not obey the London equation. If such exotic states exist (and they may indeed occur in the Quantum Hall Effect) they would not be superconductors in the sense in which we are using that word here.

The most important consequence of the London equation is to explain the Meissner–Ochsenfeld effect. In fact one can easily show that any external magnetic field is screened out inside the superconductor, as

$$B = B_0 e^{-x/\lambda}, \tag{3.52}$$

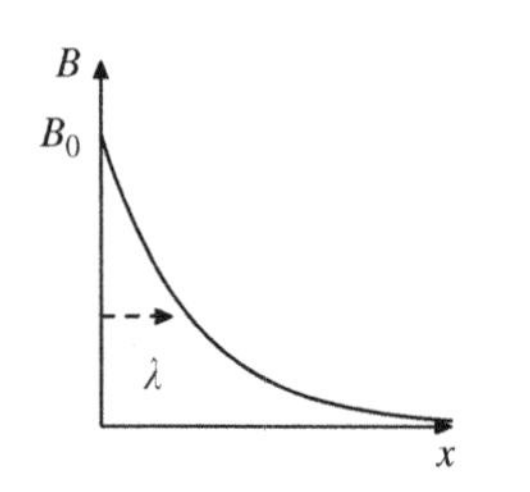

Fig. 3.12 The magnetic field near a surface of a superconductor in the Meissner state. The field decays exponentially on a length scale given by the penetration depth λ.

where x is the depth inside the surface of the superconductor. This is illustrated in Fig. 3.12. The derivation of this expression from the London equation is very straightforward, and is left to Exercise 3.1 at the end of this chapter. The implication of this result is that magnetic fields only penetrate a small distance, λ, inside the surface of a superconductor, and thus the field is equal to zero far inside the bulk of a large sample.

A modified form of the London equation was later proposed by Pippard. This form generalizes the London equation by relating the current at a point $\mathbf{r}$ in the solid, $\mathbf{j}(\mathbf{r})$, to the vector potential at nearby points $\mathbf{r}'$. The expression he proposed was

$$\mathbf{j}(\mathbf{r}) = -\frac{n_s e^2}{m_e}\frac{3}{4\pi\xi_0}\int \frac{\mathbf{R}(\mathbf{R}.\mathbf{A}(\mathbf{r}'))}{R^4} e^{-R/r_0} d^3r', \tag{3.53}$$

where $\mathbf{R} = \mathbf{r} - \mathbf{r}'$. The points which contribute to the integral are separated by distances of order r_0 or less, with r_0 defined by

$$\frac{1}{r_0} = \frac{1}{\xi_0} + \frac{1}{l}. \tag{3.54}$$

Here l is the **mean free path** of the electrons at the Fermi surface of the metal,

$$l = v_F \tau, \tag{3.55}$$

with τ the scattering time from the Drude conductivity formula, and v_F the electron band velocity at the Fermi surface. The length ξ_0 is called the **coherence length**. After the BCS theory of superconductivity was completed, it became clear that this length is closely related to the value of the energy gap, Δ, by

$$\xi_0 = \frac{\hbar v_F}{\pi \Delta}. \tag{3.56}$$

It also has the physical interpretation that it represents the physical size of the Cooper pair bound state in the BCS theory.

The existence of the Pippard coherence length implies that a superconductor is characterized by no fewer than **three different length scales**. We have the penetration depth, λ, the coherence length, ξ_0, and the mean free path, l. We shall see in the next chapter than the dimensionless ratio $\kappa = \lambda/\xi_0$ determines whether a superconductor is type I or type II. Similarly, if the mean free path is much longer than the coherence length , $l \gg \xi_0$ the superconductor is said to be in the **clean limit**, while if $l < \xi_0$ the superconductor is said to be in the **dirty limit**. It is a surprising and very important property of most superconductors that they can remain superconducting even when there are large numbers of impurities making the mean free path l very short. In fact even many alloys are superconducting despite the strongly disordered atomic structure.

3.9 The London vortex

We can use the London equation to find a simple mathematical description of a superconducting vortex, as in Fig. 3.10. The vortex will have a cylindrical core of normal material, with a radius of approximately the coherence length, ξ_0. Inside this core we will have a finite magnetic field, say B_0. Outside the vortex core we can use the London equation in the form of Eq. 3.46 to write a differential equation for the magnetic field, $\mathbf{B} = (0, 0, B_z)$. Using cylindrical polar coordinates (r, θ, z), and the expression for curl in cylndrical polars, Eq. 2.43, we obtain (Exercise 3.3)

$$\frac{d^2B_z}{dr^2} + \frac{1}{r}\frac{dB_z}{dr} - \frac{B_z}{\lambda^2} = 0. \tag{3.57}$$

This is a form of Bessel's equation (Boas 1983; Matthews and Walker 1970). The solutions to equations of this type are called modified, or hyperbolic Bessel functions, $K_\nu(z)$ and they can be found in many standard texts of mathematical physics. In this particular case the solution is $K_0(z)$. The resulting magnetic field can be written in the form,

$$B_z(r) = \frac{\Phi_0}{2\pi\lambda^2}K_0\left(\frac{r}{\lambda}\right), \tag{3.58}$$

where Φ_0 is the total magnetic flux enclosed by the vortex core,

$$\Phi_0 = \int B_z(r)d^2r. \tag{3.59}$$

We shall see in the next chapter that the magnetic flux is quantized, resulting in the universal value $\Phi_0 = h/2e$ of flux per vortex line.

For small values of z the function $K_0(z)$ becomes

$$K_0(z) \sim -\ln z$$

(Abramowitz and Stegun 1965) and so

$$B_z(r) = \frac{\Phi_0}{2\pi\lambda^2}\ln\left(\frac{\lambda}{r}\right), \tag{3.60}$$

when $r \ll \lambda$. Using $\mu_0\mathbf{j} = \nabla\times\mathbf{B}$ we find (problem 3.3) that the corresponding circulating current is irrotational,

$$\mathbf{j} \sim \frac{1}{r}\mathbf{e}_\phi \tag{3.61}$$

exactly as we found earlier for vortices in superfluid helium.

The divergence at $r = 0$ in these expressions is not physical, and is cut off by the finite coherence length of the superconductor, ξ_0. Effectively this defines a small core size for the vortex (again similar to the vortex core in superfluid ^{4}He). Superconductivity is suppressed inside the vortex core, for $r < \xi_0$, which is effectively normal material. Therefore, Eq. 3.60 is valid in the region $\xi_0 \ll r \ll \lambda$, and this simple London vortex model is only valid in superconductors where $\xi_0 \ll \lambda$.

For the case of large z the modified Bessel function becomes

$$K_0(z) \sim \sqrt{\frac{\pi}{2z}}e^{-z}$$

asymptotically (Abramowitz and Stegun 1965). Therefore the magnetic field very far from the core of a London vortex is of the form (Exercise 3.3)

$$B_z(r) = \frac{\Phi_0}{2\pi\lambda^2}\sqrt{\frac{\pi\lambda}{2r}}e^{-r/\lambda}. \tag{3.62}$$

Qualitatively this is similar to the penetration of a magnetic field near a surface as shown in Fig. 3.12.

Overall then, in this London vortex model the magnetic field has some large constant value B_0 inside the vortex core, $r < \xi$, then decreases logarithmically between $\xi_0 < r < \lambda$ and then goes to zero exponentially outside the vortex on a length scale of order λ. Clearly this picture is only useful in the limit $\lambda > \xi_0$, corresponding to a type II superconductor.

It is also instructive to calculate the energy of the rotating supercurrents in the vortex. The result[7] is that the energy of the vortex is approximately

[7] See Exercise 3.4 below for the proof.

$$E = \frac{\Phi_0^2}{4\pi\mu_0\lambda^2} \ln\left(\frac{\lambda}{\xi_0}\right) \tag{3.63}$$

per unit length.

Further reading

To review the basic concepts of band theory of metals, see *Band theory and electronic properties of solids*, Singleton (2001), a companion volume to this book in the Oxford Master Series in Condensed Matter.

There are many text books dealing with superconductivity. Probably the ones which are especially good for beginners are *Supercondctivity Today*, Ramakrishnan and Rao 1992, and *Superconductivity and Superfluidity* by Tilley and Tilley (1990).

Among the more advanced books, *Superconductivity of metals and Alloys*, de Gennes (1966), has the most extensive discussion of the topics covered in this chapter, especially vortices and the vortex lattice.

Bessel functions and their mathematical properties are described in many texts. Their definitions and properties are given in depth by Abramowitz and Stegun (1965). Good introductions are given by Boas (1983) and Matthews and Walker (1970).

Exercises

(3.1) (a) Using the London equation show that

$$\nabla \times (\nabla \times \mathbf{B}) = -\frac{1}{\lambda^2}\mathbf{B}$$

in a superconductor.

(b) In Fig. 3.12, the surface of the superconductor lies in the y–z plane. A magnetic field is applied in the z direction parallel to the surface, $\mathbf{B} = (0, 0, B_0)$. Given that inside the superconductor the magnetic field is a function of x only, $\mathbf{B} = (0, 0, B_z(x))$ show that

$$\frac{d^2 B_z(x)}{dx^2} = \frac{1}{\lambda^2} B_z(x).$$

(c) Solving the ordinary differential equation in (b) show that the magnetic field near a surface of a superconductor has the form

$$B = B_0 \exp(-x/\lambda)$$

as shown in Fig. 3.12.

(3.2) Consider a thin superconducting slab, of thickness $2L$, as shown in Fig. 3.13. If an external parallel magnetic field, B_0, is applied parallel to the slab surfaces, show that inside the slab the magnetic field becomes

$$B_z(x) = B_0 \frac{\cosh(x/\lambda)}{\cosh(L/\lambda)}.$$

(3.3) (a) A vortex in a superconductor can be modeled as having a cylindrical core of normal metal of radius ξ_0. Use $\nabla \times (\nabla \times \mathbf{B}) = -\mathbf{B}/\lambda^2$ and the expression

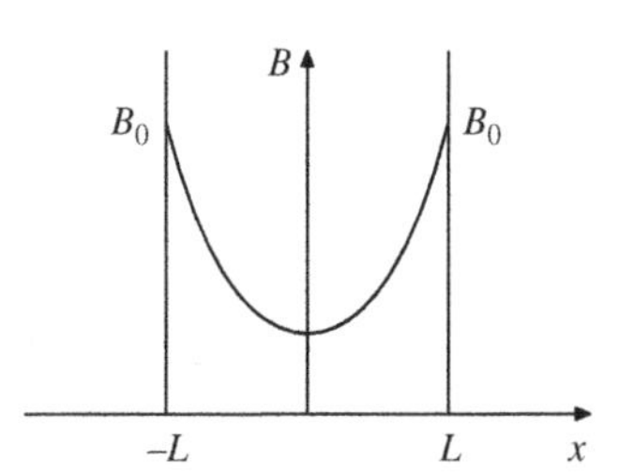

Fig. 3.13 Exercise 3.2: the magnetic field inside a superconducting slab of thickness $2L$.

for curl in cylindrical polar coordinates (Eq. 2.43) to show that the magnetic field $B_z(r)$ outside of the core obeys the Bessel equation:

$$\frac{1}{r}\frac{d}{dr}\left(r\frac{dB_z}{dr}\right) = \frac{B_z}{\lambda^2}.$$

(b) For small r, obeying $\xi_0 < r \ll \lambda$, the right-hand-side of the Bessel equation in (a) can be approximated by zero. Show that this approximation leads to

$$B_z(r) = a\ln(r) + b,$$

where a and b are unknown constants.

(c) Show that the current corresponding to the field $B_z(r)$ found in (b) is equal to

$$\mathbf{j} = -\frac{a}{\mu_0 r}\mathbf{e}_\phi$$

similar to the superfluid current in a ^{4}He vortex. Hence find the vector potential $\mathbf{A}$ and find a as a function of the magnetic flux enclosed by the vortex core, Φ.

(d) For larger values of r ($r \sim \lambda$ and above) assume that we can approximate the Bessel equation from (a) by:

$$\frac{d}{dr}\left(\frac{dB_z}{dr}\right) = \frac{B_z}{\lambda^2}.$$

Hence show that $B_z(r) \sim e^{-r/\lambda}$ for large r.

(e) The large r solution given in part (d) is not exactly the correct asymptotic form of the solution, as described in Section 3.9. For large values of r, assume that

$$B_z(r) \sim r^p e^{-r/\lambda}$$

and hence show that the correct exponent is $p = -1/2$, as described above.

(3.4) Suppose that any supercurrent flow corresponds to an effective superfluid flow velocity $\mathbf{v}$ of the electrons, where $\mathbf{j} = -en_s\mathbf{v}$. Assume that the corresponding kinetic energy is $\frac{1}{2}mv^2n_s$ per unit volume. Hence, using the results from Exercise 3.3 parts (c) and (d), show that the total energy of a vortex line is roughly of order

$$E = \frac{\Phi^2}{4\pi\mu_0\lambda^2}\ln\left(\frac{\lambda}{\xi_0}\right)$$

per unit length.

(3.5) The complex conductivity $\sigma(\omega)$ has real and imaginary parts that are related together by **Kramers–Kronig** relations

$$\mathrm{Re}[\sigma(\omega)] = \frac{1}{\pi}\mathcal{P}\int_{-\infty}^{\infty}\frac{\mathrm{Im}[\sigma(\omega')]}{\omega'-\omega}d\omega'$$

and

$$\mathrm{Im}[\sigma(\omega)] = -\frac{1}{\pi}\mathcal{P}\int_{-\infty}^{\infty}\frac{\mathrm{Re}[\sigma(\omega')]}{\omega'-\omega}d\omega',$$

where, here $\mathcal{P}\int$ means the **principal value** of the integral (Boas 1983; Matthews and Walker 1970). Therefore an experimental measurement of the real part is sufficient to determine the imaginary part, and *vice versa*.

(a) Using these expressions, and assuming that the real part of the conductivity $\mathrm{Re}\sigma(\omega)$ is a Dirac delta function

$$\mathrm{Re}[\sigma(\omega)] = \frac{\pi n_s e^2}{m_e}\delta(\omega)$$

show that the imaginary part is given by

$$\mathrm{Im}[\sigma(\omega)] = \frac{n_s e^2}{\omega m_e}$$

exactly as given in Eq. 3.41.

(b) Exercise for those who have studied analytic complex function theory. We can derive the Kramers–Kronig relations as follows. Consider the contour integral

$$I = \oint\frac{\sigma(\omega')}{\omega'-\omega}d\omega',$$

around the contour shown in Fig. 3.14. Find the poles of $\sigma(\omega')$ according to Eq. 3.38 and show that it is analytic in the upper half plane ($\mathrm{Im}[\omega'] > 0$) in Fig. 3.14.

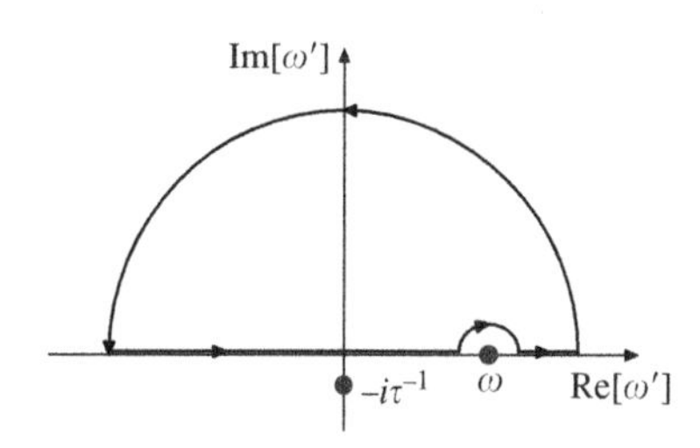

Fig. 3.14 Complex integration contour for Exercise 3.5.

(c) Use the result from (b) to show that $I = 0$, and thus prove that

$$0 = \mathcal{P}\int_{-\infty}^{\infty} \frac{\sigma(\omega')}{\omega' - \omega} d\omega' - i\pi\sigma(\omega) = 0,$$

where the integral is now just along the real ω' axis. Take the real and imaginary parts of this expression and show that this results in the Kramer's Kronig equations given above.[8]

[8] The proof is actually very general. The fact that $\sigma(\omega')$ is analytic in the upper half plane is in fact just a consequence of **causality**, that is, the applied current always responds to the applied external field. Effect follows cause, never the reverse! Therefore the Kramers–Kronig relations are always true for any such response function.

The Ginzburg–Landau model

4

4.1 Introduction

The superconducting state and the normal metallic state are separate thermodynamic phases of matter in just the same way as gas, liquid, and solid are different phases. Similarly, the normal Bose gas and Bose–Einstein Condensate (BEC), or normal liquid He^4 and superfluid He II are separated by a thermodynamic phase transitions. Each such phase transition can be characterized by the nature of the singularities in specific heat and other thermodynamic variables at the transition, T_c. We can therefore examine the problems of superfluidity and superconductivity from the point of view of the thermodynamics of phase transitions.

The theory of superconductivity introduced by Ginzburg and Landau in 1950 describes the superconducting phase transition from this thermodynamic point of view. It was originally introduced as a phenomenological theory, but later Gor'kov showed that it can be derived from full the microscopic Bardeen Cooper Schrieffer (BCS) theory in a suitable limit.[1]

In this chapter we shall first discuss the superconducting phase transition from the point of view of equilibrium thermodynamics. Then we gradually build up toward the full Ginzburg–Landau (GL) model. First we discuss spatially uniform systems, then spatially varying systems and finally systems in an external magnetic field. The Ginzburg–Landau theory makes many useful and important predictions. Here we focus on just two applications: to flux quantization, and to the Abrikosov flux lattice in type II superconductors.

The GL theory as originally applied to superconductors was *par excellence* a **mean-field theory** of the thermodynamic state. However, in fact, one of its most powerful features is that it can be used to go beyond the original mean-field limit, so as to include the effects of thermal fluctuations. We shall see below that such fluctuations are largely negligible in the case of conventional "low-T_c" superconductors, making the mean-field approximation essentially exact. However, in the newer high T_c superconductors these fluctuations lead to many important phenomena, such as flux flow, and the melting of the Abrikosov vortex lattice.

[1] In fact, the Ginzburg–Landau model is very general and has applications in many different areas of physics. It can be modified to describe many different physical systems, including magnetism, liquid crystal phases, and even the symmetry breaking phase transitions, which took place in the early universe as matter cooled following the big bang!

4.2 The condensation energy

We already have enough information about superconductivity to derive some important thermodynamic properties about the superconducting phase transition. We can analyze the phase diagram of superconductors in exactly the same manner as one would consider the well known thermodynamics of a liquid–gas phase transition problem, such as given by the van der Waals equation of state. However, for the superconductor instead of the pair of thermodynamic variables P, V (pressure and volume) we have the magnetic variables $\mathbf{H}$ and $\mathbf{M}$ as the relevant thermodynamic parameters.

Let us first briefly review the basic thermodynamics of magnetic materials. This is covered in several undergraduate text books on thermodynamics such as Mandl (1987) or Callen (1960), Blundell (2001). If we consider a long cylindrical sample in a solenoidal field, as shown in Fig. 3.6, then the magnetic field $\mathbf{H}$ inside the sample is given by

$$\mathbf{H} = \frac{N}{L} I \mathbf{e}_z, \tag{4.1}$$

where the coil has N/L turns per meter, I is the current and $\mathbf{e}_z$ is a unit vector along the axis of the cylinder. The total work done, $đW$, on increasing the current infinitesimally from I to $I + dI$ can be calculated as

$$\begin{aligned} đW &= -N\mathcal{E}I\,dt, \\ &= +N\frac{d\Phi}{dt} I\,dt, \\ &= +NI\,d\Phi, \\ &= +NAI\,dB, \\ &= +NV\mathbf{H}\cdot d\mathbf{B}, \\ &= +\mu_0 V\left(\mathbf{H}\cdot d\mathbf{M} + \mathbf{H}\cdot d\mathbf{H}\right), \end{aligned} \tag{4.2}$$

where A is the cross sectional area of the coli, $V = AL$ is its volume, $\mathcal{E} = -d\Phi/dt$ is the e.m.f. induced in the coil by the change in the total magnetic flux, Φ, through the sample. We also used the identity $\mathbf{B} = \mu_0(\mathbf{M} + \mathbf{H})$ in writing the last step in Eq. 4.2.

This analysis shows that we can divide the total work done by increasing the current in the coil into two separate parts. The first part,

$$\mu_0 \mathbf{H}\cdot d\mathbf{M}$$

per unit volume, is the **magnetic work** done on the sample. The second part,

$$\mu_0 \mathbf{H}\cdot d\mathbf{H}$$

is the work per unit volume which would have been done even if no sample had been present inside the coil; it is the work done by the **self-inductance** of the coil. If the coil is empty, $\mathbf{M} = 0$ and so $\mathbf{B} = \mu_0\mathbf{H}$ and one can easily see that the work done is exactly the change in the vacuum field energy of the electromagnetic field

$$E_B = \frac{1}{2\mu_0}\int B^2 d^3r \tag{4.3}$$

due to the change of current in the solenoid coils. By convention[2] we shall not include this vacuum field energy, as work done "on the sample." Therefore we define the magnetic work done on the sample as $\mu_0\mathbf{H}\,d\mathbf{M}$ per unit volume.

[2]Unfortunately there is no single standard convention used by all books and papers in this field. Different contributions to the total energy are either included or not, and so one must be very careful when comparing similar looking equations from different texts and research papers. Our convention follows Mandl (1987) and Callen (1960).

With this definition of magnetic work the first law of thermodynamics for a magnetic material reads

$$dU = T\,dS + \mu_0 V \mathbf{H} \cdot d\mathbf{M}, \tag{4.4}$$

where U is the total internal energy, TdS is the heat energy with T the temperature and S the entropy. We see that the magnetic work is analogous to the work, $-PdV$, in a gas. As in the usual thermodynamics of gases the internal energy, U, is most naturally thought of as a function of the entropy and volume: $U(S, V)$. The analog of the first law for a magnetic system, Eq. 4.4, shows that the internal energy of a magnetic substance is most naturally thought of as a function of S and $\mathbf{M}$, $U(S, \mathbf{M})$. In terms of this function the temperature and field $\mathbf{H}$ are given by

$$T = \frac{\partial U}{\partial S} \tag{4.5}$$

$$\mathbf{H} = \frac{1}{\mu_0 V} \frac{\partial U}{\partial \mathbf{M}}. \tag{4.6}$$

However, S and $\mathbf{M}$ are usually not the most convenient variables to work with. In a solenoidal geometry such as Fig. 3.6 it is the H-field which is directly fixed by the current, not $\mathbf{M}$. It is therefore useful to define magnetic analogs of the Helmholtz and Gibbs free energies

$$F(T, \mathbf{M}) = U - TS \tag{4.7}$$

$$G(T, \mathbf{H}) = U - TS - \mu_0 V \mathbf{H} \cdot \mathbf{M}. \tag{4.8}$$

As indicated, the Gibbs free energy G is naturally viewed as a function of T and $\mathbf{H}$ since,

$$dG = -S\,dT - \mu_0 V \mathbf{M} \cdot d\mathbf{H}. \tag{4.9}$$

In terms of G one can calculate the entropy and magnetization,

$$S = -\frac{\partial G}{\partial T}, \tag{4.10}$$

$$\mathbf{M} = -\frac{1}{\mu_0 V} \frac{\partial G}{\partial \mathbf{H}}. \tag{4.11}$$

$G(T, \mathbf{H})$ is usually the most convenient thermodynamic quantity to work with since T and $\mathbf{H}$ are the variables which are most naturally controlled experimentally. Furthermore from $G(T, \mathbf{H})$ one can also reconstruct the free energy, $F = G + \mu_0 V \mathbf{H} \cdot \mathbf{M} V$ or the internal energy $U = F + TS$.

The Gibbs free energy allows us to calculate the free energy difference between the superconducting state and the normal state. Consider the H, T phase diagram of a type I superconductor, as sketched above in Fig. 4.1. We can evaluate the change in Gibbs free energy in the superconducting state by integrating along the vertical line drawn. Along this line $dT = 0$, and so, clearly,

$$G_s(T, H_c) - G_s(T, 0) = \int dG = -\mu_0 V \int_0^{H_c} \mathbf{M} \cdot d\mathbf{H},$$

where the subscript s implies that $G(T, \mathbf{H})$ is in the superconducting state. But for a type I superconductor in the superconducting state we know from the

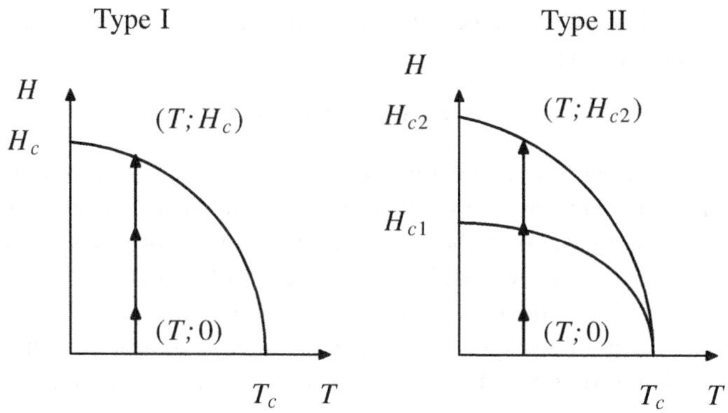

Fig. 4.1 We obtain the condensation energy for superconductors by thermodynamic integration of the Gibbs free energy along the contours in the (T, H) plane, as shown above.

Meissner–Ochsenfeld effect that $\mathbf{M} = -\mathbf{H}$ and thus,

$$G_s(T, H_c) - G_s(T, 0) = \mu_0 \frac{H_c^2}{2} V.$$

Now, at the critical field H_c in Fig. 4.1 the normal state and the superconducting state are in thermodynamic equilibrium. Equilibrium between phases implies that the two Gibbs free energies are equal:

$$G_s(T, H_c) = G_n(T, H_c).$$

Furthermore, in the normal state $M \approx 0$ (apart from the small normal metal paramagnetism or diamagnetism which we neglect). So if the normal metal state had persisted below H_c down to zero field, it would have had a Gibbs free energy of,

$$G_n(T, H_c) - G_n(T, 0) = \int dG = -\mu_0 V \int_0^{H_c} M dH \approx 0.$$

Putting these together we find the difference in Gibbs free energies of superconducting and normal states at zero field:

$$G_s(T, 0) - G_n(T, 0) = -\mu_0 V \frac{H_c^2}{2}. \tag{4.12}$$

The Gibbs potential for the superconducting state is lower, so it is the stable state.

We can also write the above results in terms of the more familiar Helmholtz free energy. Using $F = G - \mu_0 V \mathbf{H} \cdot \mathbf{M}$ and substituting $\mathbf{H} = \mathbf{M} = 0$ we can see that the difference in Helmholtz free energies $F(T, \mathbf{M})$ is the same as for the Gibbs potentials, and hence

$$F_s(T, 0) - F_n(T, 0) = -\mu_0 V \frac{H_c^2}{2}. \tag{4.13}$$

The quantity $\mu_0 H_c^2/2$ is the **condensation energy**. It is a measure of the gain in free energy per unit volume in the superconducting state compared with the normal state at the same temperature.

As an example lets consider niobium. Here $T_c = 9\,\mathrm{K}$, and $H_c = 160\,\mathrm{kA\,m^{-1}}$ ($B_c = \mu_0 H_c = 0.2T$). The condensation energy $\mu_0 H_c^2/2 = 16.5\,\mathrm{kJ\,m^{-3}}$. Given that Nb has a bcc crystal structure with a 0.33 nm lattice constant we can work out the volume per atom and find that the condensation energy is only around 2 μeV/atom! Such tiny energies were a mystery until the BCS theory, which shows that the condensation energy is of order $(k_B T_c)^2 g(E_F)$, where $g(\epsilon_F)$ is the density of states at the Fermi level. The energy is so small because $k_B T_c$ is many orders of magnitude smaller than the Fermi energy, ϵ_F.

The similar thermodynamic arguments can also be applied to calculate the condensation energy of type II superconductors. Again the magnetic work per unit volume is calculated as an integral along a countour, as shown in the right panel of Fig. 4.1,

$$G_s(T, H_{c2}) - G_s(T, 0) = \mu_0 V \int_0^{H_{c2}} \mathbf{M} \cdot d\mathbf{H}. \tag{4.14}$$

The integral is simply the area under the curve of M as a function of H drawn in Fig. 3.8 (assuming that **M** and **H** have the same vector directions). **Defining** the value of H_c for a type II superconducting from the value of the integral

$$\frac{1}{2} H_c^2 \equiv \int_0^{H_{c2}} \mathbf{M} \cdot d\mathbf{H} \tag{4.15}$$

we again can express the zero field condensation energy in terms of H_c,

$$F_s(T, 0) - F_n(T, 0) = -\mu_0 V \frac{H_c^2}{2}. \tag{4.16}$$

Here H_c is called the **thermodynamic critical field**. Note that there is no phase transition at H_c in a type II superconductor. The only real transitions are at H_{c1} and H_{c2}, and H_c is merely a convenient measure of the condensation energy.

We can also calculate the entropy of the superconducting state using the same methods. A simple calculation (Exercise 4.1) shows that in a type I superconductor there is a finite change in entropy per unit volume between the normal and superconducting states at H_c,

$$s_s(T, H_c) - s_n(T, H_c) = -\mu_0 H_c \frac{dH_c}{dT}. \tag{4.17}$$

This shows that the phase transition is generally **first-order**, that is, it has a finite latent heat. But, in zero external field, at the point $(T, H) = (T_c, 0)$ in Fig. 4.1, this entropy difference goes to zero, and so in this case the phase transition is **second-order**.

4.3 Ginzburg–Landau theory of the bulk phase transition

The GL theory of superconductivity is built upon a general approach to the theory of second-order phase transitions which Landau had developed in the 1930s. Landau had noticed that typically second-order phase transitions, such as the Curie temperature in a ferromagnet, involve some change in symmetry of the system. For example, a magnet above the Curie temperature, T_c, has no magnetic moment. But below T_c a spontaneous magnetic moment develops. In principle this could point in any one of a number of different directions, each with an equal energy, but the system spontaneously chooses one particular direction. In Landau's theory such phase transitions are characterized by an

order parameter which is zero in the disordered state above T_c, but becomes nonzero below T_c. In the case of a magnet the magnetization, $\mathbf{M}(\mathbf{r})$, is a suitable order parameter.

For superconductivity Ginzburg and Landau (GL) postulated the existence of an order parameter denoted by ψ. This characterizes the superconducting state, in the same way as the magnetization does in a ferromagnet. The order parameter is assumed to be some (unspecified) physical quantity which characterizes the state of the system. In the normal metallic state above the critical temperature T_c of the superconductor it is zero. While in the superconducting state below T_c it is nonzero. Therefore it is assumed to obey:

$$\psi = \begin{cases} 0 & T > T_c, \\ \psi(T) \neq 0 & T < T_c. \end{cases} \tag{4.18}$$

GL postulated that the order parameter ψ should be a complex number, thinking of it as a macroscopic wave function for the superconductor in analogy with superfluid ^{4}He. At the time of their original work the physical significance of this complex ψ in superconductors was not at all clear. But, as we shall see below, in the microscopic BCS theory of superconductivity there appears a parameter, Δ, which is also complex. Gor'kov was able to derive the GL theory from BCS theory, and show that ψ is essentially the same as Δ, except for some constant numerical factors. In fact, we can even identify $|\psi|^2$ as the density of BCS "Cooper pairs" present in the sample.

GL assumed that the free energy of the superconductor must depend smoothly on the parameter ψ. Since ψ is complex and the free energy must be real, the energy can only depend on $|\psi|$. Furthermore, since ψ goes to zero at the critical temperature, T_c, we can Taylor expand the free energy in powers of $|\psi|$. For temperatures close to T_c only the first two terms in the expansion should be necessary, and so the free energy density ($f = F/V$) must be of the form:

$$f_s(T) = f_n(T) + a(T)|\psi|^2 + \tfrac{1}{2}b(T)|\psi|^4 + \cdots \tag{4.19}$$

since $|\psi|$ is small. Here $f_s(T)$ and $f_n(T)$ are the superconducting state and normal state free energy densities, respectively. Clearly Eq. 4.19 is the only possible function which is real for any complex ψ near $\psi = 0$ and which is a differentiable function of ψ and ψ^* near to $\psi = 0$. The parameters $a(T)$ and $b(T)$ are, in general, temperature dependent pheonomenological parameters of the theory. However, it is assumed that they must be smooth functions of temperature. We must also assume that $b(T)$ is positive, since otherwise the free energy density would have no minimum, which would be unphysical (or we would have to extend the expansion to include higher powers such as $|\psi|^6$).

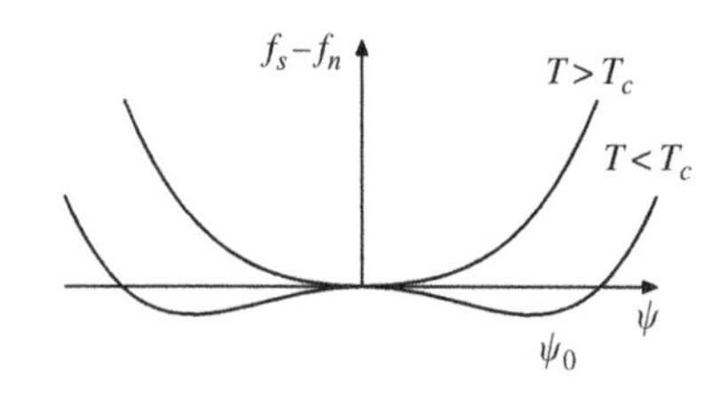

Fig. 4.2 Free energy difference between the normal and superconducting states (per unit volume) as a function of the order parameter ψ. For $T < T_c$ the free energy has a minimum at ψ_0, while for $T > T_c$ the only minimum is at $\psi = 0$.

Plotting $f_s - f_n$ as a function of ψ is easy to see that there are two possible curves, depending on the sign of the parameter $a(T)$, as shown in Fig. 4.2. In the case $a(T) > 0$, the curve has one minimum at $\psi = 0$. On the other hand, for $a(T) < 0$ there are minima wherever $|\psi|^2 = -a(T)/b(T)$. Landau and Ginzburg assumed that at high temperatures, above T_c, we have $a(T)$ positive, and hence the minimum free energy solution is one with $\psi = 0$, that is, the normal state. But if $a(T)$ gradually decreases as the temperature T is reduced, then the state of the system will change suddenly when we reach the point $a(T) = 0$. Below this temperature the minimum free energy solution changes to one with $\psi \neq 0$. Therefore we can identify the temperature where $a(T)$ becomes zero as the critical temperature T_c.

Near to this critical temperature, T_c, assuming that the coefficients $a(T)$ and $b(T)$ change smoothly with temperature, we can make a Taylor expansion,

$$\begin{aligned} a(T) &\approx \dot{a} \times (T - T_c) + \cdots, \\ b(T) &\approx b + \cdots, \end{aligned} \tag{4.20}$$

where $\dot{a}$ and b are two pheonomenological constants. Then for temperatures just above T_c, $a(T)$ will be positive, and we have the free energy minimum, $\psi = 0$. On the other hand, just below T_c we will have minimum energy solutions with nonzero $|\psi|$, as seen in Fig. 4.2. In terms of the parameters $\dot{a}$ and b it is easy to see that

$$|\psi| = \begin{cases} \left(\frac{\dot{a}}{b}\right)^{1/2} (T_c - T)^{1/2} & T < T_c, \\ 0 & T > T_c. \end{cases} \tag{4.21}$$

The corresponding curve of $|\psi|$ as a function of temperature, T, is shown in Fig. 4.3. One can see the abrupt change from zero to nonzero values at the critical temperature T_c. In fact, this curve is qualitatively similar to those obtained with other types of second-order phase transitions within Landau's general theory. For example, the behavior of the order parameter ψ near T_c in Fig. 4.3 resembles closely change in the magnetization $\mathbf{M}$ in a ferromagnet near its Curie point in the Stoner theory of magnetism (Blundell 2001).

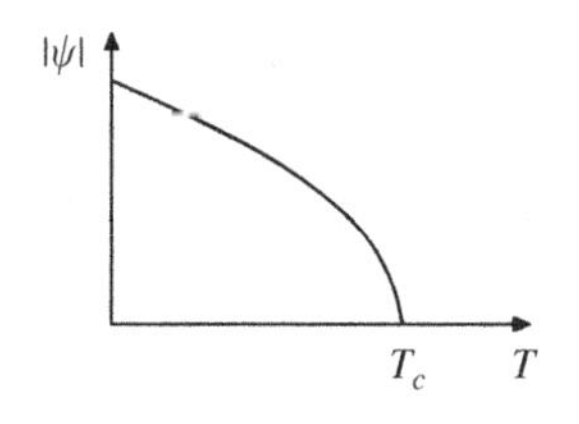

Fig. 4.3 Order parameter magnitude, $|\psi|$, as a function of temperature in the GL model.

It turns out to be very important that, because ψ is complex, there are in fact an infinite set of minima corresponding to all possible values of the complex phase θ,

$$\psi = |\psi| e^{i\theta}. \tag{4.22}$$

The phase value, θ is arbitrary, since all values lead to the same total free energy. But, just as in the case of the direction of magnetization $\mathbf{M}$ in a ferromagnet the system spontaneously chooses one particular value. A magnet heated to above T_c and then cooled again will almost certainly adopt a different random direction of magnetization, and the same would be true for the angle θ in a superconductor. In fact we have met this same concept before, in Chapter 2, when we discussed the XY symmetry of the macroscopic wave function in superfluid He II (Fig. 2.4).

The value of the minimum free energy in Fig. 4.2, is easily found to be $-a(T)^2/2b(T)$. This is the free energy difference (per unit volume) between the superconducting and non-superconducting phases of the system at temperature T. This corresponds to the condensation energy of the superconductor, and so we can write

$$f_s(T) - f_n(T) = -\frac{\dot{a}^2(T - T_c)^2}{2b} = -\mu_0 \frac{H_c^2}{2}, \tag{4.23}$$

giving the thermodynamic critical field,

$$H_c = \frac{\dot{a}}{(\mu_0 b)^{1/2}} (T_c - T) \tag{4.24}$$

near to T_c.

From this free energy we can also obtain other relevant physical quantities, such as the entropy and heat capacity. Differentiating f with respect to T gives

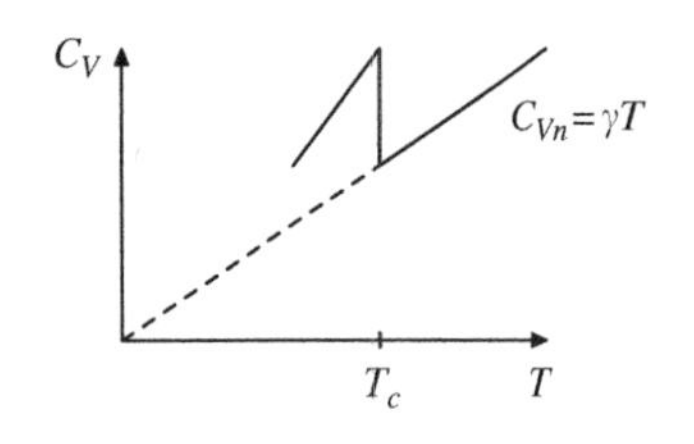

Fig. 4.4 Specific heat of a superconductor near T_c in the GL model. Above T_c the specific heat is given by the Sommerfeld theory of metals, $C_{Vn} = \gamma T$. At T_c there is a discontinuity and a change of slope.

the entropy per unit volume, $s = S/V$,

$$s_s(T) - s_n(T) = -\frac{\dot{a}^2}{b}(T_c - T), \tag{4.25}$$

below T_c. At T_c there is no discontinuity in entropy, or latent heat, confirming that the GL model corresponds to a second-order thermodynamic phase transition. But there is a sudden change in specific heat at T_c. Differentiating the entropy to find the heat capacity $C_V = Tds/dT$ per unit volume we obtain

$$C_{Vs} - C_{Vn} = \begin{cases} T\dot{a}^2/b & T < T_c, \\ 0 & T > T_c. \end{cases} \tag{4.26}$$

and so the heat capacity has a discontinuity

$$\Delta C_V = T_c \frac{\dot{a}^2}{b} \tag{4.27}$$

at T_c. The metallic normal state heat capacity is linear in T, $C_{Vn} = \gamma T$, with γ the Sommerfeld constant, and so the full heat capacity curve looks like Fig. 4.4 near to T_c.[3]

Interestingly the specific heat for superconductors shown in Fig. 4.4 is qualitatively quite very different from both the case of BEC, shown in Fig. 1.7, and the λ point of superfluid ^{4}He, Fig. 2.3.[4]

[3] The GL theory can only be reliably used at temperatures close to T_c. Therefore our calculated specific heat is only correct near to T_c, and we cannot legitimately continue the GL line in Fig. 4.4 down from T_c all the way to $T = 0$.

[4] In fact the differences are deceptive! Our theory rests on a mean-field approximation and has neglected important thermal fluctuation effects, as we shall see below. When these fluctuations are large, as in the case of high temperature superconductors, the observed specific heat near T_c appears to show exactly the same XY universality class as the lambda point in superfluid helium (Overend, Howson and Lawrie 1994).

4.4 Ginzburg–Landau theory of inhomogenous systems

The complete GL theory of superconductivity also allows for the possibility that the order parameter depends on position, $\psi(\mathbf{r})$. This, of course, now really begins to resemble the macroscopic condensate wave function introduced in Chapter 2 for the case of superfluid helium.

GL postulated that the free energy is as given above, together with a new term depending on the gradient of $\psi(\mathbf{r})$. With this term the free energy density becomes,

$$f_s(T) = f_n(T) + \frac{\hbar^2}{2m^*}|\nabla\psi(\mathbf{r})|^2 + a(T)|\psi(\mathbf{r})|^2 + \frac{b(T)}{2}|\psi(\mathbf{r})|^4 \tag{4.28}$$

at point $\mathbf{r}$ in the absence of any magnetic fields. Setting $\psi(\mathbf{r})$ to a constant value, $\psi(\mathbf{r}) = \psi$, we see that the parameters $a(T)$ and $b(T)$ are the same as for the bulk theory described in the previous section. The new parameter m^* determines the energy cost associated with gradients in $\psi(\mathbf{r})$. It has dimensions of mass, and it plays the role of an effective mass for the quantum system with macroscopic wave function $\psi(\mathbf{r})$.

In order to find the order parameter $\psi(\mathbf{r})$ we must minimize the total free energy of the system,

$$F_s(T) = F_n(T) + \int \left(\frac{\hbar^2}{2m^*}|\nabla\psi|^2 + a(T)|\psi(\mathbf{r})|^2 + \frac{b(T)}{2}|\psi(\mathbf{r})|^4\right) d^3r. \tag{4.29}$$

To find the minimum we must consider an infinitesimal variation in the function $\psi(\mathbf{r})$

$$\psi(\mathbf{r}) \to \psi(\mathbf{r}) + \delta\psi(\mathbf{r}) \tag{4.30}$$

relative to some function $\psi(\mathbf{r})$. Evaluating the change in the total free energy due to $\delta\psi$ and dropping all terms of higher than linear order in the variation $\delta\psi$ we find after some lengthy algebra

$$\delta F_s = \int \left[\frac{\hbar^2}{2m^*} (\nabla \delta\psi^*) \cdot (\nabla \psi) + \delta\psi^* (a\psi + b\psi|\psi^2|) \right] d^3r$$

$$+ \int \left[\frac{\hbar^2}{2m^*} (\nabla \psi^*) \cdot (\nabla \delta\psi) + (a\psi^* + b\psi^*|\psi^2|)\delta\psi \right] d^3r. \quad (4.31)$$

The two terms involving gradients can be integrated by parts, to obtain

$$\delta F_s = \int \delta\psi^* \left(-\frac{\hbar^2}{2m^*} \nabla^2 \psi + a\psi + b\psi|\psi^2| \right) d^3r$$

$$+ \int \left(-\frac{\hbar^2}{2m^*} \nabla^2 \psi + a\psi + b\psi|\psi^2| \right)^* \delta\psi d^3r. \quad (4.32)$$

The condition for $\psi(\mathbf{r})$ to produce a minimum in free energy is that $\delta F = 0$ for any arbitrary variation $\delta\psi(\mathbf{r})$. From Eq. 4.32 this can only be when $\psi(\mathbf{r})$ obeys

$$-\frac{\hbar^2}{2m^*} \nabla^2 \psi + a\psi + b\psi|\psi^2| = 0. \quad (4.33)$$

We can obtain this same result more formally by noting that the total free energy of the solid is a **functional** of $\psi(\mathbf{r})$, denoted $F_s[\psi]$, meaning that the scalar number F_s depends on the whole function $\psi(\mathbf{r})$ at all points in the system, $\mathbf{r}$. It will be minimized by a function $\psi(\mathbf{r})$ which satisfies

$$\frac{\partial F_s[\psi]}{\partial \psi(\mathbf{r})} = 0 \qquad \frac{\partial F_s[\psi]}{\partial \psi^*(\mathbf{r})} = 0. \quad (4.34)$$

where the derivatives are mathematically **functional derivatives**. Functional derivative can be defined by analogy with the idea of a partial derivative. For a function of many variables, $f(x_1, x_2, x_3, \ldots)$ we can express changes in the function value due to infinitesimal variations of the parameters using the standard expression

$$df = \frac{\partial f}{\partial x_1} dx_1 + \frac{\partial f}{\partial x_2} dx_2 + \frac{\partial f}{\partial x_3} dx_3 + \cdots. \quad (4.35)$$

Considering the free energy as a function of infinitely many variables, $\psi(\mathbf{r})$ and $\psi^*(\mathbf{r})$ at all possible points $\mathbf{r}$ we can write the analog of Eq. 4.35 as,

$$dF_s = \int \left(\frac{\partial F_s[\psi]}{\partial \psi(\mathbf{r})} d\psi(\mathbf{r}) + \frac{\partial F_s[\psi]}{\partial \psi^*(\mathbf{r})} d\psi^*(\mathbf{r}) \right) d^3r. \quad (4.36)$$

In comparison with Eq. 4.32 we see that

$$\frac{\partial F_s[\psi]}{\partial \psi^*(\mathbf{r})} = -\frac{\hbar^2}{2m^*} \nabla^2 \psi + a(T)\psi + b(T)\psi|\psi^2| \quad (4.37)$$

and

$$\frac{\partial F_s[\psi]}{\partial \psi(\mathbf{r})} = \left(-\frac{\hbar^2}{2m^*} \nabla^2 \psi + a(T)\psi + b(T)\psi|\psi^2| \right)^*, \quad (4.38)$$

which is just the complex conjugate of Eq. 4.37. Perhaps it seems surprising that we can effectively treat $\psi(\mathbf{r})$ and $\psi^*(\mathbf{r})$ as independent variables in the differentiation, but this is correct because there are two independent real functions, $\mathrm{Re}[\psi(\mathbf{r})]$ and $\mathrm{Im}[\psi(\mathbf{r})]$, which can be varied separately.

Thus we have found that minimizing the total free energy leads to the following Schrödinger like equation for $\psi(\mathbf{r})$,

$$-\frac{\hbar^2}{2m^*}\nabla^2\psi(\mathbf{r}) + \left(a + b|\psi(\mathbf{r})|^2\right)\psi(\mathbf{r}) = 0. \tag{4.39}$$

However, unlike the usual Schrödinger equation, this is a nonlinear equation because of the second term in the bracket. Because of this nonlinearity the quantum mechanical principle of superposition does not apply, and the normalization of ψ is different from the usual one in quantum mechanics.

4.5 Surfaces of superconductors

The effective nonlinear Schödinger equation, Eq. 4.39, has several useful applications. In particular, it can be used to study the response of the superconducting order parameter to external perturbations. Important examples of this include the properties of the surfaces and interfaces of superconductors.

Consider a simple model for the interface between a normal metal and a superconductor. Suppose that the interface lies in the yz plane separating the normal metal in the $x < 0$ region from the superconductor in the $x > 0$ region. On the normal metal side of the interface the superconducting order parameter, $\psi(\mathbf{r})$, must be zero. Assuming that $\psi(\mathbf{r})$ must be continuous, we must therefore solve the nonlinear Schrödinger equation,

$$-\frac{\hbar^2}{2m^*}\frac{d^2\psi(x)}{dx^2} + a(T)\psi(x) + b(T)\psi^3(x) = 0 \tag{4.40}$$

in the region $x > 0$ with the boundary condition at $\psi(0) = 0$. It turns out that one can solve this equation directly (Exercise 4.2) to find

$$\psi(x) = \psi_0 \tanh\left(\frac{x}{\sqrt{2}\xi(T)}\right), \tag{4.41}$$

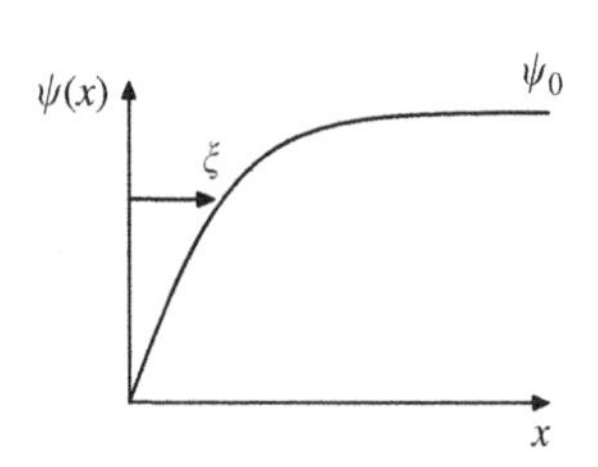

Fig. 4.5 Order parameter of a superconductor near a surface. It recovers to its bulk value ψ_0 over a length scale of the coherence length, ξ.

as shown in Fig. 4.5. Here ψ_0 is the value of the order parameter in the bulk far from the surface and the parameter $\xi(T)$ is defined by

$$\xi(T) = \left(\frac{\hbar^2}{2m^*|a(T)|}\right)^{1/2}. \tag{4.42}$$

This quantity, which has dimensions of length, is called the **Ginzburg–Landau coherence length**. It is an important physical parameter characterizing the superconductor. In Fig. 4.5 one can see that $\xi(T)$ is a measure of the distance from the surface over which the order parameter has recovered back to nearly its bulk value.

The GL coherence length arises in almost all problems of inhomogenous superconductors, including surfaces, interfaces, defects, and vortices. Using $a(T) = \dot{a}(T - T_c)$ the coherence length $\xi(T)$ can be rewritten,

$$\xi(T) = \xi(0)|t|^{-1/2}, \tag{4.43}$$

where

$$t = \frac{T - T_c}{T_c} \tag{4.44}$$

is called the reduced temperature. This expression makes it clear that the coherence length $\xi(T)$ diverges at the critical temperature T_c, and that its divergence is characterized by a critical exponent of 1/2. This exponent is typical for mean-field theories such as the GL model. The zero temperature value of ξ, $\xi(0)$, is apart from some numerical factors of order unity, essentially the same as the Pippard coherence length for superconductors, as introduced in Chapter 3. In BCS theory the coherence length relates to the physical size of a single Cooper pair.

It is also possible to calculate the contribution to the total free energy due to the surface in Fig. 4.5. The surface contribution to the total free energy is

$$\sigma = \int_0^\infty \left(\frac{\hbar^2}{2m^*} \left(\frac{d\psi}{dx} \right)^2 + a\psi^2(x) + \frac{b}{2}\psi^4(x) + \frac{1}{2}\mu_0 H_c^2 \right) dx \qquad (4.45)$$

with $\psi(x)$ given by Eq. 4.41. Here $-\mu_0 H_c^2/2 = -a^2/2b$ is the bulk free energy density. Evaluating the integral (de Gennes 1960) gives

$$\sigma = \tfrac{1}{2}\mu_0 H_c^2 \times 1.89\xi(T) \qquad (4.46)$$

free energy per unit area of the surface.

This theory can also be used to model the **proximity effect** between two superconductors. At an interface between two different superconducting materials the one with the higher T_c will become superconducting first, and will nucleate superconductivity at the surface of the second one. Superconductivity will nucleate at temperatures above the T_c for the second superconductor. If one makes the lower T_c superconductor a thin layer, of order the coherence length $\xi(T)$ in thickness, then the whole system will become superconducting at a temperature above the natural critical temperature of the lower T_c material. Effectively the order parameter $\psi(\mathbf{r})$ has been forced to become nonzero in the thin film material by its proximity to the higher T_c material.

4.6 Ginzburg–Landau theory in a magnetic field

The full power of the GL approach to superconductors only becomes apparent when we include one final term, the effect of a magnetic field. It is this which truly shows that this is a fully fledged theory of superconductivity, complete with Meissner–Ochsenfeld effect, London equation, and so on. Effectively, the GL theory as developed given in the previous sections did not include any effects of the charge of the superconducting condensate. Therefore it would be appropriate for systems of neutral particles, such as a superfliud, or for situations where there are no supercurrents. But in the presence of supercurrents of charged particles we must extend the theory to include the interaction of the current and magnetic field.

What is needed is to include the effects of magnetic fields in the free energy. GL postulated that the magnetic field enters as if $\psi(\mathbf{r})$ were the wave function for charged particles, that is, with the usual replacement in quantum mechanics

$$\frac{\hbar}{i}\nabla \to \frac{\hbar}{i}\nabla - q\mathbf{A} \qquad (4.47)$$

where q is the charge and $\mathbf{A}$ is the magnetic vector potential. For all known superconductors it turns out that the appropriate charge q is $-2e$. Why this is the case only became clear after the BCS theory was developed, and the link between the BCS theory and the GL model had been explained by Gor'kov. He showed that the correct physical interpretation of the GL order parameter $\psi(\mathbf{r})$ is that it can be understood as the wave function for the center of mass motion of Cooper pairs of electrons. Since each Cooper pair has a net charge of $-2e$, then this is the correct effective charge q.[5] Actually the sign can equally well be taken as $q = +2e$, since we can think of Cooper pairs of holes as readily as pairs of electrons. In fact no observable effects in the GL theory differ when we take a different convention for the sign.

[5] In fact, in their original paper GL assumed that the effective charge would be e, not $2e$. Reputedly, Ginzburg, then a young researcher, told his famous advisor Landau that $2e$ fitted the available experimental data better than e, but Landau overruled him and insisted that he was sure that it must be e! Ginzburg shared the 2003 Nobel Prize in Physics for his work in developing the GL model.

With this replacement the GL free energy density of the superconductor becomes,

$$f_s(T) = f_n(T) + \frac{\hbar^2}{2m^*}\left|\left(\frac{\hbar}{i}\nabla + 2e\mathbf{A}\right)\psi\right|^2 + a|\psi|^2 + \frac{b}{2}|\psi|^4. \tag{4.48}$$

To obtain the total free energy we must integrate this over the system, but we must also include an additional term, corresponding to the electromagnetic field energy of the field $\mathbf{B}(\mathbf{r}) = \nabla \times \mathbf{A}$ at each point $\mathbf{r}$. Therefore the total free energy of both the superconductor and the magnetic field is,[6]

[6] Note that in the definition of the magnetic work given in Eq. 4.4, we excluded the part of this field energy, $\mu_0 H^2/2$ per unit volume, that would be present even with no sample present inside the coil in Fig. 3.6. From now on, it will be more convenient to include this energy explicitly, making F_s the total free energy of both sample and vacuum fields.

$$F_s(T) = F_n(T) + \int \left(\frac{\hbar^2}{2m^*}\left|\left(\frac{\hbar}{i}\nabla + 2e\mathbf{A}\right)\psi\right|^2 + a|\psi|^2 + \frac{b}{2}|\psi|^4\right) d^3r$$
$$+ \frac{1}{2\mu_0}\int B(\mathbf{r})^2 d^3r. \tag{4.49}$$

The first integral is carried out over points $\mathbf{r}$ inside the sample, while the second is performed over all space.

The condition for the minimum free energy state is again found by performing a functional differentiation to minimize with respect to $\psi(\mathbf{r})$ and $\psi^*(\mathbf{r})$. The resulting equation for $\psi(\mathbf{r})$ is again a nonlinear Schrödinger equation, but now with a term containing the magnetic vector potential $\mathbf{A}$,

$$-\frac{\hbar^2}{2m^*}\left(\nabla + \frac{2ei}{\hbar}\mathbf{A}\right)^2 \psi(\mathbf{r}) + (a + b|\psi|^2)\psi(\mathbf{r}) = 0. \tag{4.50}$$

The supercurrents due to the magnetic field can be found from the functional derivative of the GL superconductor free energy with respect to the vector potential,

$$\mathbf{j}_s = -\frac{\partial F_s}{\partial \mathbf{A}(\mathbf{r})} \tag{4.51}$$

which leads to the supercurrent

$$\mathbf{j}_s = -\frac{2e\hbar i}{2m^*}(\psi^*\nabla\psi - \psi\nabla\psi^*) - \frac{(2e)^2}{m^*}|\psi|^2\mathbf{A}. \tag{4.52}$$

Note the close similarity to the superfluid current flow that we found earlier in the case of ^{4}He, Eq. 2.21. The differences from Eq. 2.21 are first the charge of the condensate particles, $-2e$, and the presence of the last term which provides the effect of the vector potential $\mathbf{A}$. Finally, the vector potential must be obtained from the magnetic field arising from both the supercurrents and any other currents, such as the external currents, $\mathbf{j}_{\text{ext}}$, in the solenoid coils of Fig. 3.6,

$$\nabla \times \mathbf{B} = \mu_0(\mathbf{j}_{\text{ext}} + \mathbf{j}_s), \tag{4.53}$$

as given by Maxwell's equations.

4.7 Gauge symmetry and symmetry breaking

The GL order parameter for superconductors has both an amplitude and a complex phase

$$\psi(\mathbf{r}) = |\psi(\mathbf{r})|e^{i\theta(\mathbf{r})}. \tag{4.54}$$

This is similar to the macroscopic wave function for superfluid He II, introduced in Chapter 2. However, unlike superfluids of neutral particles, something very interesting happens now when we consider gauge invariance.

If we make a **gauge transformation** of the magnetic vector potential

$$\mathbf{A}(\mathbf{r}) \to \mathbf{A}(\mathbf{r}) + \nabla\chi(\mathbf{r}) \tag{4.55}$$

then we must make a corresponding change in the phase of the order parameter, θ. Consider the term in the GL free energy density containing the canonical momentum operator

$$\hat{p} = \frac{\hbar}{i}\nabla + 2e\mathbf{A}.$$

If we change the phase of the order parameter by

$$\psi(\mathbf{r}) \to \psi(\mathbf{r})e^{i\theta(\mathbf{r})} \tag{4.56}$$

then we obtain

$$\begin{aligned}\hat{p}\psi(\mathbf{r})e^{i\theta(\mathbf{r})} &= e^{i\theta(\mathbf{r})}\left(\frac{\hbar}{i}\nabla + 2e\mathbf{A}\right)\psi(\mathbf{r}) + \psi(\mathbf{r})e^{i\theta(\mathbf{r})}\hbar\nabla\theta(\mathbf{r}), \\ &= e^{i\theta(\mathbf{r})}\left(\frac{\hbar}{i}\nabla + 2e\left(\mathbf{A} + \frac{\hbar}{2e}\nabla\theta\right)\right)\psi(\mathbf{r}).\end{aligned} \tag{4.57}$$

From this it follows that the free energy will be unchanged when we simultaneously change $\psi(\mathbf{r})$ to $\psi(\mathbf{r})e^{i\theta(\mathbf{r})}$ and the vector potential according to

$$\mathbf{A}(\mathbf{r}) \to \mathbf{A}(\mathbf{r}) + \frac{\hbar}{2e}\nabla\theta. \tag{4.58}$$

This shows that the theory satisfies **local gauge invariance**. Both the phase of the order parameter and the magnetic vector potential depend on the choice of gauge, but all physical observables (free energy, magnetic field **B** etc.) are gauge invariant.

So far this is all completely general. But we saw earlier that a bulk superconductor has a ground state with a constant order parameter, ψ. Therefore it must have the same θ everywhere. There must be a **phase-stiffness**, or an energy cost associated with changing θ from one part of the solid to another. If we consider a superconductor in which the order parameter has a constant magnitude, $|\psi|$, and a phase $\theta(\mathbf{r})$, which varies only slowly with position $\mathbf{r}$, then (using Eq. 4.57) we obtain the total free energy

$$F_s = F_s^0 + \rho_s \int d^3r \left(\nabla\theta + \frac{2e}{\hbar}\mathbf{A}\right)^2. \tag{4.59}$$

Here the **superfluid stiffness** is defined by,

$$\rho_s = \frac{\hbar^2}{2m^*}|\psi|^2 \tag{4.60}$$

and F_s^0 is the total free energy in the ground state ($\theta = constant$, $\mathbf{A} = 0$). Now if we choose some particular gauge for $\mathbf{A}(\mathbf{r})$, such as the London gauge, $\nabla\cdot\mathbf{A} = 0$, then within this fixed gauge there is now a free energy cost associated with further gradients in $\theta(\mathbf{r})$. To minimize the gradient energy, we must minimize the gradients, by making $\theta(\mathbf{r})$ as constant as possible throughout the system. In the case of zero applied magnetic field, we can choose $\mathbf{A} = 0$, and clearly then $\theta(\mathbf{r})$ will be constant everywhere in the system. Again we are back to the XY symmetry of Fig. 2.4. Since the system effectively chooses an (arbitrary) constant order parameter phase everywhere in the system, we can say that the system exhibits **long ranged order** in the order parameter phase, just as a ferromagnet has long ranged order in its magnetization $\mathbf{M}(\mathbf{r})$.

Because the long ranged order is in the phase variable (which is not normally a physical observable in quantum mechanics) we say that the system has **spontaneously broken global gauge symmetry**. The point is that *global* gauge symmetry refers to changing $\theta(\mathbf{r})$ by a constant amount everywhere in the whole solid (which does not require any change in $\mathbf{A}$). This is in contrast to *local* gauge symmetry in which $\theta(\mathbf{r})$ and $\mathbf{A}(\mathbf{r})$ are changed simultaneously, consistent with Eq. 4.58.

Equation 4.59 also implies the London equation, and hence the Meissner–Ochsenfeld effect, bringing us full circle back to Chapter 3. The current can be calculated from a functional derivative of the free energy

$$\begin{aligned} \mathbf{j}_s &= -\frac{\partial F_s[\mathbf{A}]}{\partial \mathbf{A}(\mathbf{r})}, \\ &= -\frac{2e}{\hbar}\rho_s\left(\nabla\theta + \frac{2e}{\hbar}\mathbf{A}\right). \end{aligned} \tag{4.61}$$

Starting in the ground state, where θ is constant, we directly find that with a small constant external vector potential, $\mathbf{A}$, the current is,

$$\mathbf{j}_s = -\rho_s\frac{(2e)^2}{\hbar^2}\mathbf{A}, \tag{4.62}$$

which is exactly the same as the London equation. The superfluid stiffness, ρ_s, is essentially just the London superfluid density, n_s, in disguise!

To make the connection between ρ_s and the London superfluid fraction, n_s, more clear, consider the London equation

$$\mathbf{j}_s = -\frac{n_s e^2}{m_e}\mathbf{A}. \tag{4.63}$$

If we rewrite Eq. 4.61 in the form

$$\mathbf{j}_s = -\frac{(2e)^2}{2m^*}|\psi|^2\mathbf{A}, \tag{4.64}$$

then clearly these are the same. It is conventional to define the constants so that the London superfluid density is $n_s = 2|\psi|^2$ and the GL effective mass is $m^* = 2m_e$ (where m_e is the bare electron mass). With this choice the equation can be interpreted physically as implying that $|\psi|^2$ is the density of pairs of electrons in the ground state. Therefore in comparison with the BCS theory of superconductivity we can interpret $|\psi|^2$ with the density of Cooper pairs in the ground state, and n_s as the density of electrons belonging to these Cooper pairs. The normal fraction, $n_n = n - n_s$ coresponds to the density of unpaired electrons. The GL parameter m^* is the mass of the Cooper pair, which is naturally twice the original electron mass.

Table 4.1 Penetration depth, $\lambda(0)$, and coherence length, $\xi(0)$, at zero temperature for some important superconductors. Data values are taken from Poole (2000).

	T_c (K)	$\lambda(0)$ (nm)	$\xi(0)$ (nm)	κ
Al	1.18	1550	45	0.03
Sn	3.72	180	42	0.23
Pb	7.20	87	39	0.48
Nb	9.25	39	52	1.3
Nb_3Ge	23.2	3	90	30
YNi_2B_2C	15	8.1	103	12.7
K_3C_{60}	19.4	2.8	240	95
$YBa_2Cu_3O_{7-\delta}$	91	1.65	156	95

In terms of the original free energy GL parameters, $\dot{a}$ and b, the superfluid density, n_s is given by

$$n_s = 2|\psi^2| = 2\frac{\dot{a}(T_c - T)}{b}. \tag{4.65}$$

Therefore the London penetration depth, $\lambda(T)$ is given by

$$\lambda(T) = \left(\frac{m_e b}{2\mu_0 e^2 \dot{a}(T_c - T)}\right)^{1/2}. \tag{4.66}$$

Clearly this will diverge at the critical temperature, T_c, since it is proportional to $(T_c - T)^{-1/2}$. We saw earlier that the GL coherence length, $\xi(T)$, also diverges with the same power of $(T_c - T)$, and so the dimensionless ratio,

$$\kappa = \frac{\lambda(T)}{\xi(T)}, \tag{4.67}$$

is independent of temperature within the GL theory. Table 4.1 summarizes the measured values of penetration depth and coherence length at zero temperature, $\lambda(0)$, $\xi(0)$, for a selection of superconductors.

4.8 Flux quantization

Let us now apply the GL theory to the case of a superconducting ring, as shown in Fig. 3.4. Describing the system using cylindrical polar coordinates, $\mathbf{r} = (r, \phi, z)$, with the z-axis perpendicular to the plane of the ring, we see that the order parameter $\psi(\mathbf{r})$ must be periodic in the angle ϕ,

$$\psi(r, \phi, z) = \psi(r, \phi + 2\pi, z). \tag{4.68}$$

We assume that the variations of $\psi(\mathbf{r})$ across the cross section of the ring are unimportant, and so we can neglect r and z dependence. Therefore the possible order parameters inside the superconductor are of the form

$$\psi(\phi) = \psi_0 e^{in\phi}, \tag{4.69}$$

where n is an integer and ψ_0 is a constant. We can interpret n as a **winding number** of the macroscopic wave function, exactly as for the case of superfluid helium in Fig. 2.9.

However, unlike the case of superfluid helium, a circulating current in a superconductor will induce magnetic fields. Assuming that there is a magnetic flux Φ through the ring, then the vector potential can be chosen to be in the tangential direction, $\mathbf{e}_\phi$ and is given by

$$A_\phi = \frac{\Phi}{2\pi R}, \tag{4.70}$$

where R is the radius of the area enclosed by the ring. This follows from

$$\Phi \equiv \int \mathbf{B} \cdot \mathbf{dS} = \int (\nabla \times \mathbf{A}) \cdot \mathbf{dS} = \oint \mathbf{A} \cdot \mathbf{dr} = 2\pi R A_\phi. \tag{4.71}$$

The free energy corresponding to this wave function and vector potential is

$$F_s(T) = F_n(T) + \int d^3r \left(\frac{\hbar^2}{2m^*} \left| \left(\nabla + \frac{2ei}{\hbar}\mathbf{A} \right) \psi \right|^2 + a|\psi|^2 + \frac{b}{2}|\psi|^4 \right) + E_B$$

$$= F_s^0(T) + V \left(\frac{\hbar^2}{2m^*} \left| \frac{in}{R} - \frac{2ei\Phi}{2\pi\hbar R} \right|^2 |\psi|^2 \right) + \frac{1}{2\mu_0} \int B^2 d^3r \tag{4.72}$$

where we have used the expression to gradient in cylindrical polar coordinates

$$\nabla X = \frac{\partial X}{\partial r}\mathbf{e}_r + \frac{1}{r}\frac{\partial X}{\partial \phi}\mathbf{e}_\phi + \frac{\partial X}{\partial z}\mathbf{e}_z \tag{4.73}$$

(Boas 1983), V is the total volume of the superconducting ring, and $F_s^0(T)$ is the ground state free energy of the ring in the absence of any currents and magnetic fluxes. The vacuum magnetic field energy $E_B = (1/2\mu_0)\int B^2 d^3r$ can be expressed in terms of the inductance, L of the ring and the current I,

$$E_B = \frac{1}{2}LI^2. \tag{4.74}$$

Clearly, it will be proportional to the square of the total flux, Φ through the ring

$$E_B \propto \Phi^2.$$

On the other hand, the energy of the superconductor contains a term depending on both the flux Φ and the winding number, n. This term can be expressed as,

$$V\frac{\hbar^2}{2m^*R^2}|\psi|^2(\Phi - n\Phi_0)^2,$$

where the **flux quantum** is $\Phi_0 = h/2e = 2.07 \times 10^{-15}$ Wb.

We therefore see that the free energy is equal to the bulk free energy plus two additional terms depending only on the winding number n and the flux Φ. The energy of the superconducting ring is therefore of the general form,

$$F_s(T) = F_s^{\text{bulk}}(T) + \text{const.}(\Phi - n\Phi_0)^2 + \text{const.}\Phi^2. \tag{4.75}$$

This energy is sketched in Fig. 4.6. We can see from the figure that the free energy is a minimum whenever the flux through the loop obeys $\Phi = n\Phi_0$. This is the phenomenon of **flux quantization** in superconductors.

Taking a ring in its normal state above T_c and cooling it to below T_c will result in the system adopting one of the metastable minima in Fig. 4.6, depending on the applied field. It will then be trapped in the minimum, and a persistent current will flow around the ring to maintain a constant flux $\Phi = n\Phi_0$. Even if any external magnetic fields are turned off, the current in the ring must maintain

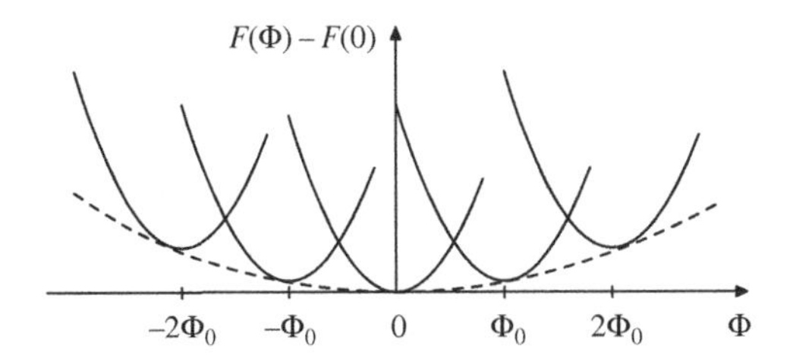

Fig. 4.6 Flux quantization in a superconducting ring. Metastable energy minima exist when the flux is an integer multiple of the flux quantum $\Phi_0 = h/2e$. There is an overall background increase with Φ corresponding to the self-inductance of the ring, making the zero flux state $\Phi = 0$ the global energy minimum. Thermal fluctuations and quantum tunnelling allow transitions between neighboring metastable energy minima.

a constant flux Φ in the ring. It is possible to directly measure the magnetic flux directly in such rings, and hence confirm that it is indeed quantized in units of Φ_0, or multiples of 2×10^{-15} Wb. Incidentally, the fact that flux quantization is observed in units of $\Phi_0 = h/2e$ and not units of h/e is clear experimental proof that the relevant charge is $2e$ not e, hence implying the existence of Cooper pairs.

Given that a system is prepared in one of the metastable minima, it can, in principle, escape over the energy barriers to move into a neighboring lower energy minimum. This would be a mechanism for the persistent current to decay, and hence for dissipation. Such an event corresponds to a change in the winding number, n, and is called a **phase-slip**. However, the rate for thermally hopping over these barriers is exponentially small, of order

$$\frac{1}{\tau} \sim e^{-E_0/k_B T} \tag{4.76}$$

where E_0 is the barrier height between minima in Fig. 4.6. Clearly this thermal hopping rate can be made negligibly small. For example, E_0 is formally proportional to the ring volume, V, and so can be made arbitrarily large in a macroscopic system. In practice persistent currents have been observed to flow for years, with essentially no decay!

Another interesting possible mechanism for a phase-slip would be a quantum tunnelling from one minimum to another. This would be possible at any temperature. But again the rate is impractically small in macroscopic systems. However, a very interesting recent development has been the direct observation of these tunnelling events in small **mesoscopic** superconducting rings. These experiments have demonstrated **macroscopic quantum coherence** and are discussed briefly in the next chapter.

4.9 The Abrikosov flux lattice

The great beauty of the GL theory is that it allows one to solve many difficult problems in superconductvity, without any reference to the underlying microscopic BCS theory. In some sense one could argue that it is more general, for example, it would almost certainly apply to exotic superconductors, such as the high T_c cuprates, even though the original BCS theory does not seem to explain these systems. The other great advantage of the GL theory is that it is considerably easier to work with than the BCS theory, especially in cases where the order parameter has complicated spatial variations. The *tour de force* example of this is the **Abrikosov flux lattice**.

Abrikosov found a solution to the GL equations in the case of a bulk superconductor in a magnetic field.[7] The result he obtained is remarkable in many respects. It is essentially an exact solution for type II superconductors, valid close to H_{c2}. Furthermore, it predicted a striking result, that just below H_{c2} the order parameter forms into a periodic structure of vortices. Each vortex carries a magnetic flux, hence explaining how magnetic flux enters superconductors in the **mixed state** between H_{c1} and H_{c2}. Abrikosov's prediction of the flux lattice was confirmed experimentally, showing not just that vortices occur, but they also tend to align in a regular triangular lattice, as predicted by the theory. This periodic lattice of vortices was perhaps the first example in physics of *emergent phenomena* in *complex systems*; the fact that sufficiently complex systems exhibit a variety of novel phenomena on different length scales. These phenomena effectively arise from *self-organization* on the macroscopic length scale.

[7] Abrikosov was awarded the 2003 Nobel Prize for Physics for this work. He shared the prize with Ginzburg and Leggett.

In type II superconductors the thermodynamic phase transition at H_{c2} is second-order (see Exercise 4.1). Therefore we can expect that the GL order parameter ψ is small in magnitude just below H_{c2} and reaches zero exactly at H_{c2}.[8] Therefore, the magnetization M will also be small at a magnetic field just below H_{c2} (since ψ is near zero, the superfluid density n_s and the screening supercurrents will also vanish at H_{c2}), as can be seen in Fig. 3.8. Therefore to a good approximation we can assume that

[8] In contrast this would not be the case for a type I superconductor, where the transition at H_c is first order. In these superconductors ψ jumps from discontinuously from zero to a finite value at H_c.

$$\mathbf{B} = \mu_0 \mathbf{H}, \tag{4.77}$$

where $\mathbf{H}$ is, as usual, the applied field given by the external apparatus as in Fig. 3.6. This also implies that sufficiently near to H_{c2} we can neglect any spatial variations in the B-field, $\mathbf{B}(\mathbf{r})$ and just treat it as a constant,

$$\mathbf{B} = (0, 0, B). \tag{4.78}$$

It will be convenient to express the corresponding vector potential $\mathbf{A}$ in the **Landau gauge** as

$$\mathbf{A}(\mathbf{r}) = (0, xB, 0). \tag{4.79}$$

In which case the GL equation, Eq. 4.50 becomes

$$-\frac{\hbar^2}{2m^*}\left(\nabla + \frac{2eBi}{\hbar}x\mathbf{e}_y\right).\left(\nabla + \frac{2eBi}{\hbar}x\mathbf{e}_y\right)\psi(\mathbf{r}) + a(T)\psi + b|\psi|^2\psi = 0, \tag{4.80}$$

where, as usual, $\mathbf{e}_y$ is the unit vector in the y direction.

Now if we are infinitesimally below H_{c2}, then ψ is essentially zero and we can drop the cubic term, $b|\psi|^2\psi$. All the other terms are linear in ψ and so we have **linearized** the equation. Expanding out the bracket (paying attention to the commutation of ∇ and $x\mathbf{e}_y$) gives

$$-\frac{\hbar^2}{2m^*}\left(\nabla^2 + \frac{4eBi}{\hbar}x\frac{\partial}{\partial y} - \frac{(2eB)^2}{\hbar^2}x^2\right)\psi(\mathbf{r}) + a(T)\psi = 0. \tag{4.81}$$

Introducing the cyclotron frequency,

$$\omega_c = \frac{2eB}{m^*}, \tag{4.82}$$

and noting that $a(T)$ is negative since we are at a temperature below the zero field T_c, the equation can be written in the form,

$$\left(-\frac{\hbar^2}{2m^*}\nabla^2 - \hbar\omega_c ix\frac{\partial}{\partial y} + \frac{m^*\omega_c^2}{2}x^2\right)\psi(\mathbf{r}) = |a|\psi(\mathbf{r}). \tag{4.83}$$

where $\xi(T)$ is the GL coherence length.

Now Eq. 4.83 has the form of an eigenvalue equation, and is well known in quantum mechanics. It is equivalent to the Scrödinger equation for the wave function of a charged particle in a magnetic field, which has well known **Landau level** solutions (Ziman 1979). The solution has the form,

$$\psi(\mathbf{r}) = e^{i(k_y y + k_z z)}f(x), \tag{4.84}$$

which is a combination of plane waves in the y and z directions, combined with an unknown function of $x, f(x)$.

To find an equation for this function, $f(x)$ we substitute the trial solution into Eq. 4.83. We find that $f(x)$ obeys

$$-\frac{\hbar^2}{2m^*}\frac{d^2f}{dx^2} + \left(\hbar\omega_c k_y x + \frac{m^*\omega_c^2}{2}x^2\right)f = \left(|a| - \frac{\hbar^2(k_y^2+k_z^2)}{2m^*}\right)f. \tag{4.85}$$

The term in brackets on the left hand side can be rearranged by "completing the square,"

$$\left(\hbar\omega_c k_y x + \frac{m^*\omega_c^2}{2}x^2\right) = \frac{m^*\omega_c^2}{2}(x-x_0)^2 - \frac{m^*\omega_c^2}{2}x_0^2, \tag{4.86}$$

where

$$x_0 = -\frac{\hbar k_y}{m\omega_c}. \tag{4.87}$$

Finally, moving all the constants over to the right-hand-side we find,

$$-\frac{\hbar^2}{2m^*}\frac{d^2f}{dx^2} + \frac{m^*\omega_c^2}{2}(x-x_0)^2 f = \left(|a| - \frac{\hbar^2k_z^2}{2m^*}\right)f. \tag{4.88}$$

Equation 4.88 is just the Schrödinger equation for a simple harmonic oscillator, except that the origin of coordinates is shifted from $x = 0$ to $x = x_0$. Therefore the term in brackets on the right is just the energy of the oscillator,

$$\left(n+\frac{1}{2}\right)\hbar\omega_c = |a| - \frac{\hbar^2k_z^2}{2m^*}, \tag{4.89}$$

or

$$\left(n+\frac{1}{2}\right)\hbar\omega_c + \frac{\hbar^2k_z^2}{2m^*} = \dot{a}(T_c - T). \tag{4.90}$$

The corresponding functions $f(x)$ are just the wave functions of a simple Harmonic oscillator for each n, shifted by x_0.

Imagine that we gradually cool a superconductor in an external field, H. At the zero field transition temperature, T_c, it will be impossible to satisfy Eq. 4.90 because of the zero point energy term $\hbar\omega_c/2$ on the left hand side. A solution will only be possible when the temperature is far enough below T_c to achieve,

$$\tfrac{1}{2}\hbar\omega_c = \dot{a}(T_c - T), \tag{4.91}$$

corresponding to the lowest possible energy solution ($n = 0$, $k_z = 0$). This equation determines the depression in transition temperature in the magnetic field,

$$T_c(H) = T_c(0) - \frac{1}{2\dot{a}}\hbar\omega_c,$$
$$= T_c(0) - \frac{2e\hbar\mu_0}{2\dot{a}m^*}H. \tag{4.92}$$

Alternatively, we can start in a large external field, H, above H_{c2} which we gradually decrease (keeping the temperature fixed) until

$$\frac{1}{2}\hbar\frac{2eB}{m^*} = \dot{a}(T_c - T). \tag{4.93}$$

Therefore, rearranging,

$$\mu_0 H_{c2} = B_{c2} = \frac{2m^*\dot{a}(T_c - T)}{\hbar^2}\frac{\hbar}{2e},$$
$$= \frac{\Phi_0}{2\pi\xi(T)^2}. \tag{4.94}$$

It is interesting to note that this result implies that at H_{c2} there is exactly one flux quantum (i.e. one vortex line), in each unit area $2\pi\xi(T)^2$. This expression also provides the simplest way to measure the GL coherence length $\xi(0)$ experimentally. Since $\xi(T) = \xi(0)t^{-1/2}$ where $t = |T - T_c|/T_c$,

$$\mu_0 H_{c2} = \frac{\Phi_0}{2\pi\xi(0)^2}\frac{T_c - T}{T_c}, \tag{4.95}$$

and so by measuring the gradient of $H_{c2}(T)$ near to T_c one can easily find the corresponding $\xi(0)$.

It is also interesting to compare this expression for H_{c2} with the corresponding result for H_c which we found earlier. Using the GL expression for the total energy in the Meissner state we had found that

$$H_c = \frac{\dot{a}}{(\mu_0 b)^{1/2}}(T_c - T),$$
$$= \frac{\Phi_0}{2\pi\mu_0\sqrt{2}\xi\lambda},$$
$$= \frac{H_{c2}}{\sqrt{2}\kappa}, \tag{4.96}$$

and therefore

$$H_{c2} = \sqrt{2}\kappa H_c. \tag{4.97}$$

From this we can deduce that for superconductors with $\kappa > 1/\sqrt{2}$ we will have $H_{c2} > H_c$, and the phase transition will be second-order with the order parameter growing continuously from zero at fields just below H_{c2}, that is, we will have a type II superconductor. On the other hand, for superconductors with $\kappa < 1/\sqrt{2}$ we will have $H_{c2} < H_c$, and the phase transition will be a first-order phase transition at the field H_c, below which the order parameter jumps discontinuously to a finite value. The Abrikosov theory therefore describes the difference between type I and type II superconductors,

$$\kappa \begin{cases} < \frac{1}{\sqrt{2}} & \text{type I,} \\ > \frac{1}{\sqrt{2}} & \text{type II.} \end{cases}$$

The linearized GL equation allows us to find at H_{c2}, but does not immediately tell us anything about the form of the solution below this field. To do this we must solve the nonlinear equation, Eq. 4.80. This is very hard to do in general, but Abrikosov made a brilliant guess and came up with essentially the exact solution! He could see from the solutions to the linearized equation, Eq. 4.80, that only the harmonic oscillator ground state solutions $n = 0$ and $k_z = 0$ will be significant. However, there are still an infinite number of degenerate states, corresponding to the different possible k_y values,

$$\psi(\mathbf{r}) = Ce^{i(k_y y)}e^{-(x-x_0)^2/\xi(T)^2}, \tag{4.98}$$

where C is a normalization constant. Here we have used the fact that the ground state wave function of a quantum harmonic oscillator is a gaussian function. The width of the gaussian solution to Eq. 4.88 turns out to be the GL coherence length $\xi(T)$.

Abrikosov's trial solution was to assume that we can combine these solutions into a **periodic** lattice. If we look for a solution which is periodic in y with period l_y, then we can restrict the values of k_y to

$$k_y = \frac{2\pi}{l_y}n \tag{4.99}$$

with n any positive or negative integer. The corresponding Landau level x shift is

$$x_0 = -\frac{2\pi\hbar}{m\omega_c l_y}n = -\frac{\Phi_0}{Bl_y}n. \tag{4.100}$$

Therefore we can try a periodic solution

$$\psi(\mathbf{r}) = \sum_{n=-\infty,\infty} C_n e^{i(2\pi ny/ly)}e^{-(x+n\Phi_0/Bl_y)^2/\xi(T)^2}. \tag{4.101}$$

In this solution we can view the parameters C_n as variational parameters which are to be chosen to minimize the total GL free energy of the system.

The solution above is periodic in y, but not necessarily periodic in x. Abrikosov noted that it can be made periodic in x provided the coefficients obey

$$C_{n+\nu} = C_n \tag{4.102}$$

for some integer ν. The period is l_x, where

$$l_x = \nu\frac{\Phi_0}{Bl_y}. \tag{4.103}$$

Abrikosov studied the simplest case, $\nu = 1$ which corresponds to a simple square lattice. Later it was shown that a slightly lower total energy is obtained for the $\nu = 2$ case and the minimum energy state corresponds to a simple triangular lattice. In each case the order parameter $\psi(\mathbf{r})$ goes to zero at one point in each unit cell, and there is exactly one flux quantum Φ_0 per unit cell. Therefore the solution is a **periodic lattice of vortices**. This is illustrated in Fig. 4.7.

Experimental evidence for the Abrikosov flux lattice comes from a variety of methods. In a "flux decoration" experiment small paramagnetic particles are dusted onto the surface of the superconductor (just like the children's experiment to see magnetic fields of a bar magnet using iron filings on paper above the magnet!). The particles concentrate at the points of highest magnetic field, that is, the vortices. Other similar methods involve scanning a small SQUID loop or Hall probe just above the surface of the superconductor to directly

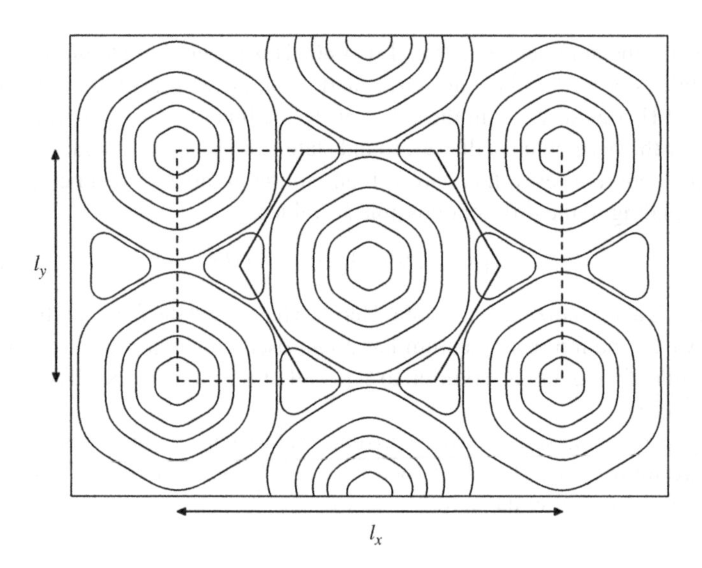

Fig. 4.7 The Abrikosov flux lattice. The figure shows the amplitude of the order parameter, $|\psi(\mathbf{r})|^2$ for the lowest energy triangular lattice. Each triangular unit cell contains one quantum, Φ_0, of magnetic flux and contains one vortex where $\psi(\mathbf{r}) = 0$. In terms of the l_x and l_y lattice periodicities used in Section 4.9, $l_y = \sqrt{3}l_x$, and the rectangular unit cell $l_x \times l_y$ (dashed line) contains two vortices and two flux quanta.

measure the variation of the local flux density $\mathbf{B}(\mathbf{r})$. For most ordinary type II low T_c superconductors, such as Pb or Nb, these experiments indeed show that the vortices form a fairly regular hexagonal lattice. The lattice can be periodic over quite long length scales, but is disrupted now and then by defects. These defects (exactly analogous to crystal dislocations or point defects) tend to concentrate near defects in the underlying crystal lattice (such as grain boundaries, impurities, etc.).

Another method of observing the order in the flux lattice is to use neutron scattering. The neutrons have a magnetic moment, and so are sensitive to the magnetic field $\mathbf{B}(\mathbf{r})$. If this is periodic, as in the Abrikosov flux lattice, then there will be diffraction. The diffraction pattern can be used to find the geometry of the flux lattice. Again the majority of systems studied are found to have triangular lattices. But, interestingly, a square lattice has now been found in a few recently discovered superconductors. These are: the "borocarbide" system $ErNi_2B_2C$, the high temperature superconductor $YBa_2Cu_3O_{7-\delta}$ and the possible p-wave superconductor strontium ruthenate Sr_2RuO_4. All of these have been found to have square vortex lattices in at least some range of external fields. It may be that this is simply because of small corrections to the original Abrikosov theory (such as terms omitted in the standard GL equations, such as higher powers of ψ, like $|\psi|^6$, or higher order gradients). In fact the square lattice and triangular lattice have energies differing by less than 1% in the Abrikosov solution. But in some cases it seems more likely to be due to the fact that the underlying form of superconductivity is "unconventional," meaning that the Cooper pairs have a different symmetry from the normal BCS case. We shall introduce these ideas briefly in Chapter 7.

Finally, we should note that while the Abrikosov solution is essentially exact just below H_{c2}, it cannot necessarily be applied far away from there, such as at H_{c1}. As we have seen, near to H_{c2} the vortices are close together, separated by distances of order the coherence length $\xi(T)$. Effectively they are so densely packed that their cores are essentially touching. On the other hand, just at H_{c1} we have very few vortices in the entire sample, and so they are well separated.

We can estimate of the lower critical field H_{c1} from the energy balance for the very first few vortices to enter a superconductor in the Meissner phase. One can show that a single London vortex has an energy of approximately (see Exercise 3.3)

$$E = \frac{\Phi_0^2}{4\pi\mu_0\lambda^2}\ln\left(\frac{\lambda}{\xi}\right) \tag{4.104}$$

per unit length. Therefore in a superconductor with N/A flux lines per unit area and thickness L, there will be a total energy cost EN/A per unit volume due to the vortices. But on the other hand, each vortex carries a flux Φ_0, and so the average magnetic induction in the sample is $B = \Phi_0 N/A$. The magnetic work gained by the presence of the vortices is $\mu_0 H dM = HdB$ (at constant H). Energy balance therefore favors the presence of the vortices when

$$E\frac{N}{A} < H\Phi_0\frac{N}{A}. \tag{4.105}$$

Thus it becomes energetically favorable for the vortices to enter the sample when $H > H_{c1}$, where

$$H_{c1} = \frac{\Phi_0}{4\pi\mu_0\lambda^2}\ln\left(\frac{\lambda}{\xi}\right). \tag{4.106}$$

This is obviously the lower critical field, and can be simply expressed as

$$H_{c1} = \frac{H_c}{\sqrt{2}\kappa}\ln(\kappa). \tag{4.107}$$

This expression is only valid when $\kappa \gg 1/\sqrt{2}$, that is, in the London vortex limit.

4.10 Thermal fluctuations

The GL theory as described above is purely a **mean-field theory**. It neglects **thermal fluctuations**. In this it is therefore similar to the Curie–Weiss or Stoner models in the theory of magnetism (Blundell 2001). But in fact it is a great strength of the GL theory that it can easily be extended so that these fluctuations can be included. It is much simpler to include these in the GL theory than in the more complex BCS theory.

In the mean-field approach, we must always find the order parameter $\psi(\mathbf{r})$ which minimizes the total free energy of the system. As discussed above, this is a functional minimization. The total free energy of the system $F[\psi]$, in Eq. 4.29, is a functional of the complex order parameter, $\psi(\mathbf{r})$, meaning that it depends on an infinite number of variables: the values of ψ all possible points $\mathbf{r}$. As we saw, the condition for minimizing the free energy is that the functional derivatives given in Eq. 4.37 are zero.

To go beyond this mean-field approach we must include fluctuations of $\psi(\mathbf{r})$ close to this minimum. For example, if we make a small variation in $\psi(\mathbf{r})$, such as $\psi(\mathbf{r}) \to \psi'(\mathbf{r}) = \psi(\mathbf{r}) + \delta\psi(\mathbf{r})$, then we expect that the energy of the system represented by $\psi'(\mathbf{r})$ would be very similar to that represented by $\psi(\mathbf{r})$. If the total energy difference is small, or no more than k_BT, then we might expect that in thermal equilibrium the system would have some probability to be in state $\psi'(\mathbf{r})$. We need to define an effective probability for each possible state.

Clearly this must be based on the usual Boltzmann probability distribution and so we expect that,

$$P[\psi] = \frac{1}{Z}e^{-\beta F[\psi]} \tag{4.108}$$

is the probability density for the system to have order parameter $\psi(\mathbf{r})$. It is again a functional of $\psi(\mathbf{r})$ as indicated by the square brackets.

The partition function Z is the normalization factor in this expression. Formally it is a **functional integral**,

$$Z = \int \mathcal{D}[\psi]\mathcal{D}[\psi^*]e^{-\beta F[\psi]}. \tag{4.109}$$

We can treat the integrals over ψ and ψ^* as formally separate for the same reason that we could view functional derivatives with respect to ψ and ψ^* as formally independent. It is really allowed because in fact we have to specify two independent real functions, the real and imaginary parts of ψ at each point $\mathbf{r}$: $\mathrm{Re}[\psi(\mathbf{r})]$ and $\mathrm{Im}[\psi(\mathbf{r})]$.

What is the meaning of the new integration symbols, like $\mathcal{D}[\psi]$ in Eq. 4.109? We are technically integrating over an infinite number of variables, the values of $\psi(\mathbf{r})$ at every point $\mathbf{r}$. It is difficult (and beyond the scope of this book!) to make this idea mathematically rigorous. But we can find an intuitive idea of what this means by supposing that we only had a discrete set of points in space, $\mathbf{r}_1, \mathbf{r}_2, \ldots, \mathbf{r}_N$. We could define values of ψ and ψ^* at each point, and then calculate a Boltzmann probability. The approximate partition function for this discrete set would be the multiple integral

$$Z(N) = \int d\psi(\mathbf{r}_1)d\psi^*(\mathbf{r}_1)\int d\psi(\mathbf{r}_2)d\psi^*(\mathbf{r}_2)\cdots \int d\psi(\mathbf{r}_N)d\psi^*(\mathbf{r}_N)e^{-\beta F[\psi]}. \tag{4.110}$$

The full functional integral is a limit in which we make the set of points infinitely dense (in fact even possibly uncountably infinite!), defining

$$Z = \lim_{N\to\infty} Z(N). \tag{4.111}$$

One way that this infinite product of integrals might be accomplished is through the Fourier transforms of $\psi(\mathbf{r})$ and $\psi^*(\mathbf{r})$. If we define $\psi_{\mathbf{k}}$ by

$$\psi(\mathbf{r}) = \sum_{\mathbf{k}} \psi_{\mathbf{k}}e^{i\mathbf{k}\cdot\mathbf{r}} \tag{4.112}$$

then specifying the parameters $\psi_{\mathbf{k}}$ and $\psi^*_{\mathbf{k}}$ at every wave vector $\mathbf{k} = (2\pi n_x/L_x, 2\pi n_y/L_y, 2\pi n_z/L_z)$ (or equivalently the real and imaginary parts), defines the full functions $\psi(\mathbf{r})$ and $\psi^*(\mathbf{r})$. In this representation we can write the partition function as

$$Z = \prod_{\mathbf{k}}\left(\int d\psi_{\mathbf{k}}d\psi^*_{\mathbf{k}}\right)e^{-\beta F[\psi]}. \tag{4.113}$$

Again there are an infinite number of integrals, two for each point $\mathbf{k}$.

As an example of the sort of thermal fluctuation effects that can be calculated within this formalism we shall consider the specific heat of a superconductor near to T_c. For a superconductor in zero magnetic field we have the free energy functional,

$$F[\psi] = \int d^3r \left(\frac{\hbar^2}{2m^*} |\nabla \psi|^2 + a|\psi|^2 + \frac{b}{2}|\psi|^4 \right) \tag{4.114}$$

(dropping the constant normal state free energy F_n, which will be irrelevant here). Writing this in terms of the Fourier coefficients $\psi_{\mathbf{k}}$ we find

$$F[\psi] = \sum_{\mathbf{k}} \left(\frac{\hbar^2 k^2}{2m^*} + a \right) \psi^*_{\mathbf{k}} \psi_{\mathbf{k}} + \frac{b}{2} \sum_{\mathbf{k}_1,\mathbf{k}_2,\mathbf{k}_3} \psi^*_{\mathbf{k}_1} \psi^*_{\mathbf{k}_2} \psi_{\mathbf{k}_3} \psi_{\mathbf{k}_1+\mathbf{k}_2-\mathbf{k}_3}, \tag{4.115}$$

which could in principle be inserted directly into Eq. 4.113. In general this would be very difficult, and requires either massive numerical Monte Carlo simulation, or some other approximation. The simplest approximation that we can make is to make the **gaussian approximation**, in which we neglect the quartic (b) term in the free energy. In this approximation we find a simple result

$$Z = \prod_{\mathbf{k}} \int d\psi_{\mathbf{k}} d\psi^*_{\mathbf{k}} \exp\left\{ -\beta \left(\frac{\hbar^2 k^2}{2m^*} + a \right) \psi^*_{\mathbf{k}} \psi_{\mathbf{k}} \right\}. \tag{4.116}$$

Changing variables to the two real functions, $\mathrm{Re}[\psi_{\mathbf{k}}]$ and $\mathrm{Im}[\psi_{\mathbf{k}}]$ gives

$$Z = \prod_{\mathbf{k}} \int d\mathrm{Re}[\psi_{\mathbf{k}}] d\mathrm{Im}[\psi_{\mathbf{k}}] \exp\left\{ -\beta \left(\frac{\hbar^2 k^2}{2m^*} + a \right) (\mathrm{Re}[\psi_{\mathbf{k}}]^2 + \mathrm{Im}[\psi_{\mathbf{k}}]^2) \right\}, \tag{4.117}$$

and so for each $\mathbf{k}$ the integral is just a two-dimensional gaussian integral. These can be done exactly, resulting in the partition function

$$Z = \prod_{\mathbf{k}} \frac{\pi}{\beta \left((\hbar^2 k^2/2m^*) + a \right)}. \tag{4.118}$$

From the partition function it is possible to calculate all thermodynamic quantities of interest. For example, the total internal energy is given by Eq. 2.57,

$$\begin{aligned} U &= -\frac{\partial \ln Z}{\partial \beta}, \\ &= +k_B T^2 \frac{\partial \ln Z}{\partial T} \\ &\sim -\sum_{\mathbf{k}} \frac{1}{\left((\hbar^2 k^2/2m^*) + a \right)} \frac{da}{dT}, \end{aligned} \tag{4.119}$$

where in the last step we have kept only the most important contribution which comes from the change of the GL parameter a with T, $da/dT = \dot{a}$.

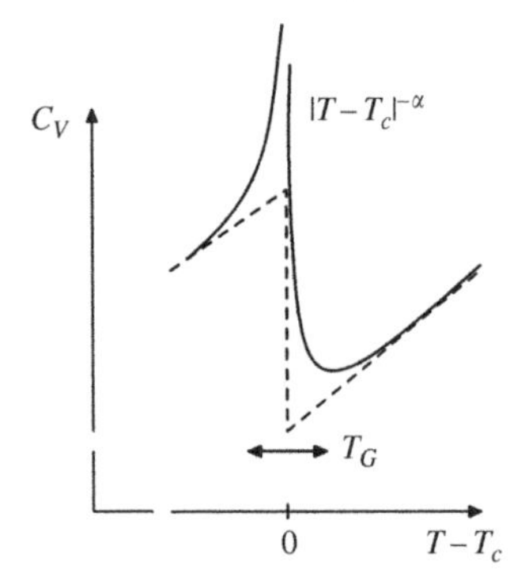

Fig. 4.8 Specific heat of a superconductor near to T_c in the gaussian approximation. The mean-field GL theory gives a discontinuity at T_c. But this is supplemented by a thermal fluctuation contribution which diverges like $|T - T_c|^{-\alpha}$ with $\alpha = 1/2$. The full renormalization group treatment (ignoring magnetic field terms) shows that α is given by the three-dimensional XY model value, exactly as in superfluid helium shown in Fig. 2.4.

The gaussian approximation for the heat capacity near to T_c is found by differentiating again, giving

$$\begin{aligned}
C_V = \frac{dU}{dT} &= \sum_{\mathbf{k}} \frac{1}{\left(\frac{\hbar^2 k^2}{2m^*} + a\right)^2} \dot{a}^2 \\
&= \frac{V}{(2\pi^3)} \frac{\dot{a}^2}{a^2} \int d^3k \frac{1}{(1 + \xi(T)^2 k^2)^2} \\
&\sim \frac{V}{(2\pi^3)} \frac{\dot{a}^2}{a^2} \frac{1}{\xi(T)^3} \\
&\sim \frac{1}{(T - T_c)^2} |T_c - T|^{3/2} \\
&\sim \frac{1}{|T - T_c|^{1/2}},
\end{aligned} \tag{4.120}$$

(where, for simplicity, we have ignored the numerical multiplying prefactors). This shows that the thermal fluctuations can make a very large contribution to the heat capacity, essentially diverging at the critical temperature T_c. If we sketch this behavior we see that the thermal fluctuations make a large difference to the original mean-field specific heat of Fig. 4.4. As can be seen in Fig. 4.8. In fact, once the fluctuations are included the heat capacity of superconductors becomes much more similar to the heat capacity of superfluid ^{4}He at T_c, as shown in Fig. 2.3.[9]

[9] In fact, the gaussian theory, as outlined above, is not truly correct since it dropped the $|\psi|^4$ terms in the GL free energy. When these terms are included the resulting theory is known in statistical physics as the XY or $O(2)$ model. Its true critical behavior near T_c can be calculated with various methods based on the renormalization group. The resulting critical exponent for specific heat α is very small and very different from the $\alpha = 1/2$ which is given by the gaussian approximation of Eq. 4.120.

Experimentally these thermal fluctuations are very difficult to see in standard "low T_c" superconductors, such as Pb or Nb. It is possible to estimate the range of temperatures near to T_c where these fluctuations are significant. This temperature range, T_G, is known as the Ginzburg critereon. Ginzburg found that this temperature range is extremely small, that is, much less than 1 μK for most low T_c superconductors. Therefore we can say that the original mean-field approach to the GL equations is perfectly well justified. However, in the high temperature superconductors, discovered in 1986, the coherence length $\xi(0)$ is very small (Table 4.1), of order just a few Angstroms. It turns out that the corresponding Ginzburg temperature range, T_G, is of order 1–2 K. Therefore it is quite possible to see such thermal fluctuation effects in these systems. The specific heat near T_c clearly shows thermal critical fluctuations, as shown in Fig. 4.9. In fact in these experimental results very good agreement is found

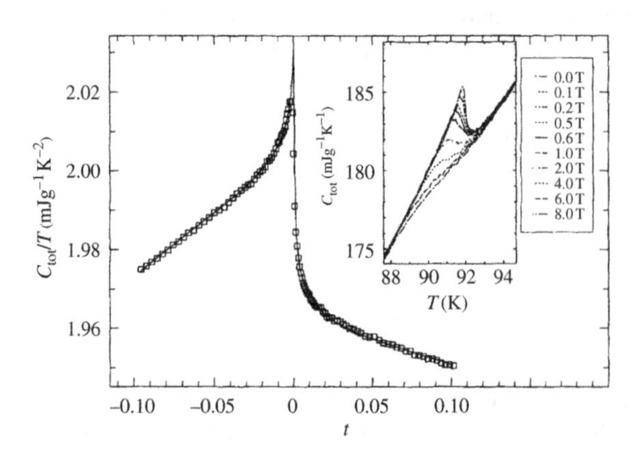

Fig. 4.9 Experimental heat capacity of the high temperature superconductor $YBa_2Cu_3O_{7-\delta}$ near to T_c. In zero magnetic field the experimental data fits very well the predictions of the three dimensional XY model. An external magnetic field (inset) removes the singularity, but does not visibly reduce T_c significantly. Reprinted figure with permission from Overend, Howson and Lawrie (1994). Copyright (1994) by the American Physical Society.

using the value of the critical exponent α given by the three-dimensional XY model predictions, exactly as is the case in superfluid ^{4}He, Fig. 2.4. The gaussian model exponent $\alpha = 1/2$ does not fit at all as well. Another example of thermal fluctuation effects can be seen in the resistivity, $\rho(T)$, just above T_c. Thermal fluctuations make $\rho(T)$ begin to bend down toward zero even at temperatures quite far above T_c. This downward bending is clearly visible in the resistivity curve of the $T_c = 135$ K superconductor $HgBa_2Ca_2Cu_3O_{8+\delta}$ shown in Fig. 3.2.

4.11 Vortex matter

Another very important consequence of thermal fluctuations occurs in the mixed state of high temperature superconductors. As we have seen, Abrikosov's flux lattice theory shows that the vortices align in a periodic lattice arrangement, essentially like a crystal lattice, either triangular or square. However, this is again a mean-field approximation! We must, in principle, again include the effects of thermal fluctuations.

The theories of the resulting **vortex matter** states show a very wide range of possibilities. It is still possible to talk about the vortices, but now they themselves form a variety of different states, including liquid and glassy (random, but frozen) states, as well as nearly perfectly ordered crystalline states. It is believed that the flux lattice never has true crystalline order, and thermal fluctuations always lead to an eventual loss of long ranged order in the periodic structure (although in practice periodicity can be quite well extended). A full discussion of these topics requires a whole book in itself (Singer and Schneider 2000), and there also are many extensive review papers (Blatter 1994).

Unfortunately these thermal fluctuations have been disastrous for commercial applications of high T_c superconductors in high current wires and electromagnets (Yeshrun 1998).[10] The problem is that thermal fluctuations lead to motion of the vortices, and this leads to a source of energy dissipation. Therefore the resistivity is not zero for high T_c superconductors in a magnetic field. The problem also occurs in low T_c superconductors, but to a much lesser extent. In these systems the energy dissipation due to motion of vortices can be reduced or eliminated by providing **pinning centers** which "pin" the vortex lattice and prevent it from moving. Typically these are just impurities, or naturally occurring crystal defects such as grain boundaries and dislocations.

[10] Perhaps this is not the only difficulty with commerical applications of high T_c superconductivity. The materials are brittle and cannot easily be made into wires. Nevertheless these problems have been gradually overcome, and now high T_c superconducting wires are beginning to make a serious entry into commercial applications. For example, at least one US city receives part of its electrical power through underground superconducting cables. Some microwave receivers, such as in some masts for mobile phones beside motorways, also use superconducting devices operating at liquid nitrogen temperatures.

To see why motion of vortices leads to energy dissipation it is necessary to see that a current density $\mathbf{j}$ flowing through the vortex lattice (perpendicular to the magnetic field) leads to a Lorentz (or Magnus) force on each vortex. The overall force is

$$\mathbf{f} = \mathbf{j} \times \mathbf{B} \tag{4.121}$$

per unit volume of the vortex lattice. This will tend to make the vortex liquid flow in the direction perpendicular to the current as shown in Fig. 4.10.

Unfortunately if the vortices flow in response to this force, work is done and there is energy dissipation. To calculate the work, consider a loop of superconducting wire, with a current flowing around the wire. Vortices will tend to drift transversely across the wire, say entering on the inner side of the wire

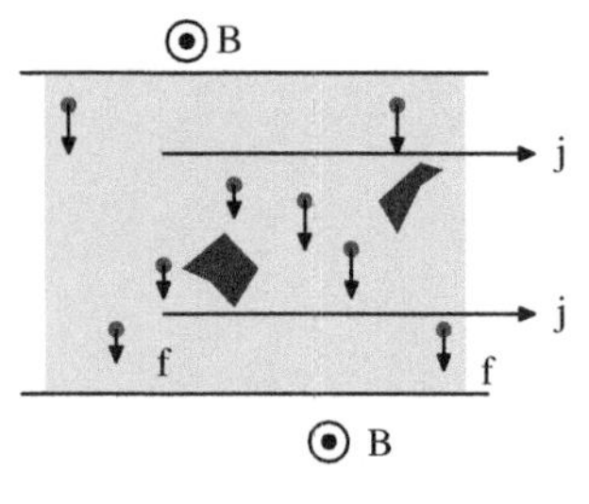

Fig. 4.10 Energy dissipation due to vortex flow in superconductors. Every vortex experiences a Lorentz (Magnus) force perpendicular to the supercurrent direction. This causes the vortices to drift sideways across the wire, unless pinned by defects (shown as black regions above). For each vortex which crosses the wire from one side to the other, a certain amount of work is done, and energy is dissipated.

and drifting over to the outer side. This is illustrated in Fig. 4.10. Each vortex carries a magnetic flux Φ_0, and so the total magnetic flux in the ring Φ changes by Φ_0 with each vortex that crosses from one side of the wire to the other. But by elementary electromagnetism there is an EMF induced in the wire given by $\mathcal{E} = -d\Phi/dt$. Power is dissipated, at a rate given by $P = \mathcal{E}I$ where I is the total current. Therefore vortex motion directly leads to **finite resistance**! In the mixed state, superconductors only have truly zero resistance when the vortices are pinned and unable to move.

In high T_c superconductors the thermal motion of vortices leads to especially bad pinning and hence a significant resistivity in the mixed state. To make matters worse, the lower critical field H_{c1} is tiny, often less than the Earth's magnetic field, and so vortices can never by truly eliminated. At high temperatures and near to H_{c2} it is believed that the vortex matter is in a liquid state, and so the vortices can move freely and pinning is essentially impossible.[11] Lowering the temperature, or going further away from H_{c2} the vortex matter appears to "freeze" into a glassy state. A glass is random spatially, but frozen in time. Since glass is effectively rigid the vortices cannot move and pinning is able to largely prevent flux motion. Therefore in this state the resistivity is quite low. Unfortunately, even in this glassy vortex state the resistivity is not fully zero, since **flux creep** can occur. The random pinning force provides a set of energy barriers to vortex motion, but thermal motions mean that from time to time the vortices can hop over the local energy barrier and find a new configuration.[12] The line in the (H, T) phase diagram where the glassy phase occurs is called the "irreversibility line," as shown in Fig. 4.11. Something approaching zero resitivity is approached only well below this line. This effectively limits the useful range of magnetic fields for applications of high T_c superconductivity to very much less than the hundreds of Tesla that one might have expected from the nominal values of $\mu_0 H_{c2} > 100\,\mathrm{T}$, such as one might expect from Table 4.1.

[11] A liquid can always flow around any impurities and so pinning centers have no effect in the vortex liquid state.

[12] The process is presumably analogous to the way that window glass in medieval cathedrals appears to have gradually flowed downward over timescales of hundreds of years.

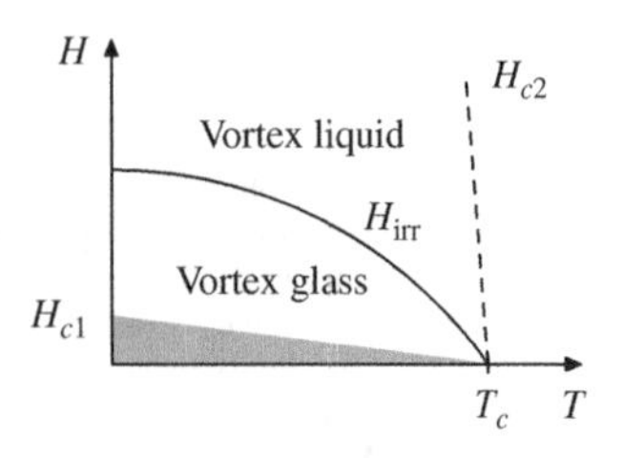

Fig. 4.11 Proposed magnetic phase diagram of some high T_c superconductors. Below H_{c2} vortices form, but are in a liquid state, leading to finite resistance of the superconductor. Below the "irreversibility line" the vortices freeze (either into a glassy or quasi periodic flux lattice). In this state resistivity is still finite, due to flux creep, but becomes negligible far below the irreversibility line. H_{c1} is extremely small.

4.12 Summary

We have seen how the GL theory provides a mathematically rather simple picture with which to describe quite complex phenomena in superconductivity. In terms of the phenomenological order parameter, $\psi(\mathbf{r})$, and four empirically determined parameters ($\dot{a}$, b, m^*, and T_c) we can can construct a theory of superconductivity which encompasses fully phenomena such as the Abrikosov flux lattice, flux quantization, and from which one can "derive" the London equation.

The power of the theory is also apparent in the way it can be modified to incorporate thermal fluctuations, including critical phenomena and vortex matter physics. It should be noted that these areas are still highly active areas of experimental and theoretical activity. Even some very simple and fundamental questions are still hotly debated, such as the various vortex phases occurring in high temperature superconductors. These also have important implications for potential commercial applications of these materials.

Further reading

Magnetic work in thermodymanics is discussed in detail in the textbooks by Mandl (1987) and Callen (1960).

The idea of order parameters and Ginzburg–Landau theory in general, including superconductors, are discussed in Chakin and Lubensky (1995) and Anderson (1984). Applications of Ginzburg–Landau theory to vortex states and other problems are covered in detail by de Gennes (1966). Other books are also useful, such as Tilley and Tilley (1990), or Tinkham (1996). In fact almost all textbooks on superconductivity have at least one chapter dealing with Ginzburg–Landau theory and its predictions.

Thermal fluctuations and critical phenomena are in themselves huge fields of study. A good introductory course is Goldenfeld (1992), while the books Amit (1984) and Ma (1976) are very comprehensive. These books discuss very general classes of theoretical models, but the Ginzburg–Landau theory we have discussed is equivalent to the model they call XY or $O(2)$.

For a modern view of thermal fluctuation phenomena and the problems of vortex matter physics, especially in its application to high temperature superconductors, see the book by Singer and Schneider (2000), or the review articles by Blatter (1994) and Yeshrun (1996).

Exercises

(4.1) (a) For a type I superconductor $H_c(T)$ is the boundary between normal metal and superconductor in the H, T phase diagram. Everywhere on this boundary thermal equilibrium requires

$$G_s(T,H) = G_n(T,H).$$

Apply this equation and $dG = -S\,dT - \mu_0\,VMdH$ to two points on the $H - T$ phase boundary (T, H) and $(T + \delta T, H + \delta H)$, as illustrated in Fig. 4.12. Hence show that:

$$-S_s\delta T - \mu_0 VM_s\delta H = -S_n\delta T - \mu_0 VM_n\delta H$$

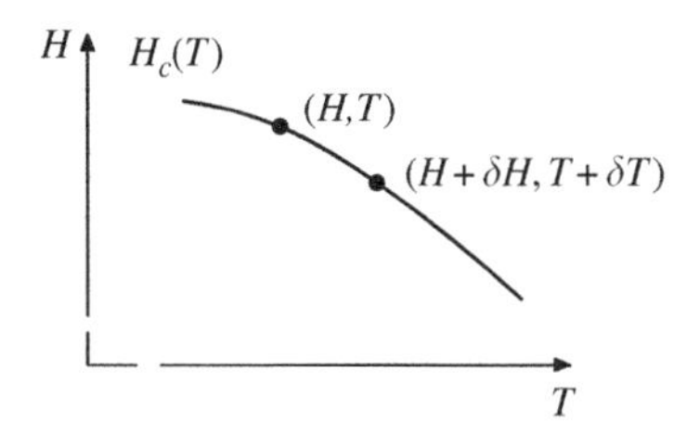

Fig. 4.12 Exercise 4.1. Consider the Gibbs free energy at the points shown on the normal-superconducting phase boundary of a type I superconductor (T, H) and $(T + \delta T, H + \delta H)$. At both points equilibrium requires that both normal and superconducting Gibbs free energies must be equal: $G_n(T,H) = G_s(T,H)$.

when δT and δH are small, and where $S_{s/n}$ $M_{s/n}$ are the superconducting and normal state entropy and magnetization, respectively.

(b) Using part (a), and $M_n = 0$, and $M_s = -H$ show that the latent heat per unit volume for the phase transition, $L = T(s_n - s_s)$, is given by

$$L = -\mu_0 TH_c\frac{dH_c(T)}{dT}$$

where the phase boundary curve is $H_c(T)$. (This is exactly analogous to the Clausius–Claperyon equation in a gas–liquid phase change except H replaces P and $-\mu_0 M$ replaces V. See Mandl 1987, p. 228.)

(4.2) Find $|\psi|^2$, the free energy $F_s - F_n$, the entropy and the heat capacity of a superconductor near T_c, using the bulk Ginzburg–Landau free energy. Sketch their variations with temperature assuming that $a = \dot{a} \times (T - T_c)$ and $\dot{a}$ and b are constant near T_c.

(4.3) (a) Show that for one-dimensional problems, such as the surface discussed in Section 4.5, the Ginzburg–Landau equations for $\psi(x)$ can be rewritten as:

$$-\frac{d^2}{dy^2}f(y) - f(y) + f(y)^3 = 0,$$

where $\psi(x) = \psi_0 f(x/\xi)$, $y = x/\xi$ and $\psi_0 = \sqrt{|a|/b}$.

(b) Verify that

$$f(y) = \tanh(y/\sqrt{2})$$

is a solution to the equation in (b) corresponding to the boundary condition $\psi(0) = 0$. Hence sketch $\psi(x)$ near the surface of a superconductor.

(c) Often the surface boundary condition of a superconductor is not $\psi(x) = 0$, but $\psi(x) = C$ where C is a numerical constant. Show that if $C < \psi_0$ we can just translate the solution from Problem 4.2 sideways to find a valid solution for any value of C in the range $0 \leq C < \psi_0$.

(d) In the proximity effect a metal (in the half-space $x > 0$) is in contact with a superconductor (occupying the region $x < 0$). Assuming that the normal metal can be described by a Ginzburg–Landau model but with $a > 0$, show that the order parameter $\psi(x)$ induced in the metal by the contact with the superconductor is approximately

$$\psi(x) = \psi(0)e^{-x/\xi(T)},$$

where $\hbar^2/2m^*\xi(T)^2 = a > 0$, and $\psi(0)$ is the order parameter of the superconductor at the interface.

(4.4) (a) In Eq. 4.120 we showed that the gaussian model gives a divergence in specific heat of the form

$$C_V \sim \frac{1}{|T - T_c|^\alpha}$$

with $\alpha = 1/2$. Repeat the steps given in Eq. 4.120 for the case of a two-dimensional system, and show that in this case the gaussian model predicts $\alpha = 1$.

(b) This argument can also be extended easily to the general case of d-dimensions. By replacing the **k** sum in Eq. 4.120 by an integral of the form

$$\sum_{\mathbf{k}} \rightarrow \frac{1}{(2\pi^d)} \int d^d k$$

show that in d dimensions we have the critical exponent

$$\alpha = 2 - \frac{d}{2}.$$

The macroscopic coherent state

5

5.1 Introduction

We have seen in the previous chapters that the concept of the macroscopic wave function,$\psi(\mathbf{r})$, is central to understanding atomic Bose–Einstein condensates (BEC), superfluid ^{4}He, and even superconductivity within the Ginzburg–Landau (GL) theory. But the connection between these ideas is not at all clear, since the atom condensates and ^{4}He are bosonic systems, while superconductivity is associated with the conduction electrons in metals which are fermions. The physical meaning of the GL order parameter was not explained until after 1957 when Bardeen Cooper and Schrieffer (BCS) published the first truly microscopic theory of superconductivity. Soon afterward the connection was finally established by Gor'kov. He was able to show that, at least in the range of temperatures near T_c, the GL theory can indeed be derived from the BCS theory. Furthermore this provides a physical interpretation of the nature of the order parameter. Essentially it is describing a macroscopic wave function, or condensate, of Cooper pairs.

The purpose of this chapter is to clarify the concept of a macroscopic wave function, and show how it arises naturally from the physics of **coherent states**. Coherent states were first developed in the field of quantum optics, and were especially useful in the theory of the laser. The laser is, of course, yet another type of macroscopic coherent state, with close similarities to atomic BEC. In this chapter we shall first introduce the concept and mathematical properties of coherent states before applying them to bosonic systems. Using this approach we shall rederive the Gross–Pitaevskii equations for the weakly interacting bose gas, as originally introduced above in Chapter 1.

Coherent states can be defined for fermions as well as for bosons. But single fermion coherent states, while very useful in other contexts, are not directly useful in the theory of superconductivity. What is needed is a coherent state of **fermion pairs**. Such coherent states are exactly the type of many-body quantum state first written down in the theory of BCS in their 1957 theory of superconductivity. We shall postpone a full discussion of this BCS theory until the following chapter. Here we focus specifically on the physics of the BCS coherent state, without worrying, for example, about why it is a stable ground state. This separation has the advantage that the key concepts can be presented more clearly and they can be seen to be very general. Indeed it is not necessary to rely on every detail of the BCS theory in order to understand physically the properties of the quantum coherent state. In this approach we can also see very generally the connection between the BCS state and the GL theory, since the coherent state of electron pairs provides a direct connection to the order parameter $\psi(\mathbf{r})$.

This logical separation between the full detail of the BCS theory and the physical origin of the order parameter is not just an educational device; it also has a useful purpose more generally. For example, there are several superconductors where we do not know if the BCS theory is applicable, the high T_c superconductors being the most obvious example. However, even though we do not know the **mechanism of pairing** we do know that these systems do have Cooper pairs. For example, they have flux quantization in the usual units of $\Phi_0 = h/2e$, showing that the fundamental charge unit is $2e$. We can also assume that there will be a GL order parameter whatever the actual pairing mechanism, and this knowledge will provide a sound basis for many theories of the superconducting state (e.g. theories of the vortex matter states in high T_c superconductors). We can therefore separate the actual mechanism of pairing from its main consequence: the existence of the order parameter.

5.2 Coherent states

To start with, let us just go back to some elementary undergraduate quantum mechanics: the quantum harmonic oscillator. The Hamiltonian operator is

$$\hat{H} = \frac{\hat{p}^2}{2m} + \frac{m\omega_c^2}{2}x^2 \tag{5.1}$$

where $\hat{p} = -i\hbar(d/dx)$ is the one-dimensional momentum operator, m is the particle mass, and ω_c is the classical oscillator angular frequency. The eigenstates of the oscillator are given by

$$\hat{H}\psi_n(x) = E_n\psi_n(x), \tag{5.2}$$

with energy levels E_n.

The most elegant method of solving this classic problem, to find E_n and $\psi_n(x)$ is to introduce the **ladder operators**,

$$\begin{aligned} \hat{a} &= \frac{1}{(\hbar\omega_c)^{1/2}}\left(\frac{\hat{p}}{(2m)^{1/2}} - i\frac{(m\omega_c^2)^{1/2}x}{(2)^{1/2}}\right), \\ \hat{a}^+ &= \frac{1}{(\hbar\omega_c)^{1/2}}\left(\frac{\hat{p}}{(2m)^{1/2}} + i\frac{(m\omega_c^2)^{1/2}x}{(2)^{1/2}}\right). \end{aligned} \tag{5.3}$$

These operators have a number of very useful and easily derived properties,[1] which can be summarized as follows:

[1] See Exercise 5.1.

$$\hat{a}^+\psi_n(x) = (n+1)^{1/2}\psi_{n+1}(x), \tag{5.4}$$

$$\hat{a}\psi_n(x) = (n)^{1/2}\psi_{n-1}(x), \tag{5.5}$$

$$\hat{a}^+\hat{a}\psi_n(x) = n\psi_n(x), \tag{5.6}$$

$$\left[\hat{a}, \hat{a}^+\right] = 1. \tag{5.7}$$

The first of these relations shows that the operator $\hat{a}^+$ changes any state to the next one higher up the "ladder" of the possible n values. Similarly the second shows that $\hat{a}$ moves down the ladder. From the third relation, the combination $\hat{a}^+\hat{a}$ results in no change of n. Therefore Eq. 5.6 shows that we can identify the combination $\hat{n} = \hat{a}^+\hat{a}$ as the **number operator**, which gives the quantum number n of any state,

$$\hat{n}\psi_n(x) = n\psi_n(x). \tag{5.8}$$

The commutator relation, Eq. 5.7,

$$[\hat{a}, \hat{a}^+] = \hat{a}\hat{a}^+ - \hat{a}^+\hat{a} = 1 \tag{5.9}$$

is fundamental to the quantum mechanics of bosonic systems, as we shall see below.

In terms of the latter operators the oscillator Hamiltonian is given by,

$$\hat{H} = \hbar\omega_c \left(\hat{a}^+\hat{a} + \tfrac{1}{2}\right). \tag{5.10}$$

Combined with Eq. 5.6 it immediately shows that the energy levels are

$$E_n = \hbar\omega_c \left(n + \tfrac{1}{2}\right). \tag{5.11}$$

exactly as expected.

Using the ladder raising operator, $\hat{a}^+$ repeatedly, Eq. 5.4 shows that one can construct all of the eigenvectors, $\psi_n(x)$, iteratively by acting repeatedly on an initial ground state $\psi_0(x)$,

$$\psi_n(x) = \frac{1}{(n!)^{1/2}}(\hat{a}^+)^n\psi_0(x). \tag{5.12}$$

Therefore to find the complete set of states it is only necessary to find $\psi_0(x)$ (which elementary quantum mechanics tells us is a simple gaussian function), and then all of the remaining quantum states can be generated essentially automatically.

But the ladder operators also have many other uses. In particular let us define a **coherent state** by,

$$|\alpha\rangle = C\left(\psi_0(x) + \frac{\alpha}{1!^{1/2}}\psi_1(x) + \frac{\alpha^2}{2!^{1/2}}\psi_2(x) + \frac{\alpha^3}{3!^{1/2}}\psi_3(x) + \cdots\right) \tag{5.13}$$

where α is any arbitrary complex number, and C is a normalization constant. This constant C can be found easily from the normalization condition

$$\begin{aligned} 1 &= \langle\alpha|\alpha\rangle \\ &= |C|^2\left(1 + \frac{|\alpha|^2}{1!} + \frac{(|\alpha|^2)^2}{2!} + \frac{(|\alpha|^2)^3}{3!} + \cdots\right) \\ &= |C|^2 e^{|\alpha|^2} \end{aligned} \tag{5.14}$$

and so we can take $C = e^{-|\alpha|^2/2}$.

Coherent states have many interesting properties. The following expression is a particularly useful relation

$$|\alpha\rangle = e^{-|\alpha|^2/2}\left(1 + \frac{\alpha\hat{a}^+}{1!} + \frac{(\alpha\hat{a}^+)^2}{2!} + \frac{(\alpha\hat{a}^+)^3}{3!} + \cdots\right)|0\rangle \tag{5.15}$$

where $|0\rangle = \psi_0(x)$ is the ground state and is also the coherent state with $\alpha = 0$. This expression can be written very compactly as,

$$|\alpha\rangle = e^{-|\alpha|^2/2}\exp(\alpha\hat{a}^+)|0\rangle. \tag{5.16}$$

Note that the exponential of any operator, $\hat{X}$, is just defined by the usual series expansion of exponential

$$\exp(\hat{X}) = 1 + \frac{\hat{X}}{1!} + \frac{\hat{X}^2}{2!} + \frac{\hat{X}^3}{3!} + \cdots \tag{5.17}$$

Another interesting relation which can be obtained from Eq. 5.13 is

$$\hat{a}|\alpha\rangle = \alpha|\alpha\rangle \tag{5.18}$$

therefore they are eigenstates of the ladder operator $\hat{a}$. To prove this we can write the state $\hat{a}|\alpha\rangle$ explicitly

$$\hat{a}|\alpha\rangle = e^{-|\alpha|^2/2}\hat{a}\left(\psi_0(x) + \frac{\alpha}{1!^{1/2}}\psi_1(x) + \frac{\alpha^2}{2!^{1/2}}\psi_2(x) + \frac{\alpha^3}{3!^{1/2}}\psi_3(x) + \cdots\right). \tag{5.19}$$

But $\hat{a}\psi_n(x) = n^{1/2}\psi_{n-1}(x)$ and so this gives

$$\hat{a}|\alpha\rangle = e^{-|\alpha|^2/2}\left(0 + \frac{\alpha 1^{1/2}}{1!^{1/2}}\psi_0(x) + \frac{\alpha^2 2^{1/2}}{2!^{1/2}}\psi_1(x) + \frac{\alpha^3 3^{1/2}}{3!^{1/2}}\psi_2(x) + \cdots\right), \tag{5.20}$$

which is clearly equal to $\alpha|\alpha\rangle$.

Finally, two further nice properties are also simple consequences of Eq. 5.18,

$$\langle\alpha|\hat{a}|\alpha\rangle = \alpha \tag{5.21}$$

$$\langle\alpha|\hat{a}^+\hat{a}|\alpha\rangle = |\alpha|^2. \tag{5.22}$$

Therefore the value of $|\alpha|^2$ gives the mean number operator $\langle\hat{n}\rangle$ of the quantum state. Extending this to $\hat{n}^2$ we can find the number uncertainty Δn,

$$\begin{aligned}\langle\hat{n}^2\rangle &= \langle\alpha|\hat{a}^+\hat{a}\hat{a}^+\hat{a}|\alpha\rangle \\ &= \langle\alpha|\hat{a}^+(\hat{a}^+\hat{a} + 1)\hat{a}|\alpha\rangle \\ &= |\alpha|^4 + |\alpha|^2 \end{aligned} \tag{5.23}$$

$$\begin{aligned}\Delta n &= \sqrt{\langle\hat{n}^2\rangle - \langle\hat{n}\rangle^2} \\ &= |\alpha|. \end{aligned} \tag{5.24}$$

Coherent states do not have a definite value of the quantum number n, simply because they are not eigenstates of the number operator. In fact the probability of observing the value of n in a quantum measurement of state $|\alpha\rangle$ is actually a

Poisson distribution shown in Fig. 5.1

$$P_n = \frac{|\alpha|^{2n}}{n!} e^{-|\alpha|^2} \tag{5.25}$$

as can easily be seen from Eq. 5.13. In this distribution, as given by Eq. 5.24, the standard deviation in the number n is

$$\Delta n = \sqrt{\langle \hat{n} \rangle} \tag{5.26}$$

or

$$\frac{\Delta n}{\langle n \rangle} \sim \frac{1}{\sqrt{\langle n \rangle}}. \tag{5.27}$$

We shall be mostly interested in **macroscopic** coherent states, where $\langle n \rangle$ is essentially infinite. For such states, one can see that the standard deviation Δn becomes essentially negligible compared with $\langle n \rangle$. Therefore to a good approximation we can approximate many operator expectation values by their **mean-field** values derived from the replacement $\hat{n} \approx \langle n \rangle$. In Fig. 5.1 one can see that the distribution becomes strongly peaked about its mean value of $\langle n \rangle$ even for quite small values of $\langle n \rangle$.

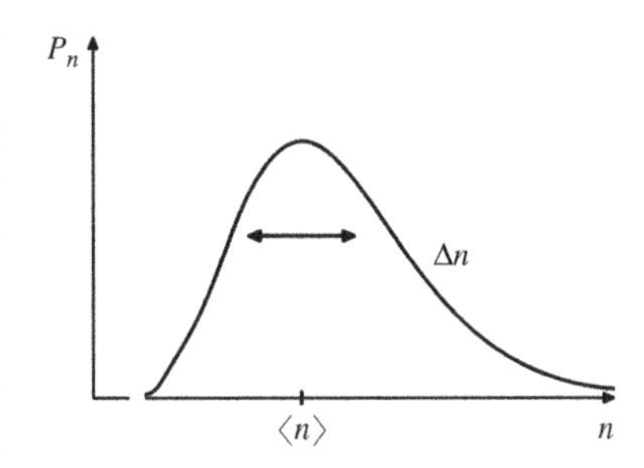

Fig. 5.1 The probability of the coherent state containing quantum number n is a Poisson distribution. The width Δn is of order $\sqrt{\langle n \rangle}$, and so $\Delta n / \langle n \rangle \to 0$ for large $\langle n \rangle$. Therefore Δn becomes negligible for coherent states with macroscopic numbers of particles.

Importantly, the coherent state $|\alpha\rangle$ does have a definite **phase**, θ, though it does not have a definite quantum number n. The coherent state can be defined for any complex number α,

$$\alpha = |\alpha| e^{i\theta}. \tag{5.28}$$

Rewriting Eq. 5.13 in terms of these variables we see that

$$|\alpha\rangle = C\left(\psi_0(x) + e^{i\theta}\frac{|\alpha|}{1!^{1/2}}\psi_1(x) + e^{2i\theta}\frac{|\alpha|^2}{2!^{1/2}}\psi_2(x) + \cdots\right). \tag{5.29}$$

We see that the term containing ψ_n depends on $e^{in\theta}$. Differentiating with respect to θ one can see that

$$\frac{1}{i}\frac{\partial}{\partial\theta}|\alpha\rangle = \hat{n}|\alpha\rangle. \tag{5.30}$$

But since the states $|\alpha\rangle$ are a complete set (actually an overcomplete set!), we can make the operator identification

$$\frac{1}{i}\frac{\partial}{\partial\theta} = \hat{n}. \tag{5.31}$$

So the phase θ and the number n are **conjugate operators**, in a similar way to momentum and position in standard quantum mechanics. It is possible to state a form of the uncertainty principle for these operators[2]

$$\Delta n \Delta\theta \geq \frac{1}{2}. \tag{5.32}$$

[2] A precise proof of this uncertainty relation is a little more tricky than the usual Heisenberg momentum-position uncertainty principle, since θ is only strictly defined in the range between 0 and 2π.

Coherent states have a fixed phase, but do not have definite values of number n. In contrast the energy eigenstates, $\psi_n(x)$, have a well defined value of n, but have an arbitrary (or ill-defined) phase.

Coherent states have many other beautiful mathematical properties, which we will not have time to explore in detail. In particular they are an overcomplete set, since they can be defined for any point in the complex α plane, and so there are uncountably infinite many such states. They are also not orthogonal, and it is easy to show that (Exercise 5.2)

$$|\langle\alpha|\beta\rangle|^2 = e^{-|\alpha-\beta|^2}. \tag{5.33}$$

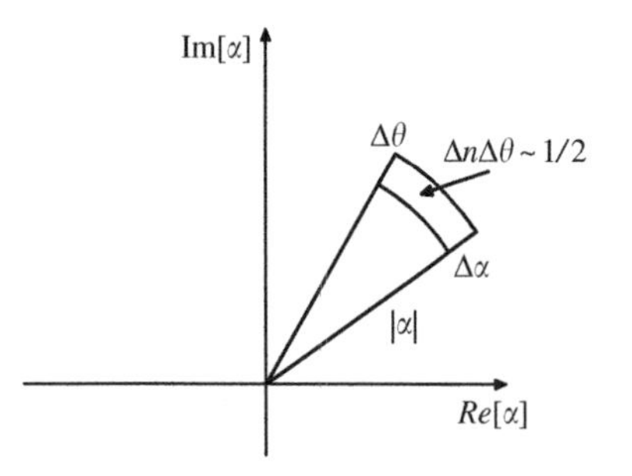

Fig. 5.2 Complex plane of coherent states, $|\alpha\rangle$, with $\alpha = |\alpha|e^{i\theta}$. The area $\Delta\theta\Delta n$ contains approximately one independent orthogonal quantum state.

We can interpret this in terms of the Argand diagram for the complex number α, shown in Fig. 5.2. Every point in the plane represents a valid coherent state. Coherent states at neighboring points are not orthogonal, but their overlap dies off when $|\alpha - \beta| \sim 1$. Therefore there is effectively one "independent" orthogonal quantum state per unit area of the complex plane. Using polar coordinates, $|\alpha|$, θ we see that an element of area contains exactly one quantum state if

$$1 \sim |\alpha|\Delta\theta\Delta|\alpha| \tag{5.34}$$

which is of order $2\Delta n\Delta\theta$, since $\langle n\rangle = |\alpha|^2$. Therefore the number-phase uncertainty principle gives essentially the minimum area per quantum state in the complex α plane of Fig. 5.2.

5.3 Coherent states and the laser

Coherent states were first studied extensively in the theory of the laser. In quantum optics the use of the operators $a^+_{\mathbf{k}s}$ and $a_{\mathbf{k}s}$ correspond directly to **creation** and **annihilation** of photons in a particular mode of electromagnetic radiation with wave number $\mathbf{k}$ and polarization s (i.e. left or right circularly polarized). A general quantum state of the system can be represented in the **occupation number representation**

$$|n_{\mathbf{k}_0 s_0}, n_{\mathbf{k}_1 s_1}, n_{\mathbf{k}_2 s_2}, n_{\mathbf{k}_3 s_3}, \ldots\rangle \tag{5.35}$$

where $\mathbf{k}_0$, $\mathbf{k}_1$, $\mathbf{k}_2$, etc. count all the different plane wave states of the system, e.g. the cavity of the laser.

In the case of light, the creation and annihilation operators arise quite naturally when one quantizes the electromagnetic radiation field. Each specific mode of the classical radiation field $\mathbf{k}$, s obeys Maxwell's equations. When these equations are quantized each mode becomes an independent quantum harmonic oscillator. The quantum states, $n_{\mathbf{k}s}$, of each oscillator are interpreted as the number of "photons" present. The creation operator adds a photon while the annihilation operator destroys one,[3]

$$a^+_{\mathbf{k}_s}|\ldots n_{\mathbf{k}_s}\ldots\rangle = (n_{\mathbf{k}s}+1)^{1/2}|\ldots n_{\mathbf{k}_s}+1\ldots\rangle, \tag{5.36}$$

$$a_{\mathbf{k}_s}|\ldots n_{\mathbf{k}_s}\ldots\rangle = (n_{\mathbf{k}_s})^{1/2}|\ldots n_{\mathbf{k}_s}\ldots\rangle, \tag{5.37}$$

[3]For convenience we will no longer write these as $\hat{a}^+$ and $\hat{a}$, but just as a^+ and a. No ambiguity will arise from this simplified notation, but one must not forget that these are operators and do not commute.

in complete analogy with the harmonic oscillator ladder operators. From these one can deduce that the same commutation law as the ladder operators must apply. But the operators for independent radiation field modes must commute, and therefore we can write,

$$\left[a_{\mathbf{k}_s}, a^+_{\mathbf{k}'s'}\right] = \delta_{\mathbf{k}s,\mathbf{k}'s'}, \tag{5.38}$$

$$[a_{\mathbf{k}s}, a_{\mathbf{k}'s'}] = 0, \tag{5.39}$$

$$\left[a^+_{\mathbf{k}s}, a^+_{\mathbf{k}'s'}\right] = 0. \tag{5.40}$$

In the case of the laser we can naturally shift from the occupation number representation to a coherent state representation. A general coherent state is of the form

$$|\alpha_{\mathbf{k}_0 s_0}, \alpha_{\mathbf{k}_1 s_1}, \alpha_{\mathbf{k}_2 s_2}, \alpha_{\mathbf{k}_3 s_3}, \dots\rangle \equiv e^{-\sum |\alpha_{\mathbf{k}s}|^2/2} \exp\left(\sum_{\mathbf{k}s} \alpha_{\mathbf{k}s} a^+_{\mathbf{k}_s}\right)|0\rangle. \qquad (5.41)$$

Here $|0\rangle$ is the vacuum state, with no photons present. An ideal coherent laser source is one in which just one of these modes has a macroscopic occupation, $\langle \hat{n}_{\mathbf{k}s}\rangle = |\alpha_{\mathbf{k}s}|^2$ while the others have essentially zero occupation.

In practical lasers, usually a few closely spaced **k** modes become macroscopically excited, and the system can randomly jump from locking onto one mode to another due to the nonlinear optical pumping which maintains the macroscopic mode occupation. It is the frequency of these jumps and the finite range of **k** values which limits the otherwise prefect optical coherence of a typical laser light source. See Loudon (1979) for a more detailed discussion of optical coherent states and their application to the laser.

5.4 Bosonic quantum fields

In this section we shall introduce **quantum field operators** for the case of boson particles. This will allow us to consider the quantum states of atomic BEC and superfluid ^{4}He. The BEC is a **weakly interacting** Bose system, while ^{4}He, as discussed in chapter 2, is a strongly interacting liquid of boson particles. At the same time we shall also see how the coherent state concept can also be applied to boson particles. This will enable us to define the macroscopic wave function $\psi(\mathbf{r})$ which we need in order to describe the condensate of particles. In this way we can generalize the simple intuitive approach of Chapters 1 and 2 into something which is both systematic and rigorous.

Understanding BEC and superfluids obviously requires us to work with many-particle quantum states for very large numbers of particles. In elementary quantum mechanics we would write a Schrödinger equation for a wave function of N particles, in order to obtain an N body wave function

$$\Psi(\mathbf{r}_1, \mathbf{r}_2, \dots, \mathbf{r}_N). \qquad (5.42)$$

As discussed in Chapters 1 and 2, if we consider a system of N interacting Bose atoms then we could in principle write a wave function $\psi(\mathbf{r}_1, \dots, \mathbf{r}_N)$ obeying the $3N$-dimensional Schrödinger equation

$$\hat{H}\Psi(\mathbf{r}_1, \dots, \mathbf{r}_N) = E\Psi(\mathbf{r}_1, \dots, \mathbf{r}_N), \qquad (5.43)$$

where

$$\hat{H} = \sum_{i=1,N}\left(-\frac{\hbar^2}{2m}\nabla_i^2 + V_1(\mathbf{r})\right) + \frac{1}{2}\sum_{i,j=1,N} V(\mathbf{r}_i - \mathbf{r}_j). \qquad (5.44)$$

Here $V_1(\mathbf{r})$ is the external potential, and $V(\mathbf{r})$ is the particle–particle interaction. In the case of ^{4}He this interaction could be taken as the Lennard–Jones potential of two helium atoms, Eq. 2.1 or Eq. 2.2, while for an atomic BEC we would use the delta function interaction of Eq. 1.62. The fact that the particles are bosons is expressed by the fact that the wave function must be symmetric under permutation of any two of the particle coordinates

$$\psi(\dots \mathbf{r}_i, \dots, \mathbf{r}_j \dots) = \psi(\dots \mathbf{r}_j, \dots, \mathbf{r}_i \dots) \qquad (5.45)$$

representing an exchange of the identical particles at $\mathbf{r}_i$ and $\mathbf{r}_j$.

This approach is feasible for systems with a few particles, but it quickly becomes impractical in larger systems. A much more useful method is to adopt the methods of quantum field theory and introduce **field operators** which add or remove particles to the system. If we consider a single particle in a box we know that the wave functions are plane wave states

$$\psi_{\mathbf{k}}(\mathbf{r}) = \frac{1}{(V)^{1/2}} e^{i\mathbf{k}\cdot\mathbf{r}} \tag{5.46}$$

where V is the volume. We saw in Chapter 1 that each of these single particle states can be occupied by 0, 1, 2, 3 or any other finite number of Bose particles. We denote these possibility by the occupation number, $n_{\mathbf{k}}$. A general quantum many body state of the system will be a superposition of different N-body plane wave states. The complete basis of all possible states can be represented by the set of all possible occupation numbers of each plane wave. In exact analogy with the case of the laser we can define creation and annihilation operators $a^+_{\mathbf{k}}$ and $a_{\mathbf{k}}$, which increase or decrease these occupation numbers. In order to satisfy the Bose symmetry condition on the wave function Eq. 5.45, it turns out that it is necessary that

$$[a_{\mathbf{k}}, a^+_{\mathbf{k}}] = 1.$$

For two different plane wave states the occupation numbers are independent, and hence the creation operators must commute. Therefore the complete set of commutation relations are exactly as given in Eqs 5.38–5.40. Similarly the occupation operator is given by the number operator,

$$\hat{n}_{\mathbf{k}} = a^+_{\mathbf{k}} a_{\mathbf{k}}. \tag{5.47}$$

These relations completely define the boson quantum field operators.

The set of many-particle states with all possible number operators, $\{n_{\mathbf{k}}\}$, is a complete set of wave functions. But, just as in the case of the harmonic oscillator, we can generate any $n_{\mathbf{k}}$ by successive actions of the operators $a^+_{\mathbf{k}}$, starting with a ground state $|0\rangle$. However, the interpretation is different now. $|0\rangle$ is the **vacuum state**, that is, a state with no particles present. Successive actions of $a^+_{\mathbf{k}}$ add more particles to the system. Any many-particle quantum state can be represented by a superposition of states generated in this way.

These field operators can also be cast into a real space form. We can define the quantum field operators $\hat{\psi}^+(\mathbf{r})$ and $\hat{\psi}(\mathbf{r})$ which create and annihilate particles at point $\mathbf{r}$. These can be defined by a Fourier transform of the $\mathbf{k}$ space operators,

$$\hat{\psi}(\mathbf{r}) = \frac{1}{\sqrt{V}} \sum_{\mathbf{k}} e^{i\mathbf{k}\cdot\mathbf{r}} a_{\mathbf{k}}, \tag{5.48}$$

$$\hat{\psi}^+(\mathbf{r}) = \frac{1}{\sqrt{V}} \sum_{\mathbf{k}} e^{-i\mathbf{k}\cdot\mathbf{r}} a^+_{\mathbf{k}}. \tag{5.49}$$

The function $e^{i\mathbf{k}\cdot\mathbf{r}}/\sqrt{V}$ here is obviously just a free particle plane wave in quantum state, $\mathbf{k}$. The inverse Fourier transforms are

$$a_{\mathbf{k}} = \frac{1}{\sqrt{V}} \int e^{-i\mathbf{k}\cdot\mathbf{r}} \hat{\psi}(\mathbf{r})\, d^3r, \tag{5.50}$$

$$a^+_{\mathbf{k}} = \frac{1}{\sqrt{V}} \int e^{i\mathbf{k}\cdot\mathbf{r}} \hat{\psi}^+(\mathbf{r})\, d^3r. \tag{5.51}$$

Using these definitions and the Bose commutation law, one can show (Exercise 5.4) that these real-space field operators have the commutation laws,

$$\left[\hat{\psi}(\mathbf{r}), \hat{\psi}^{+}(\mathbf{r}')\right] = \delta(\mathbf{r} - \mathbf{r}'), \tag{5.52}$$

$$\left[\hat{\psi}(\mathbf{r}), \hat{\psi}(\mathbf{r}')\right] = 0, \tag{5.53}$$

$$\left[\hat{\psi}^{+}(\mathbf{r}), \hat{\psi}^{+}(\mathbf{r}')\right] = 0. \tag{5.54}$$

We can also represent any operator in terms of its actions on quantum states described in terms of these operators. In particular the Hamiltonian, Eq. 5.44 becomes

$$\hat{H} = \int \left(\hat{\psi}^{+}(\mathbf{r}) \left[-\frac{\hbar^2}{2m}\nabla^2 + V_1(\mathbf{r}) \right] \hat{\psi}(\mathbf{r}) \right) d^3r$$
$$+ \frac{1}{2} \int V(\mathbf{r} - \mathbf{r}')\hat{\psi}^{+}(\mathbf{r})\hat{\psi}(\mathbf{r})\hat{\psi}^{+}(\mathbf{r}')\hat{\psi}(\mathbf{r}')\, d^3r\, d^3r'. \tag{5.55}$$

where the combination $\hat{\psi}^{+}(\mathbf{r})\hat{\psi}(\mathbf{r})$ is obviously the density operator of particles at $\mathbf{r}$.

It turns out to be convenient to always work in "normal order," in which all the creation operators are on the left and all the annihilation operators are on the left. Commuting two of the field operators above we obtain

$$\hat{H} = \int \left(\hat{\psi}^{+}(\mathbf{r}) \left[-\frac{\hbar^2}{2m}\nabla^2 + V_1(\mathbf{r}) \right] \hat{\psi}(\mathbf{r}) \right) d^3r$$
$$+ \frac{1}{2} \int V(\mathbf{r} - \mathbf{r}')\hat{\psi}^{+}(\mathbf{r})\hat{\psi}^{+}(\mathbf{r}')\hat{\psi}(\mathbf{r})\hat{\psi}(\mathbf{r}')\, d^3r\, d^3r'$$
$$+ \frac{1}{2} \int V(\mathbf{r} - \mathbf{r}')\hat{\psi}^{+}(\mathbf{r})\delta(\mathbf{r} - \mathbf{r}')\hat{\psi}(\mathbf{r}')\, d^3r\, d^3r'. \tag{5.56}$$

The final term here arises from the commutator of $[\hat{\psi}(\mathbf{r}), \hat{\psi}^{+}(\mathbf{r}')]$, and reduces to

$$V(0) \int d^3r \hat{\psi}^{+}(\mathbf{r})\hat{\psi}(\mathbf{r}) = V(0)\hat{N}, \tag{5.57}$$

where

$$\hat{N} = \int d^3r \hat{\psi}^{+}(\mathbf{r})\hat{\psi}(\mathbf{r}) \tag{5.58}$$

is obviously just the operator for the total number of particles in the system. The $\hat{N}V(0)$ term is a constant that can be absorbed into the definition of the chemical potential, μ, and so we shall drop it from now on.

For a bulk fluid we can assume translational invariance, and ignore the external potential $V_1(\mathbf{r})$. Going back to $\mathbf{k}$-space and we can use the Fourier transforms Eqs 5.48 and 5.49 to represent the Hamiltonian in terms of $a_{\mathbf{k}}^{+}$ and $a_{\mathbf{k}}$. The kinetic energy term is

$$\hat{T} = -\int \left(\hat{\psi}^+(\mathbf{r}) \frac{\hbar^2}{2m} \nabla^2 \hat{\psi}(\mathbf{r}) \right) d^3r,$$

$$= \frac{1}{V} \sum_{\mathbf{k}\mathbf{k}'} \int \left(a^+_{\mathbf{k}'} e^{-i\mathbf{k}'.\mathbf{r}} \frac{\hbar^2 k^2}{2m} a_{\mathbf{k}} e^{i\mathbf{k}\cdot\mathbf{r}} \right) d^3r,$$

$$= \sum_{\mathbf{k}} \frac{\hbar^2 k^2}{2m} a^+_{\mathbf{k}} a_{\mathbf{k}}. \tag{5.59}$$

The potential energy term is

$$\hat{V} = \frac{1}{2} \int V(\mathbf{r} - \mathbf{r}') \hat{\psi}^+(\mathbf{r}) \hat{\psi}^+(\mathbf{r}') \hat{\psi}(\mathbf{r}) \hat{\psi}(\mathbf{r}') d^3r d^3r',$$

$$= \frac{1}{2V^2} \sum_{\mathbf{k}_1\mathbf{k}_2\mathbf{k}_3\mathbf{k}_4} \int V(\mathbf{r} - \mathbf{r}') a^+_{\mathbf{k}_1} a^+_{\mathbf{k}_2} a_{\mathbf{k}_3} a_{\mathbf{k}_4},$$

$$\times\, e^{i(-\mathbf{k}_1\cdot\mathbf{r} - \mathbf{k}_2\cdot\mathbf{r}' + \mathbf{k}_3\cdot\mathbf{r}' + \mathbf{k}_4\cdot\mathbf{r})} d^3r d^3r',$$

$$= \frac{1}{2V} \sum_{\mathbf{k}_1\mathbf{k}_2\mathbf{k}_3\mathbf{k}_4} a^+_{\mathbf{k}_1} a^+_{\mathbf{k}_2} a_{\mathbf{k}_3} a_{\mathbf{k}_4} \delta_{\mathbf{k}_3+\mathbf{k}_4, \mathbf{k}_1+\mathbf{k}_2} \int V(\mathbf{r}) e^{i(\mathbf{k}_4 - \mathbf{k}_1)\cdot\mathbf{r}} d^3r.$$

Introducing the Fourier transform of the interaction

$$V_{\mathbf{q}} = \frac{1}{V} \int V(r) e^{i\mathbf{q}\cdot\mathbf{r}} d^3r \tag{5.60}$$

and making the replacements $\mathbf{k}_1 \to \mathbf{k}+\mathbf{q}$, $\mathbf{k}_2 \to \mathbf{k}'-\mathbf{q}$, $\mathbf{k}_3 \to \mathbf{k}'$, and $\mathbf{k}_4 \to \mathbf{k}$ we can express the full Hamiltonian as

$$\hat{H} = \sum_{\mathbf{k}} \frac{\hbar^2 k^2}{2m} a^+_{\mathbf{k}} a_{\mathbf{k}} + \frac{1}{2} \sum_{\mathbf{k}\mathbf{k}'\mathbf{q}} V_{\mathbf{q}} a^+_{\mathbf{k}+\mathbf{q}} a^+_{\mathbf{k}'-\mathbf{q}} a_{\mathbf{k}'} a_{\mathbf{k}}. \tag{5.61}$$

We can interpret the interaction term simply as a process in which a pair of particles are scattered from initial states $\mathbf{k}$, $\mathbf{k}'$ to final states $\mathbf{k} + \mathbf{q}$, $\mathbf{k}' - \mathbf{q}$. The momentum transferred between the particles is $\mathbf{q}$, and the matrix element for the process is $V_{\mathbf{q}}$.

5.5 Off-diagonal long ranged order

The field operators introduced in the previous section provide a general way to discuss quantum coherence in condensates and superfluids. Even though Eq. 5.61 is still too difficult to solve in general, we can still use it to explore the consequences of macroscopic quantum coherence in bosonic systems.

First, let us revisit the idea of the macroscopic quantum state, as introduced for BEC and superfluid ^{4}He in Chapters 1 and 2. Using the field operators we can redefine the one-particle density matrix as

$$\rho_1(\mathbf{r}-\mathbf{r}') \equiv \langle \hat{\psi}^+(\mathbf{r})\hat{\psi}(\mathbf{r}')\rangle. \tag{5.62}$$

This definition is clearly more compact than the equivalent one given in Chapter 2. Using the Fourier transformations Eqs 5.48 and 5.49 we find

$$\rho_1(\mathbf{r}-\mathbf{r}') = \frac{1}{V}\sum_{\mathbf{k}\mathbf{k}'} e^{i(\mathbf{k}'\cdot\mathbf{r}'-\mathbf{k}\cdot\mathbf{r})}\langle a^+_{\mathbf{k}} a_{\mathbf{k}'}\rangle,$$

$$= \frac{1}{V}\sum_{\mathbf{k}} e^{-i\mathbf{k}\cdot(\mathbf{r}-\mathbf{r}')}\langle a^+_{\mathbf{k}} a_{\mathbf{k}}\rangle, \tag{5.63}$$

which is just the Fourier transform of the momentum distribution

$$n_{\mathbf{k}} \equiv \langle a^+_{\mathbf{k}} a_{\mathbf{k}}\rangle \tag{5.64}$$

exactly as found in Chapter 2.

Now let us consider the consequences of these definitions in the case of a quantum coherent many-particle state. Just as in the case of the laser, we can define a coherent state

$$|\alpha_{\mathbf{k}_1}, \alpha_{\mathbf{k}_2}, \alpha_{\mathbf{k}_3}, \ldots\rangle$$

for any set of complex numbers $\alpha_{\mathbf{k}_i}$. Using the standard properties of coherent states we find

$$n_{\mathbf{k}} \equiv |\alpha_{\mathbf{k}}|^2 \tag{5.65}$$

and hence

$$\rho_1(\mathbf{r}-\mathbf{r}') = \frac{1}{V}\sum_{\mathbf{k}} e^{-i\mathbf{k}\cdot(\mathbf{r}-\mathbf{r}')}|\alpha_{\mathbf{k}}|^2. \tag{5.66}$$

Typically we will be interested in quantum states where only one of the $\mathbf{k}$ states is macroscopically occupied (usually but not always $\mathbf{k} = 0$). So suppose that state $\mathbf{k}_0$ has occupation $N_0 = |\alpha_{\mathbf{k}_0}|^2$ where N_0 is a macroscopic number (a finite fraction of the total particle number N) and all the other $|\alpha_{\mathbf{k}_i}|^2$ are small. For such a state we will have the momentum distribution

$$n_{\mathbf{k}} = N_0\delta_{\mathbf{k},\mathbf{k}_0} + f(\mathbf{k}), \tag{5.67}$$

where $f(\mathbf{k})$ is a smooth function of $\mathbf{k}$. The corresponding density matrix is

$$\rho_1(\mathbf{r}-\mathbf{r}') = n_0 + \frac{2}{(2\pi)^3}\int d^3k e^{-i\mathbf{k}\cdot(\mathbf{r}-\mathbf{r}')}f(\mathbf{k}), \tag{5.68}$$

where $n_0 = N_0/V$.

These results are exactly the same as we found in Chapter 2 by more elementary methods. The presence of the condensate is shown by the constant contribution, n_0 to the density matrix. If the function $f(\mathbf{k})$ is sufficiently smooth, then its Fourier transform will vanish for large $|\mathbf{r}-\mathbf{r}'|$ leaving just the constant contribution,

$$\langle \hat{\psi}^+(\mathbf{r})\hat{\psi}(\mathbf{r}')\rangle \to n_0 \tag{5.69}$$

as $|\mathbf{r}-\mathbf{r}'| \to \infty$. This is what is meant by the term **off diagonal long ranged order** (ODLRO), introduced by Oliver Penrose.

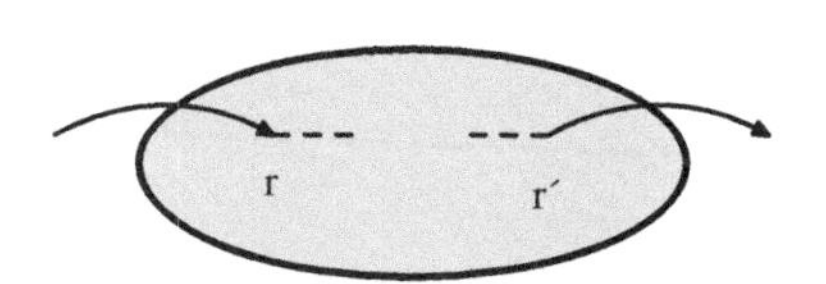

Fig. 5.3 Schematic illustrating the interpretation of ODLRO in the one particle density matrix $\rho_1(\mathbf{r}-\mathbf{r}')$. A particle is inserted into the condensate at $\mathbf{r}$, and a particle is removed from it at $\mathbf{r}'$. In a condensate, this process has a coherent quantum amplitude and phase, however great the separation between $\mathbf{r}$ and $\mathbf{r}'$.

Figure 5.3 shows the physical interpretation of the ODLRO in superfluids. A particle can be annihilated at $\mathbf{r}$, and absorbed into the condensate, while a second particle is created at $\mathbf{r}'$ out of the condensate. This process has a quantum mechanical amplitude because of the quantum coherence of the condensate, even when the points $\mathbf{r}$ and $\mathbf{r}'$ are separated arbitrarily far apart. In contrast, for a normal liquid (even a normal quantum liquid) these processes would be incoherent except when $\mathbf{r}$ and $\mathbf{r}'$ are close together.

Using the coherent state concept, there is one more step which we can take. If the density matrix

$$\langle \hat{\psi}^+(\mathbf{r})\hat{\psi}(\mathbf{r}')\rangle \tag{5.70}$$

is a constant, however far apart points $\mathbf{r}$ and $\mathbf{r}'$ are, then it seems plausible that we can treat the points as independent statistically. Then we can view the above as an average of a product of independent random variables and hence write it as a product of the two averages computed separately

$$\langle \hat{\psi}^+(\mathbf{r})\hat{\psi}(\mathbf{r}')\rangle \rightarrow \langle \hat{\psi}^+(\mathbf{r})\rangle\langle \hat{\psi}(\mathbf{r}')\rangle \tag{5.71}$$

for $|\mathbf{r}-\mathbf{r}'| \rightarrow \infty$.

If we were to work in the standard fixed particle number, N, many-body formulation of quantum mechanics then averages such as $\langle \hat{\psi}^+(\mathbf{r})\rangle$ would be automatically zero, since acting on any N particle state $\hat{\psi}^+(\mathbf{r})|N\rangle$ is an $N+1$ particle state, and is necessarily orthogonal to $\langle N|$. But if we are working on the basis of coherent states, then there is no such problem. The coherent states have definite phase, not definite N, and this type of expectation value is perfectly well allowed.

Therefore we can say that there is an **order parameter** or **macroscopic wave function**, defined by

$$\psi_0(\mathbf{r}) = \langle \hat{\psi}(\mathbf{r})\rangle. \tag{5.72}$$

In terms of this function we see that[4]

$$\rho_1(\mathbf{r}-\mathbf{r}') = \psi_0^*(\mathbf{r})\psi_0(\mathbf{r}'), \tag{5.73}$$

for $|\mathbf{r}-\mathbf{r}'| \rightarrow \infty$. In a translationally invariant system (the condensation occurs in the $\mathbf{k}=0$ state), we must therefore have

$$\psi_0(\mathbf{r}) = \sqrt{n_0}e^{i\theta}, \tag{5.74}$$

where θ is an arbitrary constant phase angle.

[4]The creation operator $\hat{\psi}^+(\mathbf{r})$ is just the Hermitian conjugate of $\hat{\psi}(\mathbf{r})$ and so $\psi_0^*(\mathbf{r}) = \langle \hat{\psi}^+(\mathbf{r})\rangle$.

Of course this phase θ is nothing more than the XY model phase angle introduced in Chapter 2, Fig. 2.4. But now we can see that its true meaning is that we have a coherent quantum state in which the $\mathbf{k}=0$ state has a macroscopic occupation.

Since we have not yet made any connection to the Hamiltonian, Eq. 5.61, it is impossible to prove from these arguments that such a coherent quantum state will be stable. But at least we can see how to construct coherent many-particle wave functions in which a definite order parameter phase θ is possible. In the case of the ideal Bose condensate discussed in Chapter 1, one can still work in the fixed particle number representation, and so there is no advantage to explicitly introduce a coherent state formalism. But as soon as there are any interactions, however weak, the coherent state approach becomes advantageous. In the next section we shall consider the weakly interacting bose gas, in which the advantages of the coherent state approach can be seen explicitly.

5.6 The weakly interacting Bose gas

The theory of the **weakly interacting Bose gas** was originally developed by Bogoliubov in the late 1940s. It was developed as a theory of superfluid helium, although as we have seen, for ^{4}He the interatomic interactions are very strong. In this case the theory has some qualitative features which agree with experimental properties of ^{4}He, most notably the linear phonon like quasiparticle excitation spectrum, $\epsilon_{\mathbf{k}} = ck$, at small wave vectors in Fig. 2.16. But it fails to reproduce other important features, such as the roton minimum in the spectrum. On the other hand, the theory is believed to be close to exact for the case of atomic BEC, since the conditions under which it is derived are close to the experimental ones.

First of all we shall assume that we are at zero temperature, or close to zero, so that the system is close to its ground state. We assume that the system is in a coherent many-particle state, characterized by a macroscopic wave function $\psi_0(\mathbf{r})$, as in Eq. 5.72. Suppose that the many-particle quantum state ,$|\psi\rangle$, is an ideal coherent state at zero temperature. Then it is an eigenstate of the annihilation operator,

$$\hat{\psi}(\mathbf{r})|\psi\rangle = \psi_0(\mathbf{r})|\psi\rangle. \tag{5.75}$$

We can view this as a trial many-particle wave function, and we will vary the parameter $\psi_0(\mathbf{r})$ so as to variationally minimize the total energy. The variational energy is found by taking the expectation value of the Hamiltonian,

$$\begin{aligned}\hat{H} &= \int \hat{\psi}^{+}(\mathbf{r})\left(-\frac{\hbar^2\nabla^2}{2m} + V_1(\mathbf{r})\right)\hat{\psi}(\mathbf{r})\,d^3r \\ &+ \frac{1}{2}\int V(\mathbf{r}-\mathbf{r}')\hat{\psi}^{+}(\mathbf{r})\hat{\psi}^{+}(\mathbf{r}')\hat{\psi}(\mathbf{r})\hat{\psi}(\mathbf{r}')\,d^3r\,d^3r'.\end{aligned} \tag{5.76}$$

Here the single particle potential $V_1(\mathbf{r})$ is the effective external potential of the atom trap. In the case of bulk superfluids this is obviously zero.

Using the definition of the coherent state $|\psi\rangle$ from Eq. 5.75 we can find immediately that the variational energy is

$$\begin{aligned}E_0 &= \langle\psi|\hat{H}|\psi\rangle \\ &= \int \psi_0^*(\mathbf{r})\left(-\frac{\hbar^2\nabla^2}{2m} + V_1(\mathbf{r})\right)\psi_0(\mathbf{r})\,d^3r \\ &+ \frac{1}{2}\int V(\mathbf{r}-\mathbf{r}')\psi_0^*(\mathbf{r})\psi_0^*(\mathbf{r}')\psi_0(\mathbf{r})\psi_0(\mathbf{r}')\,d^3r\,d^3r'.\end{aligned} \tag{5.77}$$

We can find the minimum by functional differentiation, exactly as for the GL equation. Setting

$$\frac{\partial E_0}{\partial \psi_0^*(\mathbf{r})} = 0$$

yields,

$$\left(-\frac{\hbar^2\nabla^2}{2m} + V_1(\mathbf{r}) - \mu\right)\psi_0(\mathbf{r}) + \int V(\mathbf{r}-\mathbf{r}')\psi_0(\mathbf{r})\psi_0^*(\mathbf{r}')\psi_0(\mathbf{r}')\,d^3r' = 0. \tag{5.78}$$

The parameter μ is a Lagrange multiplier, necessary to maintain a constant normalization of the macroscopic wave function

$$N_0 = \int |\psi_0(\mathbf{r})|^2 d^3r. \tag{5.79}$$

Clearly Eq. 5.78 is of the form of an effective Schrödinger equation

$$\left(-\frac{\hbar^2\nabla^2}{2m} + V_1(\mathbf{r}) + V_{\text{eff}}(\mathbf{r}) - \mu\right)\psi_0(\mathbf{r}) = 0, \tag{5.80}$$

where μ is the chemical potential, and the effective potential is

$$V_{\text{eff}}(\mathbf{r}) = \int V(\mathbf{r}-\mathbf{r}')|\psi_0(\mathbf{r}')|^2 \cdot d^3r'.$$

This Schrödinger equation is exactly the Gross–Pitaevskii equation again, which we derived by a different method in Chapter 2.

To examine the accuracy of this ground state, and to examine the low energy excited states, we need to consider possible quantum states which are close to our trial ground state $|\psi\rangle$, but which do not deviate from it too much. We need to consider many-particle states which do not exactly obey the coherent state condition, Eq. 5.75, but for which it is nearly obeyed. Bogoliubov introduced an elegant method to achieve this. He assumed that the field operators can be expressed approximately as their constant coherent state value, plus a small deviation,

$$\hat{\psi}(\mathbf{r}) = \psi_0(\mathbf{r}) + \delta\hat{\psi}(\mathbf{r}). \tag{5.81}$$

From the commutation relations for the field operators, it is easy to see that

$$[\delta\hat{\psi}(\mathbf{r}), \delta\hat{\psi}^+(\mathbf{r}')] = \delta(\mathbf{r}-\mathbf{r}'), \tag{5.82}$$

and so the deviation operators $\delta\hat{\psi}(\mathbf{r})$ and $\delta\hat{\psi}^+(\mathbf{r})$ are also bosonic quantum fields.[5] We can then rewrite the hamiltonian in terms of $\psi_0(\mathbf{r})$ and $\delta\hat{\psi}(\mathbf{r})$. We can group terms according to whether $\delta\hat{\psi}(\mathbf{r})$ occurs never, once, twice, three times, or four,

$$\hat{H} = \hat{H}_0 + \hat{H}_1 + \hat{H}_2 + \cdots \tag{5.83}$$

and we assume that it is valid to simply ignore terms higher than second order.

[5] A different way to see this is to imagine that we simply translate the origin of the coherent state complex plane, Fig. 5.2. Shifting the origin from $\alpha = 0$ to $\alpha = \psi_0(\mathbf{r})$ we can then describe states which are near to $|\psi\rangle$ in terms of the coherent states α which are in the vicinity of the point $\psi_0(\mathbf{r})$.

In this expansion, the first term, $\hat{H}_0$ is simply the original coherent state energy, Eq. 5.77, which we can minimize using the Gross–Pitaevskii equations. Furthermore, if we have variationally minimized the energy there will be no corrections to the energy to linear order in the deviation operators $\delta\hat{\psi}(\mathbf{r})$. The first significant correction term is quadratic in the deviation operators. Several terms contribute, but the net result is that

$$\begin{aligned}
\hat{H}_2 = &\int \delta\hat{\psi}^+(\mathbf{r})\left(-\frac{\hbar^2\nabla^2}{2m} + V_1(\mathbf{r})\right)\delta\hat{\psi}(\mathbf{r})\,d^3r \\
&+\frac{1}{2}\int V(\mathbf{r}-\mathbf{r}')d^3r d^3r'\Big(\delta\hat{\psi}^+(\mathbf{r})\delta\hat{\psi}^+(\mathbf{r}')\psi_0(\mathbf{r})\psi_0(\mathbf{r}') \\
&+2\delta\hat{\psi}^+(\mathbf{r})\psi_0^*(\mathbf{r}')\delta\hat{\psi}(\mathbf{r})\psi_0(\mathbf{r}') + 2\psi_0^*(\mathbf{r})\delta\hat{\psi}^+(\mathbf{r}')\delta\hat{\psi}(\mathbf{r})\psi_0(\mathbf{r}') \\
&+\psi_0^*(\mathbf{r})\psi_0^*(\mathbf{r}')\delta\hat{\psi}(\mathbf{r})\delta\hat{\psi}(\mathbf{r}')\Big).
\end{aligned} \tag{5.84}$$

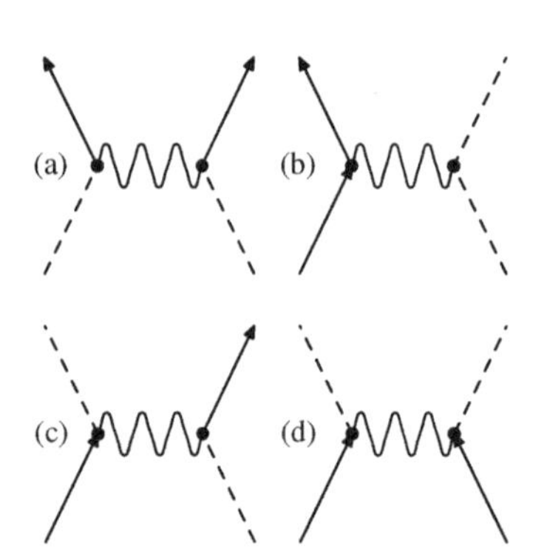

Fig. 5.4 Four types of interactions between quasiparticles and a Bose condensate. The quasiparticles are denoted by the solid lines, the condensate particles by the dashed line, and the interaction $V(\mathbf{r})$ by the wavy line. (a) Two particles are excited out of the condensate. (b) An existing quasiparticle interacts with the condensate. (c) An existing quasiparticle is absorbed into the condensate, while simultaneously a second quasiparticle becomes excited out of it. (d) Two quasiparticles are absorbed into the condensate.

One can visualize the meaning of these various terms in terms of the simple diagrams shown in Fig. 5.4. There are four distinct terms. The first, shown in panel (a), corresponds to the creation of two particles, one at $\mathbf{r}$ and one at $\mathbf{r}'$ under the action of the potential $V(\mathbf{r}-\mathbf{r}')$. Of course they are not really created, but scattered out of the condensate. The second term corresponds to the scattering of an existing quasiparticle by interaction with particles in the condensate. It has an extra factor of 2 since we can find an identical diagram with $\mathbf{r}$ and $\mathbf{r}'$ interchanged. The third diagram (c) is also a scattering of an existing quasiparticle, but now the quasiparticle at $\mathbf{r}$ is absorbed into the condensate, while at the same time a second quasiparticle appears at $\mathbf{r}'$. Again $\mathbf{r}$ and $\mathbf{r}'$ can be interchanged leading to an extra factor of 2. The final diagram (d), shows two quasiparticles being absorbed into the condensate.

To keep the algebra manageable let us specialize to the case (relevant for the atomic BEC) of a pure contact interaction,

$$V(\mathbf{r}-\mathbf{r}') = g\delta(\mathbf{r}-\mathbf{r}'). \tag{5.85}$$

For convenience we will also assume that $\psi_0(\mathbf{r})$ is real, and equal to $\psi_0(\mathbf{r}) = \sqrt{n_0(\mathbf{r})}$, where $n_0(\mathbf{r})$ the spatially varying condensate density in the atom trap. The quadratic Hamiltonian $\hat{H}_2$ simplifies to

$$\hat{H}_2 = \int \delta\hat{\psi}^+(\mathbf{r})\left(-\frac{\hbar^2\nabla^2}{2m} + V_1(\mathbf{r}) - \mu\right)\delta\hat{\psi}(\mathbf{r})d^3r \tag{5.86}$$
$$+\frac{g}{2}\int n_0(\mathbf{r})\left(\delta\hat{\psi}^+(\mathbf{r})\delta\hat{\psi}^+(\mathbf{r}) + 4\delta\hat{\psi}^+(\mathbf{r})\delta\hat{\psi}(\mathbf{r}) + \delta\hat{\psi}(\mathbf{r})\delta\hat{\psi}(\mathbf{r})\right)d^3r.$$

This Hamiltonian is a quadratic form in the operators, and it turns out that all such quadratic Hamiltonians can be diagonalized exactly. The procedure makes use of the **Bogoliubov** transformation to eliminate the "anomalous" terms $\delta\hat{\psi}(\mathbf{r})\delta\hat{\psi}(\mathbf{r})$ and $\delta\hat{\psi}^+(\mathbf{r})\delta\hat{\psi}^+(\mathbf{r})$. Define a new pair of operators by

$$\hat{\varphi}(\mathbf{r}) = u(\mathbf{r})\delta\hat{\psi}(\mathbf{r}) + v(\mathbf{r})\delta\hat{\psi}^+(\mathbf{r}) \tag{5.87}$$

$$\hat{\varphi}^+(\mathbf{r}) = u^*(\mathbf{r})\delta\hat{\psi}^+(\mathbf{r}) + v^*(\mathbf{r})\delta\hat{\psi}(\mathbf{r}). \tag{5.88}$$

These are again bosonic quantum field operators if

$$\left[\hat{\varphi}(\mathbf{r}), \hat{\varphi}^+(\mathbf{r})\right] = \delta(\mathbf{r}-\mathbf{r}'), \tag{5.89}$$

which is true provided that the functions $u(\mathbf{r})$ and $v(\mathbf{r})$ are chosen to obey

$$|u(\mathbf{r})|^2 - |v(\mathbf{r})|^2 = 1. \tag{5.90}$$

The general solution becomes quite complicated, so let us further specialize to the case of a uniform Bose liquid, without the atom trap potential $V_1(\mathbf{r})$. Assuming that the macroscopic wave function is also just a constant, $\psi_0 = \sqrt{n_0}$, and going to $\mathbf{k}$-space Eq. 5.86 becomes

$$\hat{H}_2 = \sum_{\mathbf{k}}\left(\left(\frac{\hbar^2k^2}{2m} - \mu\right)a^+_{\mathbf{k}}a_{\mathbf{k}} + \frac{n_0g}{2}(a^+_{\mathbf{k}}a^+_{-\mathbf{k}} + 4a^+_{\mathbf{k}}a_{\mathbf{k}} + a_{-\mathbf{k}}a_{\mathbf{k}})\right). \tag{5.91}$$

Here the chemical potential is $\mu = n_0g$, which follows from Eq. 5.80. The Bogoliubov transformation in $\mathbf{k}$-space gives the new operators

$$b_{\mathbf{k}} = u_{\mathbf{k}}a_{\mathbf{k}} + v_{\mathbf{k}}a^+_{-\mathbf{k}} \tag{5.92}$$

$$b^+_{-\mathbf{k}} = u_{\mathbf{k}}a^+_{\mathbf{k}} + v_{\mathbf{k}}a_{\mathbf{k}}, \tag{5.93}$$

where $u_{\mathbf{k}}$ and $v_{\mathbf{k}}$ are real and obey

$$u^2_{\mathbf{k}} - v^2_{\mathbf{k}} = 1. \tag{5.94}$$

The idea is to rewrite the Hamiltonian in terms of these new operators and then to vary the parameters $u_{\mathbf{k}}$ and $v_{\mathbf{k}}$ to make it diagonal. In particular it is necessary to make the coefficients of the anomalous terms, $b^+_{\mathbf{k}} b^+_{-\mathbf{k}}$ and $b_{-\mathbf{k}} b_{\mathbf{k}}$ equal to zero. Since the calculation is lengthy we shall just quote the results. It turns out that when the anomalous terms are eliminated, the new quasiparticle excitations created by the $b^+_{\mathbf{k}}$ operators have the energy spectrum

$$E_{\mathbf{k}} = \left(\frac{\hbar^2 k^2}{2m}\right)^{1/2} \left(\frac{\hbar^2 k^2}{2m} + 2n_0 g\right)^{1/2}. \tag{5.95}$$

For small $|\mathbf{k}|$ (less than $\sim \sqrt{4n_0 g m}/\hbar$) the spectrum is linear

$$E_{\mathbf{k}} \sim ck \tag{5.96}$$

where the "phonon" velocity is

$$c = \left(\frac{\hbar^2 n_0 g}{m}\right)^{1/2}. \tag{5.97}$$

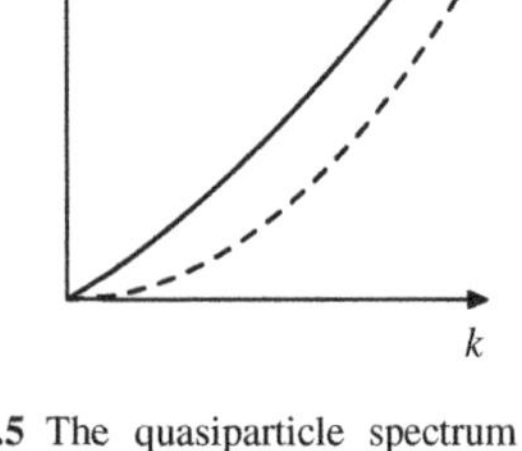

Fig. 5.5 The quasiparticle spectrum of a weakly interacting Bose gas, as found by Bogoliubov. The spectrum is linear at small $\mathbf{k}$, and approaches the free particle energy $\hbar^2 k^2/2m$ (dashed line) for large $\mathbf{k}$. Unlike the case of superfluid ^{4}He, Fig. 2.16 there is no roton minimum, and there is a slight upward curvature near to $\mathbf{k} = 0$.

This Bogoliubov quasiparticle spectrum is sketched in Fig. 5.5. The spectrum is linear at small $\mathbf{k}$, and joins smoothly onto the independent particle energy $\hbar^2 k^2/2m$ for large $\mathbf{k}$. The success of this Bogoliubov theory is that it explains the linear dispersion in the excitation spectrum near to $\mathbf{k} = 0$, as we saw in the case of superfluid ^{4}He, Fig. 2.16. In Chapter 2 we have also seen that a linear spectrum is necessary to prevent quasiparticle scattering from the container walls, and hence to maintain a dissipationless superflow. Therefore we can conclude that a very weakly interacting Bose gas is a superfliud, even when the ideal noninteracting Bose gas is not. The critical velocity for the superfluid will be less than c, and so by Eq. 5.97 the critical velocity will approach zero in the limits of weak interaction $g \to 0$, or low density $n_0 \to 0$.

Of course there are still huge differences between the Bogoliubov quasiparticle spectrum in Fig. 5.5 and the experimental helium quasiparticle spectrum of Fig. 2.16. Most importantly, there is no roton minimum. The Bogoliubov spectrum is linear at small $\mathbf{k}$ and becomes equal to the free particle spectrum $\hbar^2 k^2/2m$ for large $\mathbf{k}$, but it has no minimum at intermediate $\mathbf{k}$. The roton minimum is therefore an effect emerging only in a strongly interacting Bose liquid. This difference has another effect, namely that the Bogoliubov spectrum $E_{\mathbf{k}}$ has a slight upward curvature at small $\mathbf{k}$, while the true spectrum has a slight downward curvature. This curvature means that a quasiparticle of momentum $\hbar\mathbf{k}$ and energy $E_{\mathbf{k}}$ has a nonzero cross section to decay into three quasiparticles of lower energy and momentum. Therefore the Bogoliubov quasiparticle is not an exact eigenstate, but only a long lived resonance.[6]

[6] It makes perfect sense here to borrow the concepts of elementary particle physics, and to talk about one particle decaying into a set of others. Just as in particle physics we can call such "particles" resonances. In this context the Bogoliubov quasiparticles are just the "elementary particles" of the Bose gas, and the background condensate is the analog of the vacuum. Indeed in particle physics language our Bogoliubov quasiparticles are the "Goldstone bosons" of the system.

5.7 Coherence and ODLRO in superconductors

The above ideas were originally introduced to explain the strongly interacting Bose superfluid ^{4}He. But they also apply to superconductors. However, in this case the argument must be modified in order to take account of the fact that the electrons in a superconductor are fermions. Although it is perfectly possible to define coherent states for fermions, they are not immediately useful for superconductivity. This is because a single fermion state can only ever be occupied by either 0 or 1 fermions, due to the exclusion principle. Therefore it is not possible to have a macroscopic number of fermions in a single plane wave state.

It was Robert Schrieffer who first managed to write down a coherent many-particle wave function for fermions. His colleauges Bardeen and Cooper had already realized that electrons bind into pairs in a superconductor. There was even an earlier theory by Schafroth Blatt and Butler in which superconductivity was seen as a Bose condensate of electron pairs.[7] But BCS knew that the pairs of electrons in superconductors could not be simply treated as bosons. The problem was to write down a valid many-body wave function for the electrons in which each electron participates in the pairing. The brilliant solution Schrieffer discovered was effectively another type of coherent state, similar to those we have already seen. But the key point is to have a coherent state in which a macroscopic number of **pairs** are all in the same state.

[7] Unfortunately this theory was not able to make quantitative predictions for the superconductors which were known at that time, and it was generally discarded in favor of the BCS theory which was very much more successful in making quantitative predictions. The electron pairs in the BCS theory **are not bosons**. In general a pair of fermions is not equivalent to a boson. In the BCS case the bound electron pairs are very large, and so different pairs strongly overlap each other. In this limit it is not possible to describe the pair as a boson, and so the BCS theory is not normally expressed in terms of bose condensation. However, in recent years theories based upon the ideas of the Schafroth, Blatt, and Butler theory have been somewhat revived as possible models of high T_c superconductors.

Nevertheless even in the BCS theory we have a form of ORLRO and an order parameter, as we shall discover in this section. Their appearance is a quite general phenomenon. Here we concentrate on the most general statements about the ODLRO, and leave the actual details of the BCS theory, and its specific predictions until the next chapter.

First we must define the correct quantum field operators to describe the conduction electrons in a solid. In single-particle quantum mechanics of solids we know that the wave functions are Bloch waves

$$\psi_{n\mathbf{k}}(\mathbf{r}) = e^{i\mathbf{k}\cdot\mathbf{r}} u_{n\mathbf{k}}(\mathbf{r}). \tag{5.98}$$

Here the crystal wave vector $\mathbf{k}$ must lie within the first Brillouin zone. For simplicity of notation we shall assume that only one of the energy bands, n, is relevant (i.e. the one at the Fermi surface) and so we will drop the index n from now on.

A particular Bloch state of a given spin σ, $\psi_{\mathbf{k}\sigma}(\mathbf{r})$, can either be empty or occupied by an electron. A quantum state of N particles can be specified by saying for each individual state whether it is occupied or not. This is **the occupation number representation** of quantum mechanics. We can introduce operators which change these occupation numbers. Suppose we label the state as $|0\rangle$ or $|1\rangle$ if the given Bloch state $\psi_{\mathbf{k}\sigma}(\mathbf{r})$ is empty or occupied. Then we can define operators which change the occupation numbers

$$c^+|0\rangle = |1\rangle,$$
$$c|1\rangle = |0\rangle.$$

These are obviously similar to the similar Bose field operators, or the harmonic oscillator ladder operators, $\hat{a}^+$ and $\hat{a}$. The only difference is that the state of a harmonic oscillator can have $n = 0, 1, 2, \ldots$, while for fermions the occupation number is only 0 or 1. The operators c^+ and c are again called the creation and annihilation operators. Since electrons are neither being created or annihilated in solids these names may seem a bit misleading, however the operators work the same way in particle physics where particles are indeed being created, or annihilated back into the vacuum (e.g. electron–positron pairs annihilating each other). In a solid one can imagine adding electrons to the solid (e.g. with an external current source into the surface), or removing them from the solid (e.g. in photo-emission).

These operators have several important properties. First, just as for bosons, the combination c^+c measures the occupation number $|n\rangle$ since

$$c^+c|0\rangle = 0,$$
$$c^+c|1\rangle = |1\rangle,$$

that is, $c^+c|n\rangle = n|n\rangle$. Second the exclusion principle implies that we cannot occupy a state with more than one fermion, and hence $c^+c^+|n\rangle = 0$ and $cc|n\rangle = 0$. This fermion nature of the particles means that

$$\{c, c^+\} \equiv cc^+ + c^+c = 1 \tag{5.99}$$

where $\{A, B\} = AB + BA$ is the **anti-commutator** of operators A and B. The antisymmetry of the many-particle wave function for fermions means that the operators for different Bloch states or spin states **anti-commute**, and so in general we have the anti-commutation relations,

$$\{c_{\mathbf{k}\sigma}, c^+_{\mathbf{k}'\sigma'}\} = \delta_{\mathbf{k}\sigma,\mathbf{k}'\sigma'} \tag{5.100}$$
$$\{c_{\mathbf{k}\sigma}, c_{\mathbf{k}'\sigma'}\} = 0 \tag{5.101}$$
$$\{c^+_{\mathbf{k}\sigma}, c^+_{\mathbf{k}'\sigma'}\} = 0, \tag{5.102}$$

where $\sigma = \pm 1$ denotes the two different spin states. Naturally, just as for bosons, we can also represent these operators in real space, by a Fourier transformation, obtaining

$$\{\hat{\psi}_\sigma(\mathbf{r}), \hat{\psi}^+_{\sigma'}(\mathbf{r}')\} = \delta(\mathbf{r} - \mathbf{r}')\delta_{\sigma\sigma'} \tag{5.103}$$
$$\{\hat{\psi}_\sigma(\mathbf{r}), \hat{\psi}_{\sigma'}(\mathbf{r}')\} = 0 \tag{5.104}$$
$$\{\hat{\psi}^+_\sigma(\mathbf{r}), \hat{\psi}^+_{\sigma'}(\mathbf{r}')\} = 0. \tag{5.105}$$

Now the mathematical challenge which BCS needed to solve was to write down a many-particle wave function in which every electron near the Fermi surface participates in the pairing. They knew that a single pair of electrons would bind into a spin singlet state with two body wave function (as shown below in Section 6.3)

$$\Psi(\mathbf{r}_1\sigma_1, \mathbf{r}_2\sigma_2) = \varphi(\mathbf{r}_1 - \mathbf{r}_2)\frac{1}{\sqrt{2}}\left(|\uparrow\downarrow\rangle - |\downarrow\uparrow\rangle\right). \tag{5.106}$$

They first wrote down a many-particle wave function in which every particle is paired,

$$\Psi(\mathbf{r}_1\sigma_1, \ldots, \mathbf{r}_N\sigma_N) = \frac{1}{\sqrt{N!}}\sum_P (-1)^P \Psi(\mathbf{r}_1\sigma_1, \mathbf{r}_2\sigma_2)\Psi(\mathbf{r}_3\sigma_3, \mathbf{r}_4\sigma_4)\cdots$$
$$\cdots\Psi(\mathbf{r}_{N-1}\sigma_{N-1}, \mathbf{r}_N\sigma_N). \tag{5.107}$$

Here the sum over P denotes the sum over all the $N!$ permutations of the N particle labels $\mathbf{r}_1\sigma_1$, $\mathbf{r}_2\sigma_2$, etc. The sign $(-1)^P$ is positive for an even permutation and -1 for odd permutations. This alternating sign is necessary so that the many-body wave function has the correct fermion antisymmetry

$$\Psi(\ldots, \mathbf{r}_i\sigma_i, \ldots, \mathbf{r}_j\sigma_j, \ldots) = -\Psi(\ldots, \mathbf{r}_j\sigma_j, \ldots, \mathbf{r}_i\sigma_i, \ldots). \tag{5.108}$$

But this fixed N many-body quantum mechanics is unwieldy. By fixing N we cannot have a definite overall phase, unlike in a coherent state representation. But if we use coherent states then it is possible to describe a condensate with a definite phase, and the key step achieved by Schrieffer was to find a way to write down a coherent state of fermion pairs. First we need to write down operators which create or annihilate pairs of electrons. Defining,

$$\hat{\varphi}^{+}(\mathbf{R}) \equiv \int d^3r \varphi(\mathbf{r}) \hat{\psi}^{+}_{\uparrow}(\mathbf{R} + \mathbf{r}/2) \hat{\psi}^{+}_{\downarrow}(\mathbf{R} - \mathbf{r}/2). \tag{5.109}$$

we can see that acting with this operator on a quantum state with N electrons gives a new quantum state with $N+2$ electrons. It creates a spin singlet electron pair, where the electrons are separated by $\mathbf{r}$ and with center of mass at $\mathbf{R}$.[8]

[8] Note that we assume here that the wave function for the bound electron pair obeys $\varphi(\mathbf{r}) = \varphi(-\mathbf{r})$. This is simply because a spin singlet bound state wave function must be even under exchange of particle coordinates, because the spin singlet state is odd under exchange.

Naively one might regard such a fermion pair as a boson. But this is not correct. If we try to evaluate the commutator we find that

$$\left[\hat{\varphi}(\mathbf{R}), \hat{\varphi}^{+}(\mathbf{R}')\right] \neq \delta(\mathbf{R} - \mathbf{R}'), \tag{5.110}$$

$$\left[\hat{\varphi}(\mathbf{R}), \hat{\varphi}(\mathbf{R}')\right] = 0, \tag{5.111}$$

$$\left[\hat{\varphi}^{+}(\mathbf{R}), \hat{\varphi}^{+}(\mathbf{R}')\right] = 0. \tag{5.112}$$

The operators only commute when $\mathbf{R}$ and $\mathbf{R}'$ are far apart, corresponding to non-overlapping pairs. For this reason we cannot simply make a Bose condensate out of these pairs.

But even though these are not true boson operators, we can still define the analog of ODLRO corresponding to Bose condensation. Now it **is** possible to find a state of ODLRO of **Cooper pairs**. We can define a new density matrix by,

$$\rho_1(\mathbf{R} - \mathbf{R}') = \langle \hat{\varphi}^{+}(\mathbf{R}) \hat{\varphi}(\mathbf{R}') \rangle. \tag{5.113}$$

This is a one particle density matrix for pairs, and so it is related to the two particle density matrix for the electrons

$$\rho_2(\mathbf{r}_1\sigma_1, \mathbf{r}_2\sigma_2, \mathbf{r}_3\sigma_3, \mathbf{r}_4\sigma_4) = \langle \hat{\psi}^{+}_{\sigma_1}(\mathbf{r}_1) \hat{\psi}^{+}_{\sigma_2}(\mathbf{r}_2) \hat{\psi}_{\sigma_3}(\mathbf{r}_3) \hat{\psi}_{\sigma_4}(\mathbf{r}_4) \rangle. \tag{5.114}$$

Inserting the definition of the pair operator gives the pair density matrix in terms of the electron two particle density matrix

$$\rho_1(\mathbf{R}-\mathbf{R}') = \int \varphi(\mathbf{r})\varphi(\mathbf{r}')\rho_2\left(\mathbf{R}+\frac{\mathbf{r}}{2}\uparrow, \mathbf{R}-\frac{\mathbf{r}}{2}\downarrow, \mathbf{R}'-\frac{\mathbf{r}'}{2}\downarrow, \mathbf{R}'+\frac{\mathbf{r}'}{2}\uparrow\right) d^3r\, d^3r'. \tag{5.115}$$

The pair wave function $\varphi(\mathbf{r})$ is a quantum mechanical bound state, and so it will become zero for large $|\mathbf{r}|$. If the scale of its range is defined by a length, ξ_0 (which turns out to be the BCS coherence length of the superconductor), then the main contributions to the pair density matrix comes from the parts of electron density matrix where $\mathbf{r}_1$ and $\mathbf{r}_2$ are separated by less than ξ_0, and similarly $\mathbf{r}_3$ and $\mathbf{r}_4$ are separated by less than ξ_0. But the pair $\mathbf{r}_1$,$\mathbf{r}_2$ and the pair $\mathbf{r}_3$,$\mathbf{r}_4$ can be separated by any arbitrarily large distance.

Now we can have ODLRO in the pair density matrix, provided that

$$\rho_1(\mathbf{R} - \mathbf{R}') \to \text{const.} \tag{5.116}$$

as $|\mathbf{R} - \mathbf{R}'| \to \infty$. The BCS theory does therefore correspond to a similar macroscopic quantum coherence to the ordinary theory of ODLRO in superfluids. However, it is an ODLRO of Cooper pairs, not single electrons.

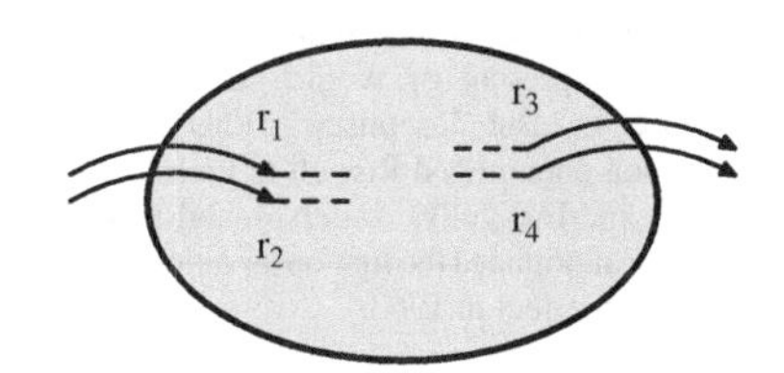

Fig. 5.6 The two body density matrix for electrons in a metal, $\rho_2(\mathbf{r}_1\sigma_1, \mathbf{r}_2\sigma_2, \mathbf{r}_3\sigma_3, \mathbf{r}_4\sigma_4)$. Off diagonal long ranged order (ODLRO) for electron pairs appears when this has a finite value however far apart the pair $\mathbf{r}_1$ and $\mathbf{r}_2$, is from $\mathbf{r}_3$ and $\mathbf{r}_4$. In contrast, the points within each pair must be no more than a coherence length, ξ_0, apart.

In terms of the electron density matrix $\rho_2(\mathbf{r}_1\sigma_1, \mathbf{r}_2\sigma_2, \mathbf{r}_3\sigma_3, \mathbf{r}_4\sigma_4)$ this ODLRO corresponds to the density matrix approaching a constant value in a limit where the two coordinates $\mathbf{r}_1$ and $\mathbf{r}_2$ are close to each other, and $\mathbf{r}_3$ and $\mathbf{r}_4$ are close, but these two pairs are separated by an arbitrarily large distance. Figure 5.6 illustrates this concept.

Following the same approach to ODLRO as in the weakly interacting Bose gas, we can make the assumption that very distant points $\mathbf{R}$ and $\mathbf{R}'$ should behave independently. Therefore we should be able to write

$$\rho_1(\mathbf{R} - \mathbf{R}') \sim \langle \hat{\varphi}^+(\mathbf{R}) \rangle \langle \hat{\varphi}(\mathbf{R}') \rangle \tag{5.117}$$

for $|\mathbf{R} - \mathbf{R}'| \to \infty$. We can therefore also define a **macroscopic wave function** by

$$\psi(\mathbf{R}) = \langle \hat{\varphi}(\mathbf{R}) \rangle. \tag{5.118}$$

Effectively this is the **Ginzburg–Landau order parameter** for the superconductor.

Of course, we have still not actually shown how to construct a many-electron quantum state which would allow this type of Cooper pair ODLRO. We leave that until the next chapter. Nevertheless we can view the above discussion as stating the requirements for the sort of quantum state which could exhibit superconductivity. It was possibly the most significant achievement of the BCS theory that it was possible to explicitly construct such a nontrivial many-body quantum state. In fact we can see immediately from Eq. 5.118 that we must have a coherent state with a definite quantum mechanical phase θ, and conversely that we should not work with fixed particle number N. At the time of the original publication of the BCS theory in 1957 this aspect was one of the most controversial of the whole theory. The explicit appearance of the phase θ also caused concern, since it appeared to violate the principle of gauge invariance. It was only the near perfect agreement between the predictions of the BCS theory and experiments, as well as clarification of the gauge invariance issue by Anderson and others, which led to the final acceptance of the BCS theory.

5.8 The Josephson effect

The Josephson effect is a direct physical test of the quantum coherence implied by superconducting ODLRO. Soon after the BCS theory was published Brian Josephson, then a PhD student at Cambridge, considered the effect of electrons tunnelling between two different superconductors.[9]

[9] Brian Josephson received the Nobel prize in 1973 for this discovery, possibly one of the few Nobel prizes to have arisen from a PhD project. After Josephon's first prediction of the effect in 1962, established experts in the superconductivity field at first objected to the theory, believing that the effect was either not present or would be too weak to observe. But Josephson's PhD adviser Anderson encouraged Rowell to look for the effect experimentally. Anderson and Rowell together announced the first observation of the Josephson effect in 1963.

Consider two superconductors, separated by a thin insulating layer, as shown in Fig. 5.7. If each superconductor has a macroscopic wave function as defined

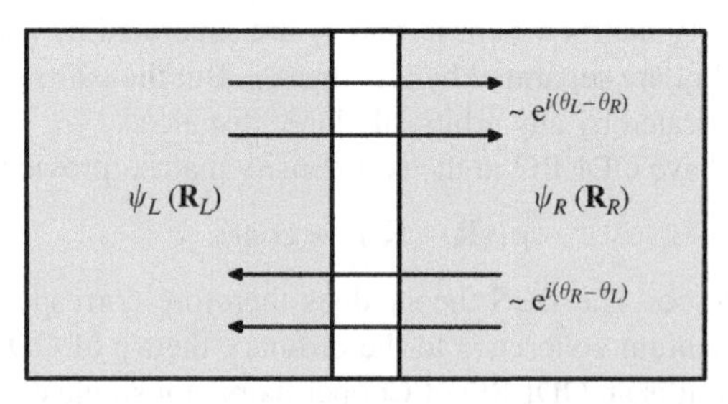

Fig. 5.7 A schematic Josephson tunnel junction between two superconductors. To second order in the tunnelling Hamiltonian there are two possible processes for a Cooper pair to tunnel from one side to another. If θ_L and θ_R denote the order parameter phases on either the left and right hand sides of the junction, then the amplitudes for processes depend on $e^{i(\theta_L-\theta_R)}$ and $e^{i(\theta_R-\theta_L)}$. When these are added together the net tunnel current is proportional to $\sin(\theta_L - \theta_R)$.

in, 5.118, then we can assign definite values to the wave functions on either side of the tunnel barrier, say ψ_L and ψ_R for the left- and right-hand side, respectively,

$$\psi_L(\mathbf{R}_L) = \langle L|\hat{\varphi}(\mathbf{R}_L)|L\rangle,$$
$$\psi_R(\mathbf{R}_R) = \langle R|\hat{\varphi}(\mathbf{R}_R)|R\rangle,$$

where $\mathbf{R}_L$ and $\mathbf{R}_R$ are points on either side of the tunnel barrier, and $|L\rangle$ and $|R\rangle$ are the many-particle quantum states of the superconductors on the left and right side respectively. Josephson assumed that electron tunnelling takes place for electrons crossing the barrier. In terms of electron field operators we can write

$$\hat{H} = \sum_\sigma \int T(\mathbf{r}_L, \mathbf{r}_R)(\hat{\psi}_\sigma^+(\mathbf{r}_L)\hat{\psi}_\sigma(\mathbf{r}_R) + \hat{\psi}_\sigma^+(\mathbf{r}_R)\hat{\psi}_\sigma(\mathbf{r}_L))\, d^3r_L\, d^3r_R \quad (5.119)$$

as the operator which tunnels electrons of spin σ from points $\mathbf{r}_L$ and $\mathbf{r}_R$ on either side of the barrier. Here $T(\mathbf{r}_L, \mathbf{r}_R)$ is the transmission matrix element for electrons to tunnel across the barrier. Using the BCS many-body wave functions on either side of the junction and second-order perturbation theory in the tunnelling Hamiltonian $\hat{H}$, Josephson was able to find the remarkable result that a current flows in the junction, given by

$$I = I_c \sin(\theta_L - \theta_R) \quad (5.120)$$

where θ_L and θ_R are the phases of the macroscopic wave functions on either side of the junction.

The details of how Josephson found this result will not be important here. But it is worth noting very roughly how this might come about. If we consider effects to second order in the tunnelling Hamiltonian $\hat{H}$ we can see that $\hat{H}^2$ contains many terms, but includes some four fermion terms of the following form,

$$\hat{H}^2 \sim T^2(\hat{\psi}_\sigma^+(\mathbf{r}_L)\hat{\psi}_{\sigma'}^+(\mathbf{r}'_L)\hat{\psi}_\sigma(\mathbf{r}_R)\hat{\psi}_{\sigma'}(\mathbf{r}'_R)$$
$$+ \hat{\psi}_\sigma^+(\mathbf{r}_R)\hat{\psi}_{\sigma'}^+(\mathbf{r}'_R)\hat{\psi}_\sigma(\mathbf{r}_L)\hat{\psi}_{\sigma'}(\mathbf{r}'_L) + \cdots). \quad (5.121)$$

The net effect of the first of these terms is to transfer a pair of electrons from the right hand superconductor to the left. Conversely the second term transfers a pair from left to right. Because the many-body states on both sides of the junction are coherent pair states, these operators will have nonzero expectation values consistent with the ODLRO

$$\langle L|\hat{\psi}_\sigma^+(\mathbf{r}_L)\hat{\psi}_{\sigma'}^+(\mathbf{r}'_L)|L\rangle \neq 0, \quad (5.122)$$
$$\langle R|\hat{\psi}_\sigma(\mathbf{r}_R)\hat{\psi}_{\sigma'}(\mathbf{r}'_R)|R\rangle \neq 0. \quad (5.123)$$

In fact we expect the first of these expectation values to be proportional to $e^{-i\theta_L}$ and the second to $e^{i\theta_R}$. The overall quantum mechanical amplitude for tunnelling a pair from right to left is thus dependent upon a phase $e^{i(\theta_R-\theta_L)}$. The reverse process, tunnelling from left to right has the opposite phase. When they are added together the net current is proportional to $\sin(\theta_L - \theta_R)$ as given by Eq. 5.120.

Equation 5.120 shows that the current flows in response to the phase difference $\theta_L - \theta_R$. Therefore it is in some sense a direct proof of the existence of such coherent state phases in superconductors. The proportionality constant

I_c is the maximum Josephson current that can flow, and is called the **critical current** of the junction.

For currents I below I_c the Josephson current is perfectly dissipationless, that is, it is a supercurrent. But if current is driven to a higher value, $I > I_c$, a finite voltage drop V develops across the junction. Therefore the typical I–V characteristic of the junction is as shown in Fig. 5.8.

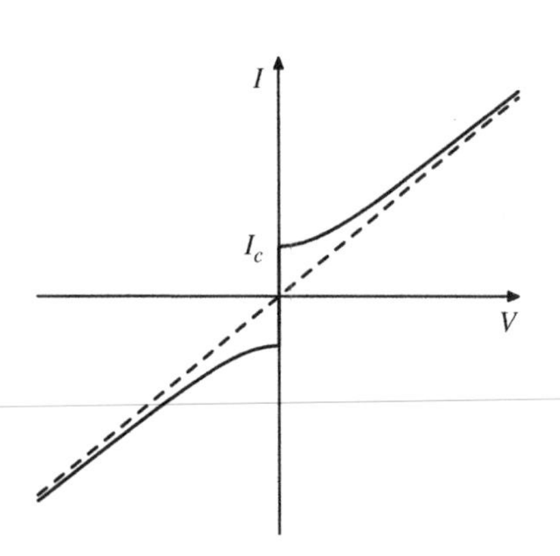

Fig. 5.8 I–V characteristic of a Josephson junction. There is no voltage drop, $V = 0$, provided that the junction current is less than the critical current I_c. Above this value a finite voltage drop occurs. This approaches the normal state Ohm's law result $I = V/R$ for large currents. The a.c. Joesphson effect occurs in the $V \neq 0$ regime above I_c.

In the case that $I > I_c$ Josephson found a second surprising consequence of this tunnel current. The finite voltage difference V between the superconductors, means that the macroscopic wave functions become time dependent. Using a version of the Heisenbeg equation of motion for the left- and right-hand side macroscopic wave functions,

$$i\hbar\frac{\partial\psi_L(t)}{\partial t} = -2eV_L\psi_L(t),$$
$$i\hbar\frac{\partial\psi_R(t)}{\partial t} = -2eV_R\psi_R(t), \tag{5.124}$$

Josephson was able to show that the finite voltage drop $V = V_L - V_R$ leads to a steadily increasing phase difference,

$$\Delta\theta(t) = \Delta\theta(0) + \frac{2eV}{\hbar}t \tag{5.125}$$

and hence the Josephson current,

$$I = I_c \sin\left(\Delta\theta(0) + \frac{2eV}{\hbar}t\right), \tag{5.126}$$

oscillates at a frequency

$$\nu = \frac{2eV}{\hbar}. \tag{5.127}$$

This surprising effect is called the a.c. Joesphson effect (in contrast to the d.c. Josephson effect for $I < I_c$, $V = 0$).

The experimental observation of the a.c. Josephson effect not only confirmed the theory and the validity of the BCS macroscopic quantum coherent state, but also provided another direct empirical confirmation that the relevant particle charge is $2e$ and not e. Thus it again confirmed the Cooper pairing hypothesis. Even more surprising is that the Josephson frequency appears to be **exactly** given by Eq. 5.127. In fact, it is so accurate that the Josephson effect has been incorporated into the standard set of measurements used to define the SI unit system. By measuring the frequency (which can be measured with accuracies of one part in 10^{12} or better) and the voltage V one can obtain the ratio of fundamental constants $e/\hbar$ with high precision. Alternatively one can use the given values of $e/\hbar$ and the Josephson effect to define a reliable and portable voltage standard.

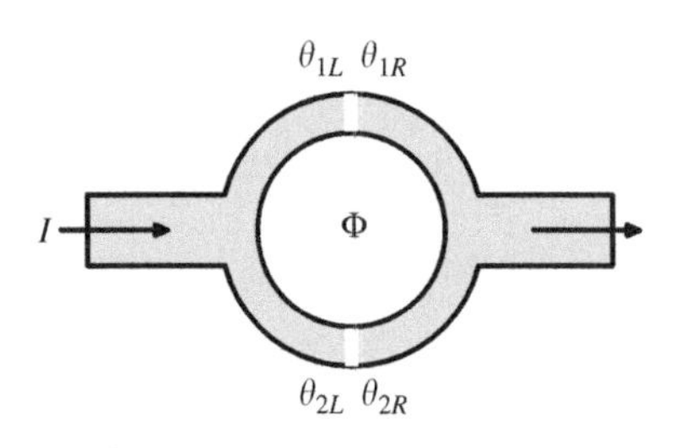

Fig. 5.9 Schematic geometry of a SQUID ring. The two Josephson junctions have currents governed by the phase differences $\Delta\theta_1 = \theta_{1L} - \theta_{1R}$ and $\Delta\theta_2 = \theta_{2L} - \theta_{2R}$. The total critical current of the whole device is modulated by the total magnetic flux through the ring Φ.

The Josephson effect is also at the heart of many different practical applications of superconductivity. One of the simplest devices to make is a SQUID ring; where SQUID stands for Superconducting Quantum Interference Device. This is simply a small (or large) superconducting ring in which there are two weak links. Each half of the ring is then connected to external leads, as shown in Fig. 5.9. By "weak link" one can mean either a tunnel barrier, such as Fig. 5.7 (an SIS junction), or a thin normal metallic spacer (an SNS junction). The current through each junction depends on the phase difference across it and so,

$$I = I_{c1} \sin(\Delta\theta_1) + I_{c2} \sin(\Delta\theta_2) \tag{5.128}$$

is the sum of the currents of each Josephson junction. The phases differences $\Delta\theta_1$, $\Delta\theta_2$ correspond to the macroscopic wave function phases differences at the points to the left and right of each junction in Fig. 5.9.

If the junctions are perfectly balanced, so $I_{c1} = I_{c2}$ and a small external current I ($< I_c$) is applied to the SQUID, then we would expect an equal steady state current to flow in both halves of the ring, and a constant phase difference equal to $\Delta\theta = \sin^{-1}(I/2I_c)$ will develop across both junctions. But this is only true if there is no magnetic flux through the ring. Using the principle of gauge invariance we find that a flux Φ implies that the phase differences are no longer equal.

$$\begin{aligned}
\Phi &= \int \mathbf{B} \cdot d\mathbf{S} \\
&= \oint \mathbf{A} \cdot d\mathbf{r} \\
&= \frac{2e}{\hbar} \oint (\nabla\theta) \cdot d\mathbf{r} \\
&= \frac{2e}{\hbar} (\Delta\theta_1 - \Delta\theta_2) \\
&= 2\pi \Phi_0 (\Delta\theta_1 - \Delta\theta_2).
\end{aligned} \tag{5.129}$$

Therefore the magnetic flux through the ring leads to a difference between the two phases. For a balanced SQUID ring system we can assume that

$$\begin{aligned}
\Delta\theta_1 &= \Delta\theta + \frac{\pi\Phi}{\Phi_0}, \\
\Delta\theta_2 &= \Delta\theta - \frac{\pi\Phi}{\Phi_0}.
\end{aligned} \tag{5.130}$$

therefore the total current in the SQUID is

$$\begin{aligned}
I &= I_c \sin(\Delta\theta_1) + I_c \sin(\Delta\theta_2) \\
&= I_c \sin\left(\Delta\theta + \frac{\pi\Phi}{\Phi_0}\right) + I_c \sin\left(\Delta\theta - \frac{\pi\Phi}{\Phi_0}\right) \\
&= 2I_c \sin(\Delta\theta) \cos\left(\frac{\pi\Phi}{\Phi_0}\right).
\end{aligned} \tag{5.131}$$

The critical current is therefore modulated by a factor depending on the net flux through the ring,[10]

$$I_c(\Phi) = I_0 \left|\cos\left(\frac{\pi\Phi}{\Phi_0}\right)\right|. \tag{5.132}$$

[10] The SQUID ring critical current will always have the same sign as the driving current and hence the modulus signs appearing in Eq. 5.132.

This modulation of the observed SQUID ring critical current is shown in Fig. 5.10. This current is essentially an ideal Fraunhoffer interference pattern, exactly analogous to the interference pattern one observes in optics with Young's two slit experiment. Here the two Josephson junctions are playing the role of the two slits, and the interference is between the supercurrents passing through the two halves of the ring. The supercurrents acquire different phases due to the

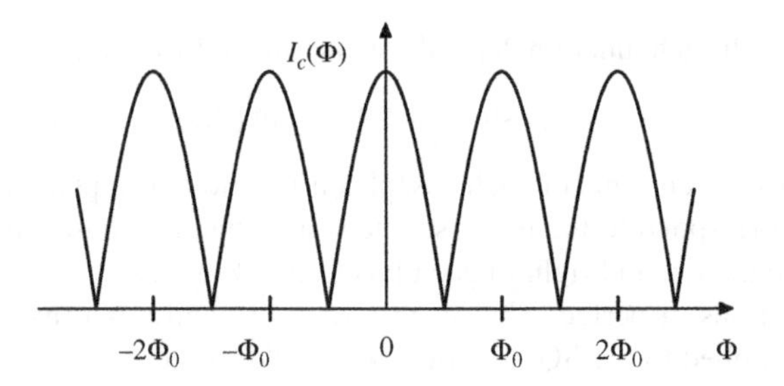

Fig. 5.10 Modulation of critical current in a SQUID ring. This is effectively equivalent to the Fraunhoffer interference pattern of a two slit optical interference pattern. Effectively there is interference between the two currents flowing through the opposite sides of the SQUID ring in Fig. 5.9.

magnetic field. One can say that this effect is also analogous to the Aharanov Bohm effect (Feynman 1964), in which a single electron passes on either side of a solenoid of magnetic flux.

The SQUID device provides a simple, but highly accurate, system for measuring magnetic flux. Since the flux quantum Φ_0 is only about 2×10^{-15} Wb in the SI unit system, and one can make SQUID devices approaching 1 cm^2 in area, it is in principle possible to measure magnetic fields to an accuracy below $B \sim 10^{-10} T$. In particular it is easy to measure **changes** in field to this accuracy by simply counting the number of minima in the SQUID critical current.

5.9 Macroscopic quantum coherence

To what extent does the Joesphson effect or the SQUID represent true evidence quantum coherence? Even though it is very cold, operating at a temperature below T_c, say at one or two degrees Kelvin, a SQUID ring is hardly isolated from its environment. In fact one can make perfectly good SQUIDs using high temperature superconductors. These operate at temperatures of over 100 K. SQUID rings are usually fabricated on some sort of insulating substrate, and are usually subject to normal external electromagnetic noise in the laboratory, unless well shielded.

Given this relatively noisy thermal environment, the SQUID shows a remarkable insensitivity to these effects. This is fundamentally because the macroscopic wave function $\psi(\mathbf{r})$ which we have defined above, and its phase θ, is not a true wave function in the sense of elementary quantum mechanics. In particular it does not obey the fundamental **principle of superposition**, and one cannot apply the usual **quantum theory of measurement** or Copenhagen interpretation to it. The macroscopic wave function behaves much more like a thermodynamic variable, such as the magnetization in a ferromagnet, than a pure wave function, even though it has a phase and obeys local gauge invariance.

But since the early 1980s there have been attempts to observe true quantum superpositions in superconductors (Leggett 1980), that is, can one construct a "Schrödinger cat" like quantum state? For example, is a state such as

$$|\psi\rangle = \frac{1}{\sqrt{2}}\left(|\psi_1\rangle + |\psi_2\rangle\right), \tag{5.133}$$

meaningful in a SQUID ring? If $|\psi_1\rangle$ and $|\psi_2\rangle$ are two pure quantum states, then the general principle of linearity of quantum mechanics, implies that any superposition such as $|\psi\rangle$ must also be a valid quantum state. Only by measuring some **physical observable** can one "collapse" the wave function and find out whether the system was in $|\psi_1\rangle$ or $|\psi_2\rangle$. For small systems, such as single atoms or photons, such superpositions are a standard part of quantum mechanics. But in his famous 1935 paper Schrödinger showed that this fundamental principle leads to paradoxes with our everyday understanding of the world when we apply it to macroscopic systems such as the famous cat in a box!

Even since Schrödinger's paper it has not been clear where to put the dividing line between the "macrosopic" world (governed by classical physics and without superposition) and the microscopic (governed by quantum mechanics). The ideas of **decoherence** provide one possible route by which quantum systems can acquire classical behavior. Interactions with the environment lead to entanglement between the quantum states of the system and those of the environment, and (in some of the most modern approaches) "quantum information" is lost.

In this context one can say that a large SQUID ring, say 1 cm (or 100 m!) in diameter will be subject to decoherence from its environment, and hence will be effectively in the classical realm. But if one makes the ring smaller, or operates at lower temperatures, is there a regime where true quantum superpositions occur? In fact the answer to this question is yes! Indeed strong evidence for quantum superposition states has now been seen in three different systems.

The first system where quantum superposition states were observed was in BEC, in 1996 (Ketterle 2002). Since these exist in a very low temperature state (a micro Kelvin or less) and are isolated from most external thermal noise sources (since they are trapped in vacuum) one could expect a high degree of quantum coherence to occur. Indeed it has proved possible to "split" a single condensate into two halves, in a similar way to which a beam splitter separates photons. When the two halves of the condensate are subsequently brought back together again, then one observes an interference pattern. The experiment is effectively the exact analog of the Young's slit interference experiment with light. Figure 5.11 illustrates the interference fringes obtained after two BEC collide.

The second type of experiment which showed true quantum interference was done using a superconducting island or "Cooper pair box." This consists of a small island of superconductor (Al was used) of order 0.1 μm on a side, as shown in Fig. 5.12. Operating at temperatures of a few millikelvin, well below the superconducting T_c, the quantum states of the box can be characterized entirely by the number of Cooper pairs present. For example, the box can have a state $|N\rangle$ of N Cooper pairs, or a state $|N+1\rangle$ etc. The energies of these

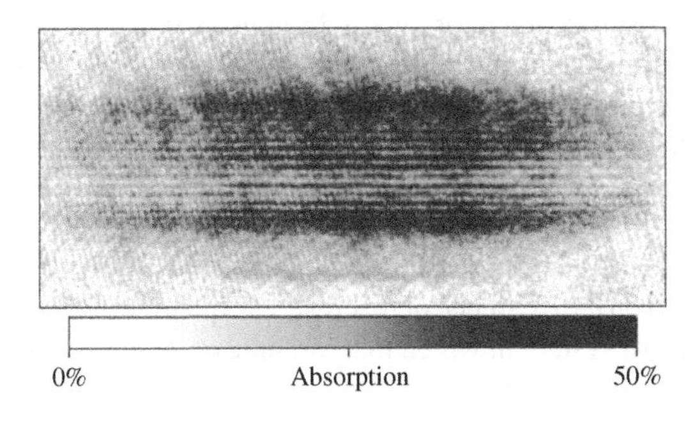

Fig. 5.11 Macroscopic quantum coherence demonstrated in a BEC. A condensate is split into two halves, which then interfere with each other, analogous to an optical beam splitter experiment. The interference fringes are clearly visible as horizontal bands of light and dark absorption, corresponding to a spatially modulated condensate density. Reprinted figure with permission from Ketterle (2002). Copyright (2002) by the American Physical Society.

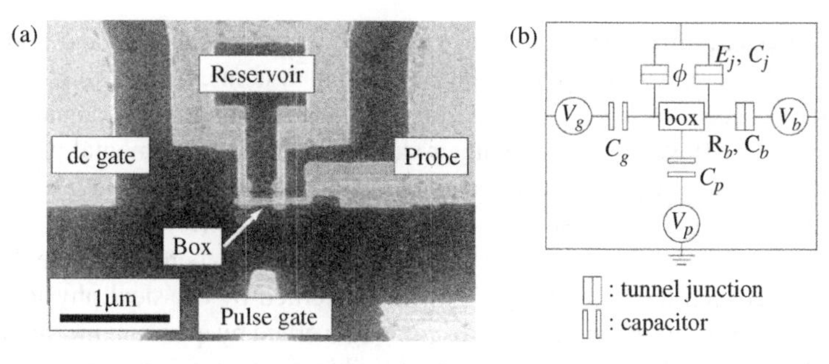

Fig. 5.12 An electron micrograph image of the Cooper pair box device used to demonstrate macroscopic quantum coherence in superconductors. The Cooper pair box, is connected to its environment via Josephson coupling to the charge reservoir, as indicated, and to a probe device, which is used to measure the number of Cooper pairs, N on the box. The device is manipulated through the two electrical gates indicated, one providing a d.c. bias, and the second delivering pico-second pulses which switch the device from one quantum state to another. Reproduced by permission from Nature (Nakamura, Pashkin and Tsai 1999). Copyright (1999) Macmillan Publishers Ltd.

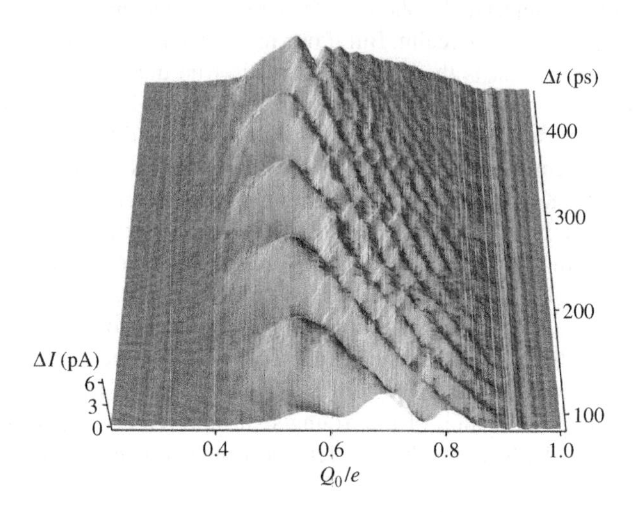

Fig. 5.13 Quantum oscillations of charge observed in the Cooper pair box of Fig. 5.12. The measured device current, I, is proportional to the Cooper pair number N on the box, and so the oscillations demonstrate quantum superpositions of states $|N\rangle$ and $|N+1\rangle$. The oscillations depend on the amplitude and duration of the gate pulses, corresponding to the two axes shown in the diagram. Reproduced by permission from Nature (Nakamura, Pashkin and Tsai 1999). Copyright (1999) Macmillan Publishers Ltd.

different states can be manipulated through external voltage gates, since they are states of different total electronic charge on the box. The analog of the Schrödinger cat state for this system is to place it in a superposition, such as

$$|\psi\rangle = \frac{1}{\sqrt{2}}\left(|N\rangle + |N+1\rangle\right). \tag{5.134}$$

Nakamura Pashkin and Tsai (1999) were able to demonstrate the presence of just such superpositions in their device. By connecting the Cooper pair box to a second superconductor, via a Josephson junction, they effectively allowed quantum mechanical transitions between these two states, as Cooper pairs tunnel onto or off the island. By pulsing external voltage gates connected to the system they observed beautiful interference fringes associated with the superposition states, as shown in Fig. 5.13. The figure shows the final charge on the box (i.e. $N+1$ or N) as a function of the voltage pulse amplitude and duration. The observed oscillations agree excellently with the theoretical predictions based on the existence of macroscopic quantum superposition states.

The third type of system in which macroscopic quantum coherence has been demonstrated are small superconducting rings, as in Fig. 4.6. As we saw in chapter 4, a superconducting ring has a set of different ground states corresponding to different winding numbers of the order parameter around the ring. These can again be represented by abstract many-particle states such as $|n\rangle$ and $|n+1\rangle$. In a large ring there will be no way for the system to tunnel from one of these states to another, but if the ring is made small enough (below $1\,\mu$m in diameter) such transitions become possible. Two experiments in the year 2000 observed direct evidence for quantum mechanical coherence in such systems (Friedman *et al.* 2000, van der Wal *et al.* 2000). These results are fascinating, since they represent a coherent quantum tunnelling of a system containing 10^{10} or more electrons. Figure 5.14 shows an electron microscope image of a small superconducting circuit, approximately $5\,\mu$m across, in which the quantum superposition states were observed. More recently Rabi oscillations were observed (Chiorescu *et al.* 2003). These oscillations demonstrate the existence of quantum superposition states in which the system is simultaneously in two macroscopically different quantum states!

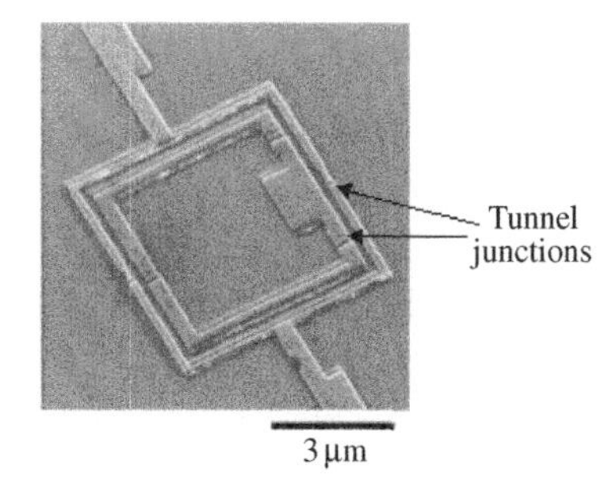

Fig. 5.14 A small superconducting SQUID type circuit, approximately $5\,\mu$m across. This device shows coherent quantum superposition states. Because the ring is macroscopic (containing or order 10^{10} superconducting electrons) this demonstrates the existence of "Schrödinger cat" quantum superposition states of macroscopically different states. Reprinted from van der Wal (2001) with permission.

Do these quantum superposition states have any practical use? In recent years there has been a huge growth in the field of **quantum information** and **quantum computation**. The idea is that "information" as used and manipulated in computer bits is actually always a physical quantity, for example, the charges on the capacitors in a RAM computer memory. Therefore it is subject to the laws of physics. But conventional computer bits are essentially based on classical physics. For example, a computer bit can be measured without disturbing its state. But if we imagine eventually making the physical computer bits smaller and smaller with each new generation of computer, then eventually we will have to use devices which are so small that quantum mechanics applies, not classical physics. For such a quantum bit, or **qubit**, information is carried by its full quantum state, not just by a classical 0 or 1. Surprisingly it turns out the computers based on manipulation of these qubits would be far more efficient than classical computers for certain types of algorithms. But whether this goal can be ever achieved depends on finding suitable physical systems in which to realize the qubit. While there are many possibilities under active investigation, superconducting devices or BEC have several possible advantages for these types of problem. At the very least, the experiments described above demonstrating macroscopic quantum superposition states show that BEC or superconducting devices are at least one viable option for a physical qubit. See Mahklin Schön and Shnirman (2001) and Annett Gyorffy and Spiller (2002) for more discussion of possible superconducting qubit devices.

5.10 Summary

In this chapter we have explored the implications of **quantum coherent states** in the theory of Bose and Fermi systems. We have seen how Bose coherent states provide an effective way to understand the laser and the weakly interacting Bose gas. The key point being that the coherent state representation allows one to discuss quantum states with definite phase θ, rather than with definite particle number N. Using the coherent state approach the ideas of an effective macroscopic quantum wave function and ODLRO also become quite natural.

For fermion systems the coherent state approach is also quite natural, provided that one deals with coherent states of Cooper pairs, not single electrons.

We have not yet seen how to explicitly construct such a coherent state (see the next chapter!), but we have already been able to see how ORLRO and the GL order parameter both arise naturally from this formalism. The Josephson effect, and its applications to SQUID devices can also be understood qualitatively even without the full description of the BCS wave function.

Finally we have seen that both BEC and superconductors do indeed exhibit **macroscopic quantum coherence**. But in the case of superconductors this is only evident when the devices are small enough and cold enough to avoid the effects of decoherence. Although the usual Josephson effect and the SQIUD interference patterns are both interference effects, they do not in themselves show the existence of quantum superposition states such as the Schrödinger cat.

Further reading

See Loudon (1979) for a more detailed discussion of optical coherent states and their application to the laser. A more advanced and general review of all applications of coherent states is given by Klauder and Skagerstam (1985).

The theory of the weakly interacting Bose gas is discussed in detail in Pines (1961), a book which also includes a reprint of the original paper by Bogoliubov (1947). More mathematically advanced approaches using many-body Green's function techniques are given by Fetter and Walecka (1971), and the Abrikosov, Gor'kov, and Dzyaloshinski (1963).

P.W. Anderson made many key contributions to the development of the ideas of ODLRO and macroscopic coherence in superconductors. His book *Basic Notions in Condensed Matter Physics* (Anderson 1984), includes several reprints of key papers in the discovery of ODLRO in superconductors, the Josephson effect and related topics. Tinkham (1996) also has a very detailed discussion about the Josephson effect and SQUID devices.

The problems and paradoxes posed by macroscopic quantum coherence are discussed by Leggett (1980), with a recent update in Leggett (2002). The possibilities of making superconducting qubit devices for quantum computation are discussed in Mahklin Schön and Shnirman (2001) and Annett Gyorffy and Spiller (2002).

Exercises

(5.1) (a) Using the definitions of the ladder operators given in Eq. 5.12 show that

$$[\hat{a}, \hat{a}^{+}] = 1$$

and

$$\hat{H} = \hbar\omega_c \left(\hat{a}^{+}\hat{a} + \tfrac{1}{2}\right).$$

(b) Show that

$$[\hat{H}, \hat{a}^{+}] = \hbar\omega_c \hat{a}^{+}.$$

Hence show that if $\psi_n(x)$ is an eigenstate of the Hamiltonian with energy E_n, then ψ_{n+1} (defined by Eq. 5.2) is also an eigenstate with energy

$$E_{n+1} = E_n + \hbar\omega_c.$$

(c) Assuming that $\psi_n(x)$ is normalized, show that ψ_{n+1} as defined by Eq. 5.2 is also a correctly normalized quantum state.

(5.2) Using the fundamental defining equation of the coherent state Eq. 5.13, show that two coherent states $|\alpha\rangle$ and $|\beta\rangle$ have the overlap,

$$\langle\alpha|\beta\rangle = e^{-|\alpha|^2/2}e^{-|\beta|^2/2}e^{\alpha^*\beta},$$

and hence derive Eq. 5.33.

(5.3) Show that for a coherent state $|\alpha\rangle$

$$\langle\alpha|(\hat{a}^+)^p\hat{a}^q|\alpha\rangle = (\alpha^*)^p\alpha^q$$

for any positive integers p and q.

(5.4) The proof that coherent states are a complete set relies on the following useful identity

$$\hat{I} = \frac{1}{4\pi}\int d^2\alpha|\alpha\rangle\langle\alpha|$$

called the *resolution of unity*. Here $\hat{I}$ is the identity operator, and $d^2\alpha$ means integrating over all possible values of α in the complex plane.

(a) Using the definition of the coherent state $|\alpha\rangle$ from Eq. 5.13, show that

$$|\alpha\rangle\langle\alpha| = e^{-|\alpha^2|}\sum_{n,m}\frac{(\alpha^*)^n\alpha^m}{(n!m!)^{1/2}}|\psi_m\rangle\langle\psi_n|,$$

where the $|\psi_n\rangle$ are the usual harmonic oscillator eigenstates of quantum number n.

(b) Transforming to polar coordinates in the complex plane, $\alpha = re^{i\theta}$, show that

$$\frac{1}{4\pi}\int d^2\alpha|\alpha\rangle\langle\alpha| = \sum_n|\psi_n\rangle\langle\psi_n|.$$

(c) Verify that the expression in (b) is the identity operator,

$$\hat{I} = \sum_n|\psi_n\rangle\langle\psi_n|,$$

by considering its action on any general quantum state

$$|\psi\rangle = \sum_m c_m|\psi_m\rangle$$

were $c_m = \langle\psi_m|\psi\rangle$.

(5.5) (a) Writing the pair of Eqs 5.92 and 5.32 in the matrix form,

$$\begin{pmatrix} b_{\mathbf{k}} \\ b^+_{-\mathbf{k}} \end{pmatrix} = \begin{pmatrix} u_{\mathbf{k}} & v_{\mathbf{k}} \\ v_{\mathbf{k}} & u_{\mathbf{k}} \end{pmatrix}\begin{pmatrix} a_{\mathbf{k}} \\ a^+_{-\mathbf{k}} \end{pmatrix}$$

show that the pair of equations can be inverted to yield

$$\begin{pmatrix} a_{\mathbf{k}} \\ a^+_{-\mathbf{k}} \end{pmatrix} = \begin{pmatrix} u_{\mathbf{k}} & -v_{\mathbf{k}} \\ -v_{\mathbf{k}} & u_{\mathbf{k}} \end{pmatrix}\begin{pmatrix} b_{\mathbf{k}} \\ b^+_{-\mathbf{k}} \end{pmatrix}$$

(b) Rewrite the Bogoliubov Hamiltonian, Eq. 5.91, in the matrix form,

$$\hat{H} = \sum_{\mathbf{k}}\begin{pmatrix} a^+_{\mathbf{k}} & a_{-\mathbf{k}} \end{pmatrix}\begin{pmatrix} \epsilon_{\mathbf{k}} + n_0 g & \frac{1}{2}n_0 g \\ \frac{1}{2}n_0 g & 0 \end{pmatrix}\begin{pmatrix} a_{\mathbf{k}} \\ a^+_{-\mathbf{k}} \end{pmatrix},$$

where $\epsilon_{\mathbf{k}} = \hbar^2k^2/2m$.

(c) Substituting the matrix expression (b) into the Hamiltonian (c), show that the Hamiltonian expressed in terms of the b operators becomes

$$\hat{H} = \sum_{\mathbf{k}}\begin{pmatrix} b^+_{\mathbf{k}} & b_{-\mathbf{k}} \end{pmatrix}\begin{pmatrix} M_{11} & M_{12} \\ M_{21} & M_{22} \end{pmatrix}\begin{pmatrix} b_{\mathbf{k}} \\ b^+_{-\mathbf{k}} \end{pmatrix},$$

where the new matrix is

$$\begin{pmatrix} M_{11} & M_{12} \\ M_{21} & M_{22} \end{pmatrix} = \begin{pmatrix} u_{\mathbf{k}} & -v_{\mathbf{k}} \\ -v_{\mathbf{k}} & u_{\mathbf{k}} \end{pmatrix} \times \begin{pmatrix} \epsilon + n_0 g & \frac{1}{2}n_0 g \\ \frac{1}{2}n_0 g & 0 \end{pmatrix} \times \begin{pmatrix} u_{\mathbf{k}} & -v_{\mathbf{k}} \\ -v_{\mathbf{k}} & u_{\mathbf{k}} \end{pmatrix}.$$

(d) Show that the condition for the transformed matrix to be diagonal is that

$$\frac{2u_{\mathbf{k}}v_{\mathbf{k}}}{u^2_{\mathbf{k}} + v^2_{\mathbf{k}}} = \frac{n_0 g}{\epsilon_{\mathbf{k}} + n_0 g}.$$

(e) Show that the sum of the diagonal elements of the M matrix (the trace) is

$$E = (\epsilon_{\mathbf{k}} + n_0 g)(u^2_{\mathbf{k}} + v^2_{\mathbf{k}}) - 2n_0 g u_{\mathbf{k}}v_{\mathbf{k}}.$$

Using the representation $u_{\mathbf{k}} = \cosh\theta$, $v_{\mathbf{k}} = \sinh\theta$, show from (d) that

$$\tanh(2\theta) = \frac{n_0 g}{\epsilon_{\mathbf{k}} + n_0 g}$$

and hence prove that

$$E = \left[\epsilon_{\mathbf{k}}(\epsilon_{\mathbf{k}} + 2n_0 g)\right]^{1/2}$$

consistent with the Bogoliubov quasiparticle energy given by Eq. 5.95.

The BCS theory of superconductivity

6

6.1 Introduction

In 1957 Bardeen Cooper and Schrieffer (BCS) published the first truly microscopic theory of superconductivity. The theory was soon recognized to be correct in all the essential aspects, and to explain a number of important experimental phenomena. For example, the theory correctly explained the isotope effect:

$$T_c \propto M^{-\alpha}, \tag{6.1}$$

in which the transition temperature changes with the mass of the crystal lattice ions, M. The original BCS theory predicts that the isotope exponent α is $1/2$. Most common superconductors agree very well with this prediction, as one can see in Table 6.1. However, it is also clear that there are exceptions to this prediction. Transition metals such as Molybdenum and Osmium (Mo, Os) show a reduced effect, and others such as Ruthenium, Ru, have essentially zero isotope effect. In these it is necessary to extend the BCS theory to include what are called strong coupling effects. In other systems, such as the high temperature superconductor, $YBa_2Cu_3O_7$, the absence of the isotope effect may indicate that the lattice phonons are not involved at all in the pairing mechanism.[1]

The second main prediction of the BCS theory is the existence of an energy gap 2Δ at the Fermi level, as shown in Fig. 6.1. In the normal metal the electron states are filled up to the Fermi energy, ϵ_F, and there is a finite density of states at the Fermi level, $g(\epsilon_F)$. But in a BCS superconductor below T_c the electron density of states acquires a small gap 2Δ separating the occupied and unoccupied states. This gap is fixed at the Fermi energy, and so (unlike a band gap in a semiconductor or insulator) it does not prevent electrical conduction.

This energy gap was discovered experimentally at essentially the same time as BCS theory was developed. Immediately after the BCS theory was published various different experimental measurements of the energy gap, 2Δ, were shown to be excellent agreement with the predictions. Perhaps most important of all of these was electron tunnelling spectroscopy. This not only showed the existence of the energy gap, 2Δ, but also showed extra features which directly showed that the gap arises from electron–phonon coupling. The gap parameter Δ also had another important role. In 1960 Gor'kov was able to use the BCS theory to derive the Ginzburg–Landau equations, and hence gave a microscopic explanation of the order parameter ψ. He not only found that ψ is directly related to the wave function for the Cooper pairs, but that it is also directly proportional to the gap parameter Δ.

[1] Even here the situation is complicated. In fact if the material is prepared with less than optimal oxygen content, for example, $YBa_2Cu_3O_{6.5}$, then there is again a substantial isotope effect, although less than the BCS prediction. The significance of these is still a matter of strong debate. Do they indicate a phonon role in the pairing mechanism, or do they just relate to variations in lattice properties, band structure, etc. which only influence T_c indirectly?

Table 6.1 Isotope effect in some selected superconductors

	T_c(K)	α
Zn	0.9	0.45
Pb	7.2	0.49
Hg	4.2	0.49
Mo	0.9	0.33
Os	0.65	0.2
Ru	0.49	0.0
Zr	0.65	0.0
Nb_3Sn	23	0.08
MgB_2	39	0.35
$YBa_2Cu_3O_7$	90	0.0

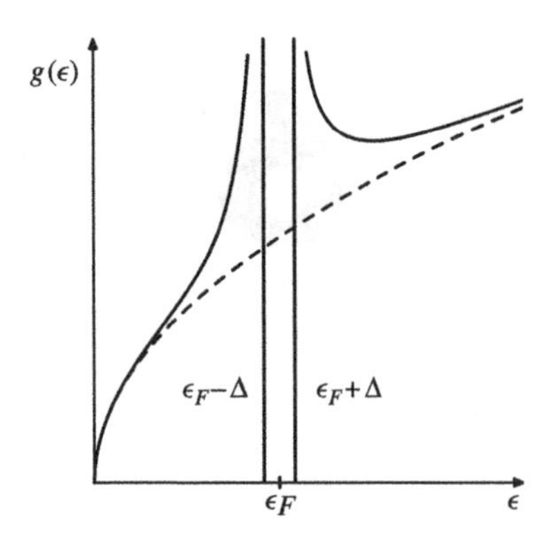

Fig. 6.1 The BCS energy gap, 2Δ in a superconductor. The gap is always pinned at the Fermi level, unlike the gap in an insulator or a semiconductor, and hence electrical conduction always remains possible.

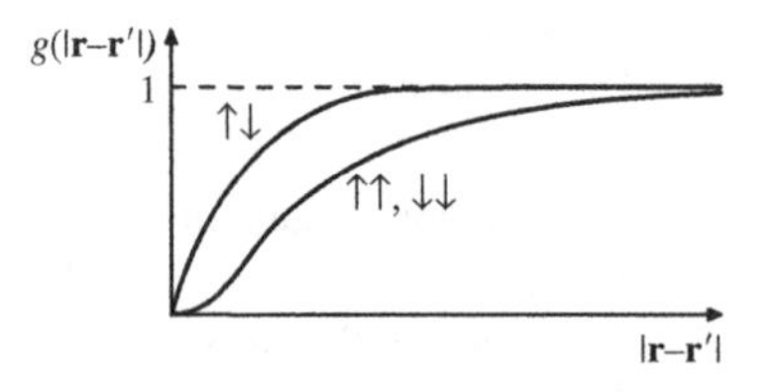

Fig. 6.2 The exchange-correlation hole for an electron moving in a metal. $g(|\mathbf{r}-\mathbf{r}'|)$ is the pair-correlation function of the electron gas. It measures the relative probability of finding an electron at $\mathbf{r}'$ given that one is at $\mathbf{r}$. By the Pauli exclusion principle this must vanish for $\mathbf{r} = \mathbf{r}'$ in the case of parallel spin particles, ↑↑ and ↓↓. This is the **exchange**-hole. But the $e^2/4\pi\epsilon_0 r$ repulsive Coulomb interaction gives an additional high energy cost for two electrons to be close together, whatever their spins. This is the **correlation** part of the exchange-correlation hole.

BCS theory built upon three major insights: (i) First it turns out that the effective forces between electrons can sometimes be **attractive** in a solid rather than repulsive. This is due to coupling between the electrons and the phonons of the underlying crystal lattice. (ii) Second, in the famous "Cooper problem," Cooper considered the simple system of just two electrons outside an occupied Fermi surface. Surprisingly, he found that the electrons form a stable pair bound state, and this is true **however weak the attractive force**! (iii) Finally Schrieffer constructed a many-particle wave function which all the electrons near to the Fermi surface are paired up. This has the form of a coherent state wave function, similar to those we have seen in the previous chapter. The BCS energy gap 2Δ comes out of this analysis, since 2Δ corresponds to the energy for breaking up a pair into two free electrons.

The full derivation of BCS theory requires more advanced methods of many-body theory than we can cover properly in this volume. For example, BCS theory can be elegantly formulated in terms of many-body Green's functions and Feynman diagrams (Abrikosov, Gor'kov, and Dzyaloshinski 1963; Fetter and Walecka 1971). But, on the other hand it is possible to at least get the main flavor of the theory with just basic quantum mechanics. Here we shall just follow this simpler approach to develop the outline of the BCS theory and to summarize the key points. Those wishing to extend their knowledge to a deeper level should consult these more advanced texts.

6.2 The electron–phonon interaction

The first key idea in BCS theory is that there is an effective attraction for electrons near the Fermi surface. This idea was first formulated by Frölich in 1950. At first it is very surprising to find an attractive force, because electrons "obviously" repel each other strongly with the electrostatic Coulomb repulsion,

$$V(\mathbf{r}-\mathbf{r}') = \frac{e^2}{4\pi\epsilon_0|\mathbf{r}-\mathbf{r}'|}. \tag{6.2}$$

While this is obviously always true for the **bare electrons**, in a metal we should properly think about **quasiparticles** not bare electrons. A quasiparticle is an excitation of a solid consisting of a moving electron together with a surrounding **exchange correlation hole**. This idea is illustrated in Fig. 6.2. The point is that when the electron moves other electrons must move out of the way. They must do this both because the exclusion principle prevents two electrons of the same spin being at the same point (this is called the exchange interaction) and because they must also try to minimize the repulsive Coulomb energy of Eq. 6.2.[2] The idea of a quasiparticle was developed by Landau, and we call such a system of strongly interacting fermions a **Landau Fermi liquid**. We shall explore the Fermi liquid idea in more depth in the next chapter.

[2]The exchange interaction arises if one treats the many-electron state of the metal using Hartree–Fock theory. But this is not adequate for metals, and modern methods of **Density Functional Theory**, DFT, include both exchange and correlation effects explicitly.

If we consider both the electron and its surrounding exchange correlation hole, then in a metal it turns out that between quasiparticles the effective Coulomb force is substantially reduced by **screening**. Using the simplest model of screening in metals, the Thomas Fermi model, we would expect an effective interaction of the form,

$$V_{\mathrm{TF}}(\mathbf{r}-\mathbf{r}') = \frac{e^2}{4\pi\epsilon_0|\mathbf{r}-\mathbf{r}'|}e^{-|\mathbf{r}-\mathbf{r}'|/r_{\mathrm{TF}}}. \tag{6.3}$$

Here r_{TF} is the Thomas Fermi screening length. One can see that the effect of screening is to substantially reduce the Coulomb repulsion. In particular the effective repulsive force is now short ranged in space, vanishing for $|\mathbf{r}-\mathbf{r}'| > r_{\mathrm{TF}}$. The overall repulsive interaction is therefore much weaker than the original $1/r$ potential.

Second the electrons interact with each other via their interaction with the phonons of the crystal lattice. In the language of Feynman diagrams an electron in Bloch state $\psi_{n\mathbf{k}}(\mathbf{r})$ can excite a phonon of crystal momentum $\hbar\mathbf{q}$, leaving the electron in a state $\psi_{n\mathbf{k}'}(\mathbf{r})$ with crystal momentum $\hbar\mathbf{k}' = \hbar\mathbf{k} - \hbar\mathbf{q}$. Later a second electron can absorb the phonon and pick up the momentum $\hbar\mathbf{q}$. This gives rise to Feynman diagrams as drawn in Fig. 6.3, which correspond to an effective interaction between the electrons.

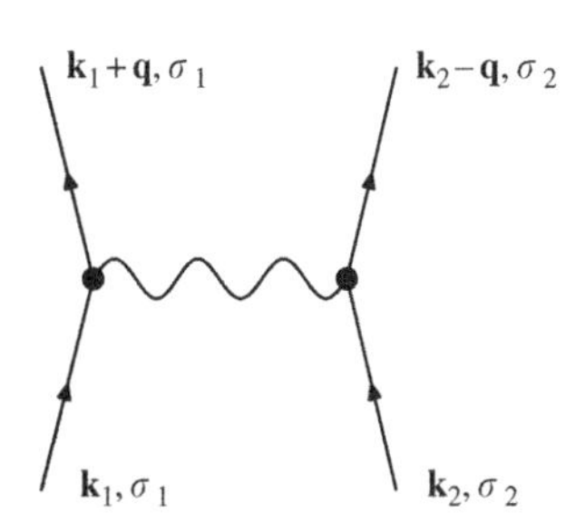

Fig. 6.3 Interaction of fermions via exchange of a boson. In particle physics such diagrams could represent interactions between quarks due to exchange of gluons, interactions between electrons by exchange of photons or by exchange of W or Z bosons. In the BCS theory the same principle gives the interaction between electrons at the Fermi surface due to exchange of crystal lattice phonons.

How does this electron–phonon interaction arise? Consider a phonon of wave vector $\mathbf{q}$ in a solid. The effective Hamiltonian for the phonons in the solid will be just a set of quantum harmonic oscillators, one for each wave vector $\mathbf{q}$ and phonon mode

$$\hat{H} = \sum_{\mathbf{q},\lambda} \hbar\omega_{\mathbf{q}\lambda}\left(a^{+}_{\mathbf{q}\lambda}a_{\mathbf{q}\lambda} + \frac{1}{2}\right), \tag{6.4}$$

where the operators $a^{+}_{\mathbf{q}\lambda}$ and $a_{\mathbf{q}\lambda}$ create or annihilate a phonon in mode λ, respectively. There are $3N_a$ phonon modes (branches) in a crystal with N_a atoms per unit cell. For simplicity let us assume that there is only one atom per unit cell, in which case there are just three phonon modes (one longitudinal mode and two transverse). Using the expressions for the ladder operators Eq. 5.12, the atoms located at $\mathbf{R}_i$ will be displaced by

$$\delta\mathbf{R}_i = \sum_{\mathbf{q}\lambda} \mathbf{e}_{\mathbf{q}\lambda}\left(\frac{\hbar}{2M\omega_{\mathbf{q}\lambda}}\right)^{1/2}(a^{+}_{\mathbf{q}\lambda} + a_{-\mathbf{q}\lambda})e^{i\mathbf{q}\cdot\mathbf{R}_i}. \tag{6.5}$$

Here $\mathbf{e}_{\mathbf{q}\lambda}$ is a unit vector in the direction of the atomic displacements for mode $\mathbf{q}\lambda$. For example, for the longitudinal mode this will be in the direction of propagation, $\mathbf{q}$.

Such a displacement of the crystal lattice will produce a modulation of the electron charge density and the effective potential for the electrons in the solid, $V_1(\mathbf{r})$. We can define the **deformation potential** by

$$\delta V_1(\mathbf{r}) = \sum_i \frac{\partial V_1(\mathbf{r})}{\partial \mathbf{R}_i}\delta\mathbf{R}_i, \tag{6.6}$$

as illustrated in Fig. 6.4.

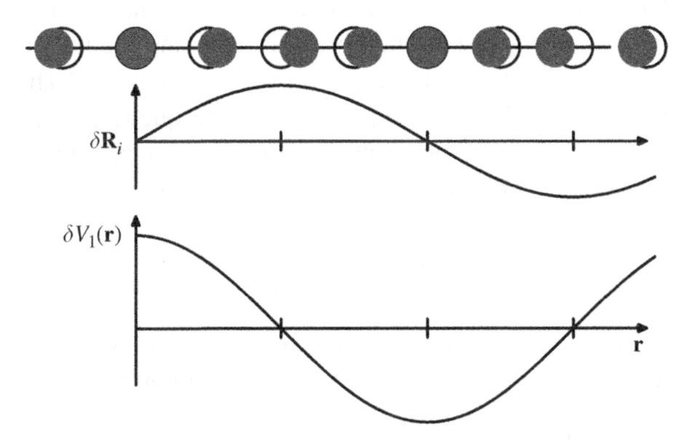

Fig. 6.4 A phonon in a solid and the resulting atomic displacements, $\delta\mathbf{R}_i$, and deformation potential, $\delta V_1(\mathbf{r})$. For example, one can see that in the diagram the atom at the origin is not displaced, and locally its neighbors are further apart than average. This leads to a locally repulsive potential for the electrons, since there is a reduced positive charge density from the ions. In contrast, in regions where the atom density is higher than average the deformation potential is attractive for electrons.

This is a periodic modulation of the potential with wavelength $2\pi/q$. An electron moving through the crystal lattice will experience this periodic potential and undergo diffraction. If it is initially in Bloch state $\psi_{n\mathbf{k}}(\mathbf{r})$, it can be diffracted to another Bloch state $\psi_{n'\mathbf{k}-\mathbf{q}}(\mathbf{r})$. The net effect of this is that an electron has been scattered from a state with crystal momentum $\mathbf{k}$ to one with momentum $\mathbf{k}-\mathbf{q}$. The extra "momentum" has been provided by the phonon. One can see that either one has created a phonon of momentum $\mathbf{q}$, or annihilated one of momentum $-\mathbf{q}$, consistent with overall conservation of crystal momentum.[3] We can draw such an interaction as a **vertex** of a Feynman diagram, as shown in Fig. 6.5. In the vertex an electron is scattered from one momentum state to another while simultaneously a phonon is created or destroyed.

[3]There are also **Umklapp** processes, where it is simultaneously scattered by a reciprocal lattice vector of the crystal from $\psi_{n\mathbf{k}}(\mathbf{r})$ to $\psi_{n'\mathbf{k}+\mathbf{q}+\mathbf{G}}(\mathbf{r})$. We shall not consider such processes here. Although they do contribute to the total electron–phonon interaction, their effect is generally less important than the direct scattering terms.

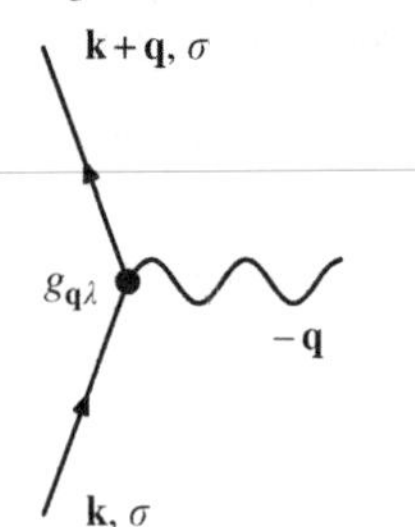

Fig. 6.5 The vertex for the electron–phonon interaction. The electron is scattered from $\mathbf{k}$ to $\mathbf{k}+\mathbf{q}$ by the annihilation of a phonon of wave vector $\mathbf{q}$, or the destruction of a phonon of wave vector $-\mathbf{q}$. The phonon can be real or virtual, depending on the available energy.

Putting together two such vertices we arrive at the diagram shown in Fig. 6.3. The meaning of this diagram is that one electron emits a phonon, it propagates for a while, and it is then absorbed by a second electron. The net effect of the process is to transfer momentum $\hbar\mathbf{q}$ from one electron to the other. Therefore it implies an effective interaction between electrons. Note that we do not have to specify which of the electrons created or destroyed the phonon. Therefore there is no need to draw an arrow on the phonon line showing which way it propagates. This effective interaction between the electrons due to exchange of phonons turns out to be of the form:

$$V_{\text{eff}}(\mathbf{q},\omega) = |g_{\mathbf{q}\lambda}|^2 \frac{1}{\omega^2 - \omega_{\mathbf{q}\lambda}^2} \tag{6.7}$$

where the virtual phonon has wave vector $\mathbf{q}$ and frequency $\omega_{\mathbf{q}\lambda}$. The parameter $g_{\mathbf{q}\lambda}$ is related to the matrix element for scattering an electron from state $\mathbf{k}$ to $\mathbf{k}+\mathbf{q}$ as shown in Fig. 6.5.

An important result due to Migdal is that the electron–phonon vertex, $g_{\mathbf{q}\lambda}$ is of order

$$g_{\mathbf{q}\lambda} \sim \sqrt{\frac{m}{M}}, \tag{6.8}$$

where m is the effective mass of the electrons at the Fermi surface and M is the mass of the ions. Since m/M is of order 10^{-4}, typically, the electrons and phonons are only weakly coupled. We are therefore justified in only using the basic electron–phonon coupling diagram Fig. 6.3 and we can neglect and higher order diagrams which would contain more vertices.

The full treatment of this effective interaction is still too complex for analytic calculations. For this reason BCS introduced a highly simplified form of the above effective interaction. They first neglected dependence of the interaction on the wave vector $\mathbf{q}$ and phonon branch, replacing the interaction by an approximate one which effectively averages over all values of $\mathbf{q}$. The frequency $\omega_{\mathbf{q}}$ is replaced by, ω_D, which is a typical phonon frequency, usually taken to be the Debye frequency of the phonons, and the $\mathbf{q}$ dependent electron–phonon interaction vertex, $g_{\mathbf{q}\lambda}$, is replaced by a constant, g_{eff}, giving

$$V_{\text{eff}}(\mathbf{q},\omega) = |g_{\text{eff}}|^2 \frac{1}{\omega^2 - \omega_D^2}. \tag{6.9}$$

This is an attractive interaction for phonon frequencies ω, which are less than ω_D, and repulsive for $\omega > \omega_D$. But BCS recognized that the repulsive part is not important. We are only interested in electrons which lie within $\pm k_B T$ of the

Fermi energy, and at the temperatures of interest to superconductivity we are in the regime $\hbar\omega_D \gg k_B T$. Therefore BCS assumed the final, simple form

$$V_{\text{eff}}(\mathbf{q}, \omega) = -|g_{\text{eff}}|^2 \quad |\omega| < \omega_D. \tag{6.10}$$

The corresponding effective Hamiltonian for the effective electron–electron interaction is

$$\hat{H}_1 = -|g_{\text{eff}}|^2 \sum c^+_{\mathbf{k}_1+\mathbf{q}\sigma_1} c^+_{\mathbf{k}_2-\mathbf{q}\sigma_1} c_{\mathbf{k}_1\sigma_1} c_{\mathbf{k}_2\sigma_2}, \tag{6.11}$$

where the sum is over all values of $\mathbf{k}_1$, σ_1, $\mathbf{k}_2$, σ_1, and $\mathbf{q}$ with the restriction that the electron energies involved are all within the range $\pm\hbar\omega_D$ of the Fermi surface,

$$|\epsilon_{\mathbf{k}_i} - \epsilon_F| < \hbar\omega_D.$$

Therefore we have interacting electrons near the Fermi surface, but the Bloch states far inside or outside the Fermi surface are unaffected, as shown in Fig. 6.6. The problem is that of electons in this thin shell of states around ϵ_F.

Note that combining the fact that the Migdal vertex is $\sim 1/M^{1/2}$ and the $1/\omega_D^2$ in the effective interaction, one finds that $|g_{\text{eff}}|^2 \sim 1/(M\omega_D^2)$. This turns out to be independent of the mass of the ions, M, since $\omega_D \sim (k/M)^{1/2}$, where k is an effective harmonic spring constant for the lattice vibrations. Therefore in the BCS model the isotope effect arises because the thickness of the energy shell around the Fermi surface is $\hbar\omega_D$, and not from changes in the coupling constant, $|g_{\text{eff}}|^2$.

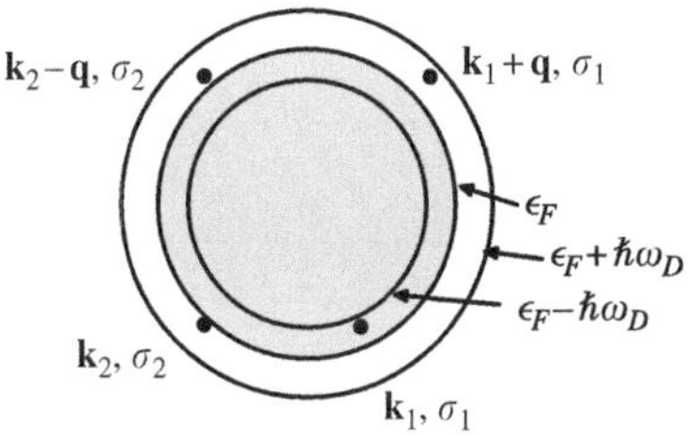

Fig. 6.6 The effective electron–electron interaction near the Fermi surface. The electrons at $\mathbf{k}_1, \sigma_1$ and $\mathbf{k}_2, \sigma_2$ are scattered to $\mathbf{k}_1 + \mathbf{q}, \sigma_1$ and $\mathbf{k}_2 - \mathbf{q}, \sigma_2$. The interaction is attractive provided that all of the wave vectors lie in the range where $\epsilon_{\mathbf{k}}$ is within energy $\pm\hbar\omega_D$ of the Fermi energy.

6.3 Cooper pairs

Having found that there is an attraction between electrons near the Fermi level is still a long way from a theory of superconductivity. The next key step was carried out by Cooper. He noted that the effective interaction is attractive only near to the Fermi surface, Fig. 6.6, and asked what the effect of this attraction would be for just a single pair of electrons outside the occupied Fermi sea. He found that they form a bound state. This result was quite unexpected, since two electrons in free space would not bind with the same weak attractive interaction. This "Cooper problem" thus shows that the Fermi liquid state (i.e. independent Bloch electrons) is unstable to even weak attractive interactions between the particles. This idea thus led the way to the full BCS state in which every electron at the Fermi surface is part of a pair.

Cooper's model is the following. Assume a spherical Fermi surface at zero temperature, where all the states with $k < k_F$ are occupied. Then place an extra two electrons outside of the Fermi surface. These interact by the electron–phonon interaction as shown in Fig. 6.7.

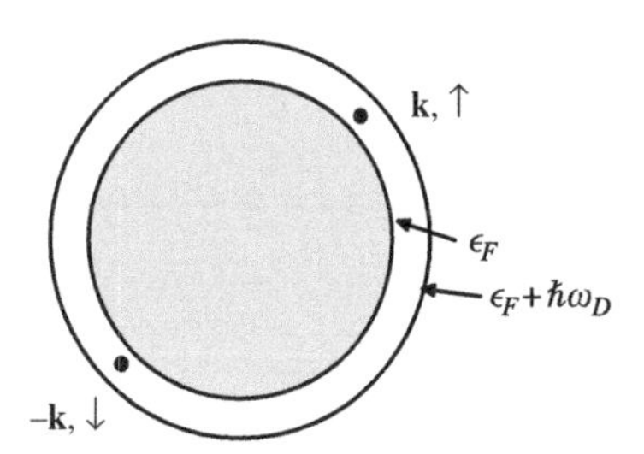

Fig. 6.7 The Cooper problem: two electrons outside a fully occupied Fermi sea. The interaction is attractive provided that the electron energies are in the range $\epsilon_F < \epsilon_{\mathbf{k}} < \epsilon_F + \hbar\omega_D$.

The two particle wave function of these extra electrons is

$$\Psi(\mathbf{r}_1, \sigma_1, \mathbf{r}_2, \sigma_2) = e^{i\mathbf{k}_{\text{cm}}\cdot\mathbf{R}_{\text{cm}}}\varphi(\mathbf{r}_1 - \mathbf{r}_2)\phi^{\text{spin}}_{\sigma_1,\sigma_2}, \tag{6.12}$$

where $\mathbf{R}_{\text{cm}}$ is the centre of mass $(\mathbf{r}_1 + \mathbf{r}_2)/2$ and $\hbar\mathbf{k}_{\text{cm}}$ is the total momentum of the pair. It turns out that the minimum energy will correspond to a pair with no center of mass motion, so in the ground state, $\mathbf{k}_{\text{cm}} = 0$, and we shall assume this is so from now on.

The spin wave function can be either spin singlet:

$$\phi^{\text{spin}}_{\sigma_1,\sigma_2} = \frac{1}{\sqrt{2}}(|\uparrow\downarrow\rangle - |\downarrow\uparrow\rangle), \tag{6.13}$$

(total spin $S = 0$) or spin triplet ($S = 1$)

$$\phi^{\text{spin}}_{\sigma_1,\sigma_2} = \begin{cases} |\uparrow\uparrow\rangle \\ \frac{1}{\sqrt{2}}(|\uparrow\downarrow\rangle + |\downarrow\uparrow\rangle) \\ |\downarrow\downarrow\rangle \end{cases}. \tag{6.14}$$

Almost all known superconductors (with a few very interesting exceptions) have singlet Cooper pairs and so we shall assume this from now on in this chapter.

Fermion antisymmetry implies that

$$\Psi(\mathbf{r}_1, \sigma_1, \mathbf{r}_2, \sigma_2) = -\Psi(\mathbf{r}_2, \sigma_2, \mathbf{r}_1, \sigma_1). \tag{6.15}$$

Since the spin singlet is an odd function of σ_1 and σ_2 the wave function $\varphi(\mathbf{r}_1 - \mathbf{r}_2)$ must be even, that is, $\varphi(\mathbf{r}_1 - \mathbf{r}_2) = +\varphi(\mathbf{r}_2 - \mathbf{r}_1)$. Conversely, for a spin triplet state bound it would have to be an odd function.

Expanding $\phi(\mathbf{r}_1 - \mathbf{r}_2)$ in terms of the Bloch waves (assumed to be simply free electron plane wave states) we have

$$\varphi(\mathbf{r}_1 - \mathbf{r}_2) = \sum_{\mathbf{k}} \varphi_{\mathbf{k}}\, e^{i\mathbf{k}(\mathbf{r}_1 - \mathbf{r}_2)} \tag{6.16}$$

where $\varphi_{\mathbf{k}}$ are some expansion coefficients to be found. $\varphi_{\mathbf{k}} = \varphi_{-\mathbf{k}}$ because the function $\varphi(\mathbf{r})$ is even. The full pair wave function is thus a sum of Slater determinants:

$$\Psi(\mathbf{r}_1, \sigma_1, \mathbf{r}_2, \sigma_2) = \sum_{\mathbf{k}} \varphi_{\mathbf{k}} \begin{vmatrix} \psi_{\mathbf{k}\uparrow}(\mathbf{r}_1) & \psi_{\mathbf{k}\downarrow}(\mathbf{r}_2) \\ \psi_{-\mathbf{k}\uparrow}(\mathbf{r}_1) & \psi_{-\mathbf{k}\downarrow}(\mathbf{r}_2) \end{vmatrix}, \tag{6.17}$$

where the single particle Bloch state is $\psi_{\mathbf{k}}(\mathbf{r}) = e^{i\mathbf{k}\cdot\mathbf{r}}$. Each Slater determinant includes an up spin and a down spin, and an electron at $\mathbf{k}$ and $-\mathbf{k}$. The state is thus a pairing of electron waves at $\mathbf{k}$ with those of opposite spin at $-\mathbf{k}$. The restriction that all the states below k_F are already filled is imposed by restricting the sum over $\mathbf{k}$ to the range $k > k_F$.

Substituting this trial wave function into the Schrödinger equation gives:

$$E\varphi_{\mathbf{k}} = 2\epsilon_{\mathbf{k}}\varphi_{\mathbf{k}} - |g_{\text{eff}}|^2 \sum_{\mathbf{k}'} \varphi_{\mathbf{k}'}, \tag{6.18}$$

where E is the total energy of the two particle state. For simplicity the energy $\epsilon_{\mathbf{k}}$ is measured relative to ϵ_F. To obtain this equation we first write

$$|\Psi\rangle = \sum_{\mathbf{k}} \varphi_{\mathbf{k}} |\Psi_{\mathbf{k}}\rangle, \tag{6.19}$$

where $|\Psi_{\mathbf{k}}\rangle$ is the two particle Slater determinant given above

$$|\Psi_{\mathbf{k}}\rangle = \begin{vmatrix} \psi_{\mathbf{k}\uparrow}(\mathbf{r}_1) & \psi_{\mathbf{k}\downarrow}(\mathbf{r}_2) \\ \psi_{-\mathbf{k}\uparrow}(\mathbf{r}_1) & \psi_{-\mathbf{k}\downarrow}(\mathbf{r}_2) \end{vmatrix}. \tag{6.20}$$

This wave function obeys the two body Schrödinger equation, which is

$$\hat{H}|\Psi\rangle = E|\Psi\rangle. \tag{6.21}$$

Multiplying this equation on the left by $\langle\Psi_{\mathbf{k}}|$ picks out the terms for a given $\mathbf{k}$. The Hamiltonian consists of the two energies of the Bloch states $\epsilon_{\mathbf{k}}$ (and note $\epsilon_{\mathbf{k}} = \epsilon_{-\mathbf{k}}$) together with the effective interaction $-|g_{\text{eff}}|^2$. The effective interaction takes a momentum $\mathbf{q} = \mathbf{k}' - \mathbf{k}$ from one of the electrons and transfers it to the other. A pair of electrons $\mathbf{k}, -\mathbf{k}$ thus becomes a pair $\mathbf{k}', -\mathbf{k}'$ with a matrix element $-|g_{\text{eff}}|^2$. The limitation that $\epsilon(\mathbf{k}) < \hbar\omega_D$ places another restriction on the possible values of $\mathbf{k}$ limiting them to a thin shell between $k = k_F$ and $k = k_F + \omega_D/v$, with v the Bloch wave group velocity at the Fermi surface.

The energy E can be found by a self-consistency argument. Let us define

$$C = \sum_{\mathbf{k}} \varphi_{\mathbf{k}}. \tag{6.22}$$

Then we can solve Eq. 6.18 for the $\varphi_{\mathbf{k}}$, giving

$$\varphi_{\mathbf{k}} = -C|g_{\text{eff}}|^2 \frac{1}{E - 2\epsilon_{\mathbf{k}}}. \tag{6.23}$$

Self-consistency requires

$$C = \sum_{\mathbf{k}} \varphi_{\mathbf{k}} = -C|g_{\text{eff}}|^2 \sum_{\mathbf{k}} \frac{1}{E - 2\epsilon_{\mathbf{k}}} \tag{6.24}$$

or

$$1 = -|g_{\text{eff}}|^2 \sum_{\mathbf{k}} \frac{1}{E - 2\epsilon(\mathbf{k})}. \tag{6.25}$$

Converting the sum over $\mathbf{k}$ into an integral over the density of states gives

$$1 = -|g_{\text{eff}}|^2 g(\epsilon_F) \int_0^{\hbar\omega_D} d\epsilon \frac{1}{E - 2\epsilon}. \tag{6.26}$$

The integration limits are present because of the restriction of $\mathbf{k}$ to the thin shell around the Fermi surface, as discussed above. The integration is easy, and the result can be rearranged to find E,

$$-E = 2\hbar\omega_D e^{-1/\lambda}. \tag{6.27}$$

where the **electron–phonon coupling parameter**, λ is defined by

$$\lambda = |g_{\text{eff}}|^2 g(\epsilon_F) \tag{6.28}$$

and is assumed to be small, $\lambda \ll 1$.

Thus a bound state does exist, and its energy is exponentially small when λ is small. As in the full BCS solution the energy scale for superconductivity is set by the Debye energy, but multiplied by a very small exponential factor. This explains why the transition temperatures T_c are so small compared with other energy scales in solids. The Debye energies of most materials usually correspond to energy scales of order 100–300 K, and it is the very small exponential factor which leads to $T_c \sim 1$ K for most metallic superconductors.

Interestingly, the bound state exists, however small the interaction constant λ is. This would not have been the case without the filled Fermi sea. In general, an attractive interaction in three dimensions does not always lead to the existence of a bound state. The presence of the filled Fermi sea is thus a key aspect of the BCS theory.

Finally, notice that obviously we could have made two particle states with different quantum numbers. For example, we could have made spin triplets instead of singlets. The relative coordinate wave function $\varphi(\mathbf{r}_1 - \mathbf{r}_2)$ we had above was independent of the direction of the vector $\mathbf{r}_1 - \mathbf{r}_2$, that is, the pair are found in an s-wave state (like the ground state of the hydrogen atom). On the other hand it might have been possible to find solutions with p or d type wave functions,

$$\varphi(\mathbf{r}_1 - \mathbf{r}_2) = f(|\mathbf{r}_1 - \mathbf{r}_2|)Y_{lm}(\theta, \phi),$$

where Y_{lm} is a spherical Harmonic function. In general these different pairing states are all quite possible, however it seems that almost all superconductors choose an s-wave singlet pairing state. In fact the BCS model electron–electron interaction we chose above only allows solutions of that type, since it is independent of the phonon wave vector $\mathbf{q}$. Fourier transforming to real space, this corresponds to a point contact interaction

$$V_{\text{eff}}(\mathbf{r}_1 - \mathbf{r}_2) = -|g_{\text{eff}}|^2\delta(\mathbf{r}_1 - \mathbf{r}_2).$$

Only s-wave spherical Harmonic functions, $l = 0$, allow pair wave function which is finite at $\mathbf{r}_1 = \mathbf{r}_2$. However, more general types of interactions, perhaps not due to electron–phonon coupling, can allow other pair types to occur. Superfluid helium-3 (^{3}He) occurs because of Cooper pairing of the (fermion) ^{3}He atoms. These Cooper pairs turn out to be *p–wave* and spin triplet. The nature of the Cooper pairs in the high T_c superconductors is still controversial, but there is now a lot of evidence suggesting spin singlet, but *d–wave* Cooper pairs. We shall return to this topic in the next chapter.

6.4 The BCS wave function

Using the insight from the Cooper problem, BCS realized that the whole Fermi surface would be unstable to the creation of such pairs. As soon as there is an effective attractive interaction essentially every electron at the Fermi surface will become bound into a Cooper pair.

The next problem was to write down a many particle wave function in which every electron is paired. At first one could try a sort of product state of the form given in Eq. 5.107. However, this function is not very convenient to work with. It also does not make clear the concept of the macroscopic quantum coherence which, as we have seen, is essential to the formation of a condensate and hence to the idea of superconductivity.

Instead, Schrieffer wrote down a **coherent state** of Cooper pairs. As discussed in the previous chapter, it is possible to construct operators which create or annihilate electron pairs centered at $\mathbf{R}$, $\hat{\varphi}^+(\mathbf{R})$ and $\hat{\varphi}(\mathbf{R})$, respectively. As we have seen, these operators **do not** obey normal Bose commutation laws, and so they cannot be regarded as creating or destroying boson particles.

We will look for a uniform translationally invariant solution, and so it is more convenient to work in $\mathbf{k}$ space. Let us define the pair creation operator by,

$$\hat{P}^+_{\mathbf{k}} = c^+_{\mathbf{k}\uparrow}c^+_{-\mathbf{k}\downarrow}. \tag{6.29}$$

This creates a pair of electrons of zero total crystal momentum, and opposite spins. In terms of this operator Schrieffer proposed the following coherent state many-body wave function,

$$|\Psi_{\rm BCS}\rangle = \text{const.} \exp\left(\sum_{\mathbf{k}} \alpha_{\mathbf{k}} \hat{P}^+_{\mathbf{k}}\right)|0\rangle, \tag{6.30}$$

where the complex numbers, $\alpha_{\mathbf{k}}$, are parameters which can be adjusted to minimize the total energy. Here the vacuum state, $|0\rangle$, is the state with no electrons at all in the band of Bloch states at the Fermi surface.

Even though these pair operators do not obey Bose commutation laws

$$\left[\hat{P}_{\mathbf{k}}, \hat{P}^+_{\mathbf{k}}\right] \neq 1 \tag{6.31}$$

they **do** commute with each other. It is easy to confirm that

$$\left[\hat{P}^+_{\mathbf{k}}, \hat{P}^+_{\mathbf{k}'}\right] = 0 \tag{6.32}$$

for different $\mathbf{k}$ points, $\mathbf{k} \neq \mathbf{k}'$. On the other hand, for the same $\mathbf{k}$ point, $\mathbf{k} = \mathbf{k}'$, the product $\hat{P}^+_{\mathbf{k}}\hat{P}^+_{\mathbf{k}}$ contains four electron creation operators for the same $\mathbf{k}$ point,

$$\hat{P}^+_{\mathbf{k}}\hat{P}^+_{\mathbf{k}} = c^+_{\mathbf{k}\uparrow}c^+_{-\mathbf{k}\downarrow}c^+_{\mathbf{k}\uparrow}c^+_{-\mathbf{k}\downarrow} = 0, \tag{6.33}$$

and it is therefore always zero because $c^+_{\mathbf{k}\uparrow}c^+_{\mathbf{k}\uparrow} = 0$. It will also be useful to note that this implies

$$\left(\hat{P}^+_{\mathbf{k}}\right)^2 = 0. \tag{6.34}$$

Using the fact that these operators commute we can rewrite the coherent state in Eq. 6.30 as a product of exponentials, one for each $\mathbf{k}$ point,

$$|\Psi_{\rm BCS}\rangle = \text{const.} \prod_{\mathbf{k}} \exp\left(\alpha_{\mathbf{k}} \hat{P}^+_{\mathbf{k}}\right)|0\rangle. \tag{6.35}$$

Then, using property, Eq. 6.34, we can also expand out each of the operator exponentials. In the expansion all terms containing $\hat{P}^+_{\mathbf{k}}$ to quadratic or higher powers are zero. Therefore we obtain

$$|\Psi_{\rm BCS}\rangle = \text{const.} \prod_{\mathbf{k}} \left(1 + \alpha_{\mathbf{k}} \hat{P}^+_{\mathbf{k}}\right)|0\rangle. \tag{6.36}$$

The normalizing constant is found from

$$1 = \langle 0| \left(1 + \alpha^*_{\mathbf{k}} \hat{P}_{\mathbf{k}}\right)\left(1 + \alpha_{\mathbf{k}} \hat{P}^+_{\mathbf{k}}\right)|0\rangle = 1 + |\alpha_{\mathbf{k}}|^2. \tag{6.37}$$

So we can finally write the normalized BCS state as

$$|\Psi_{\rm BCS}\rangle = \prod_{\mathbf{k}} \left(u^*_{\mathbf{k}} + v^*_{\mathbf{k}} \hat{P}^+_{\mathbf{k}}\right)|0\rangle, \tag{6.38}$$

where

$$u^*_{\mathbf{k}} = \frac{1}{1 + |\alpha_{\mathbf{k}}|^2}, \tag{6.39}$$

$$v^*_{\mathbf{k}} = \frac{\alpha_{\mathbf{k}}}{1 + |\alpha_{\mathbf{k}}|^2}, \tag{6.40}$$

and where

$$|u_{\mathbf{k}}|^2 + |v_{\mathbf{k}}|^2 = 1. \tag{6.41}$$

The use of complex conjugates here for $u^*_{\mathbf{k}}$ $v^*_{\mathbf{k}}$ is a matter of convention. The convention used here differs from some other textbooks, but is consistent with most of the modern research literature in the field. Notice that the constants $\alpha_{\mathbf{k}}$ can be any complex numbers, as is usual in a coherent state. Therefore we can associate a complex phase angle θ with the BCS state. On the other hand, the wave function does not have a definite particle number, N, since it is a superposition of the original vacuum, $|0\rangle$, and the vacuum plus $2, 4, 6, \ldots$ electrons. Of course this number-phase uncertainty is typical of coherent states. As BCS argued, the total number of electrons involved, N, is macroscopic and of the order the system size. For this state the uncertainty in N, ΔN, is of order $N^{1/2}$ and is therefore absolutely negligible compared with N. Nevertheless it was only several years after the original BCS paper that this became fully accepted as a valid argument.

Finally, the way the BCS state was originally written, as described above, treats electrons and holes in a relatively unsymmetrical manner. We start with a vacuum $|0\rangle$ and add pairs of electrons. But what about pairs of holes? In fact these are also included in the theory. We just have to see that by a suitable redefinition of the original reference state $|0\rangle$ we can write the BCS state in a form which treats electrons and holes more evenly,

$$|\Psi_{\text{BCS}}\rangle = \prod_{k>k_F} \left(u_{\mathbf{k}}^* + v_{\mathbf{k}}^* P_{\mathbf{k}}^+\right) \prod_{k'<k_F} \left(u_{\mathbf{k}'}^* P_{\mathbf{k}'} + v_{\mathbf{k}'}^*\right) |\psi_0\rangle. \tag{6.42}$$

Here $|\psi_0\rangle$ is the zero temperature Fermi sea, in which all states with $k < k_F$ are occupied and the rest are unoccupied. One can equally well view the BCS state as a condensate of electron pairs above a filled electron Fermi sea, of a condensate of hole pairs below an empty "hole sea." In fact electrons and holes contribute more or less equally.[4]

[4]The same duality arises in Dirac's theory of the electron sea. In Dirac's theory the positive energy electrons move above a filled Dirac sea of filled negative energy electron states. In this picture positrons are holes in this Dirac sea of electrons. But an equally valid point of view is the opposite! We could view positive energy postitrons as moving above a filled sea of negative energy positrons. Then electrons are just holes in this filled sea of postitrons! Neither point of view is any more correct than the other because they are equivalent under **particle–hole** symmetry.

6.5 The mean-field Hamiltonian

With the trial wave function given above, the next task is to find the parameters $u_{\mathbf{k}}$ and $v_{\mathbf{k}}$ which minimize the energy.

Using the BCS approximation for the effective interaction, Eq. 6.11, the relevant Hamiltonian is

$$\hat{H} = \sum_{\mathbf{k},\sigma} \epsilon_{\mathbf{k}} c_{\mathbf{k}\sigma}^+ c_{\mathbf{k}\sigma} - |g_{\text{eff}}|^2 \sum c_{\mathbf{k}_1+\mathbf{q}\sigma_1}^+ c_{\mathbf{k}_2-\mathbf{q}\sigma_2}^+ c_{\mathbf{k}_1\sigma_1} c_{\mathbf{k}_2\sigma_2}, \tag{6.43}$$

where, as discussed above, we restrict the interaction to values of $\mathbf{k}$ so that $\epsilon_{\mathbf{k}}$ is within $\pm\hbar\omega_D$ of the Fermi energy.

If we assume that the most important interactions are those involving Cooper pairs $\mathbf{k}, \uparrow$ and $-\mathbf{k}, \downarrow$, then the most important terms are those for which $\mathbf{k}_1 = -\mathbf{k}_2$ and $\sigma_1 = -\sigma_2$. Dropping all other interactions the Hamiltonian becomes

$$\hat{H} = \sum_{\mathbf{k},\sigma} \epsilon_{\mathbf{k}} c_{\mathbf{k}\sigma}^+ c_{\mathbf{k}\sigma} - |g_{\text{eff}}|^2 \sum_{\mathbf{k},\mathbf{k}'} c_{\mathbf{k}\uparrow}^+ c_{-\mathbf{k}\downarrow}^+ c_{-\mathbf{k}'\downarrow} c_{\mathbf{k}'\uparrow} \tag{6.44}$$

using the same model form of the interaction V_{eff} as we used in the Cooper problem above.

The above Hamiltonian is still an interacting electron problem and is too hard to solve exactly. But making use of the trial BCS wave function we can treat the parameters $u_{\mathbf{k}}$ and $v_{\mathbf{k}}$ as variational parameters, which are to be adjusted to minimize the total energy

$$E = \langle \Psi_{\text{BCS}} | \hat{H} | \Psi_{\text{BCS}} \rangle. \tag{6.45}$$

In this minimization the mean total number of particles $\langle \hat{N} \rangle$ must be held constant, and the parameters must obey the constraint that $|u_{\mathbf{k}}|^2 + |v_{\mathbf{k}}|^2 = 1$.

We can evaluate the variational energy as follows. First note that the average occupation of Bloch state **k** is

$$\begin{aligned}
\langle \hat{n}_{\mathbf{k}\uparrow} \rangle &= \langle \Psi_{\text{BCS}} | c^{+}_{\mathbf{k}\uparrow} c_{\mathbf{k}\uparrow} | \Psi_{\text{BCS}} \rangle \\
&= \langle 0 | (u_{\mathbf{k}} + v_{\mathbf{k}} c_{-\mathbf{k}\downarrow} c_{\mathbf{k}\uparrow}) c^{+}_{\mathbf{k}\uparrow} c_{\mathbf{k}\uparrow} (u^{*}_{\mathbf{k}} + v^{*}_{\mathbf{k}} c^{+}_{\mathbf{k}\uparrow} c^{+}_{-\mathbf{k}\downarrow}) | 0 \rangle \\
&= \langle 0 | v_{\mathbf{k}} c_{-\mathbf{k}\downarrow} v^{*}_{\mathbf{k}} c^{+}_{-\mathbf{k}\downarrow}) | 0 \rangle \\
&= |v_{\mathbf{k}}|^2,
\end{aligned} \tag{6.46}$$

and similarly one can also show that $\langle \hat{n}_{\mathbf{k}\downarrow} \rangle = |v_{\mathbf{k}}|^2$. The intermediate steps in the algebra in these expressions are obtained by repeatedly using the fermion anti-commutation relations to reorder the products of operators into **normal order**, where all the creation operators are to the left and the annihilation operators to the right. Once the expression is in normal order, then any terms with annihilation operators acting on the vacuum give zero, since $c_{\mathbf{k}\sigma}|0\rangle = 0$, and similarly any terms like $\langle 0 | c^{+}_{\mathbf{k}\sigma}$ also vanish. In this case, the only term, which is nonzero comes with the weight $|v_{\mathbf{k}}|^2$. From Eq. 6.46 it immediately follows that the total number of electrons is

$$\langle \hat{N} \rangle = 2 \sum_{\mathbf{k}} |v_{\mathbf{k}}|^2, \tag{6.47}$$

where the factor of 2 is just from the two spin states. Similarly the kinetic energy contribution to the total energy is

$$\left\langle \sum_{\mathbf{k}\sigma} \epsilon_{\mathbf{k}} c^{+}_{\mathbf{k}\sigma} c_{\mathbf{k}\sigma} \right\rangle = 2 \sum_{\mathbf{k}} \epsilon_{\mathbf{k}} |v_{\mathbf{k}}|^2. \tag{6.48}$$

The expectation value of interaction part of the BCS Hamiltonian can be evaluated using

$$\langle c^{+}_{\mathbf{k}\uparrow} c^{+}_{-\mathbf{k}\downarrow} c_{-\mathbf{k}'\downarrow} c_{\mathbf{k}'\uparrow} \rangle = v_{\mathbf{k}} v^{*}_{\mathbf{k}'} u_{\mathbf{k}'} u^{*}_{\mathbf{k}}. \tag{6.49}$$

The proof of this is left to the reader (see the Exercises at the end of this chapter). Therefore the total energy expressed in terms of the variational parameters $u_{\mathbf{k}}$ and $v_{\mathbf{k}}$ is

$$E = 2 \sum_{\mathbf{k}} \epsilon_{\mathbf{k}} |v_{\mathbf{k}}|^2 - |g_{\text{eff}}|^2 \sum_{\mathbf{k}\mathbf{k}'} v_{\mathbf{k}} v^{*}_{\mathbf{k}'} u_{\mathbf{k}'} u^{*}_{\mathbf{k}}. \tag{6.50}$$

Now we can use the method of Lagrange multipliers to find the minimum energy solution. First, it is helpful to use the normalization constraint $|u_{\mathbf{k}}|^2 + |v_{\mathbf{k}}|^2 = 1$ to rewrite the energy in the form

$$E = \sum_{\mathbf{k}} \epsilon_{\mathbf{k}} \left(|v_{\mathbf{k}}|^2 - |u_{\mathbf{k}}|^2 + 1 \right) - |g_{\text{eff}}|^2 \sum_{\mathbf{k}\mathbf{k}'} v_{\mathbf{k}} v^{*}_{\mathbf{k}'} u_{\mathbf{k}'} u^{*}_{\mathbf{k}}. \tag{6.51}$$

Similarly the total particle number is

$$N = \sum_{\mathbf{k}} \left(|v_{\mathbf{k}}|^2 - |u_{\mathbf{k}}|^2 + 1 \right). \tag{6.52}$$

Differentiating with respect to $u^*_{\mathbf{k}}$ and $v^*_{\mathbf{k}}$ (and treating them as independent variables from $u_{\mathbf{k}}$ and $v_{\mathbf{k}}$) the condition for a minimum is,

$$0 = \frac{\partial E}{\partial u^*_{\mathbf{k}}} - \mu \frac{\partial N}{\partial u^*_{\mathbf{k}}} + E_{\mathbf{k}} u_{\mathbf{k}}$$

$$0 = \frac{\partial E}{\partial v^*_{\mathbf{k}}} - \mu \frac{\partial N}{\partial v^*_{\mathbf{k}}} + E_{\mathbf{k}} v_{\mathbf{k}}.$$

Here the chemical potential μ enters as the Lagrange multiplier, which enforces the constant total particle number, and $E_{\mathbf{k}}$ is the Lagrange multiplier associated with the constraint $|u_{\mathbf{k}}|^2 + |v_{\mathbf{k}}|^2 = 1$.

Evaluating the derivatives and rearranging gives the pair of linear equations

$$(\epsilon_{\mathbf{k}} - \mu) u_{\mathbf{k}} + \Delta v_{\mathbf{k}} = E_{\mathbf{k}} u_{\mathbf{k}} \tag{6.53}$$

$$\Delta^* u_{\mathbf{k}} - (\epsilon_{\mathbf{k}} - \mu) v_{\mathbf{k}} = E_{\mathbf{k}} v_{\mathbf{k}}, \tag{6.54}$$

where we introduce (finally!) the **BCS gap parameter** Δ, which is defined by

$$\Delta = |g_{\text{eff}}|^2 \sum_{\mathbf{k}} u_{\mathbf{k}} v^*_{\mathbf{k}}. \tag{6.55}$$

Using $|\Psi_{\text{BCS}}\rangle$ this can also be expressed elegantly in the form

$$\Delta = |g_{\text{eff}}|^2 \sum_{\mathbf{k}} \langle c_{-\mathbf{k}\downarrow} c_{\mathbf{k}\uparrow} \rangle. \tag{6.56}$$

This expectation value is the expectation value of the Cooper pair operator $P_{\mathbf{k}}$, $\langle P_{\mathbf{k}} \rangle$. It is nonzero only because the BCS ground state wave function is a coherent state with components of different particle numbers $N, N+2, N+4, \ldots,$ etc. But, just as in the earlier discussion of coherent states in Chapter 5, it is possible to show that this expectation value can be finite because we are interested in macroscopic systems with essentially infinite values of $\langle N \rangle$. For such large values of $\langle N \rangle$ the fact that N has fluctuations about this mean value can be safely ignored.

In order to determine the parameter Δ, it is necessary to solve the coupled equations for $u_{\mathbf{k}}$ and $v_{\mathbf{k}}$. This pair of equations, Eqs 6.53 and 6.54, can most elegantly be written as a single matrix equation

$$\begin{pmatrix} \epsilon_{\mathbf{k}} - \mu & \Delta \\ \Delta^* & -(\epsilon_{\mathbf{k}} - \mu) \end{pmatrix} \begin{pmatrix} u_{\mathbf{k}} \\ v_{\mathbf{k}} \end{pmatrix} = E_{\mathbf{k}} \begin{pmatrix} u_{\mathbf{k}} \\ v_{\mathbf{k}} \end{pmatrix}. \tag{6.57}$$

In this form, it is now clear that $(u_{\mathbf{k}}, v_{\mathbf{k}})$ is just the eigenvector of the two by two matrix, and the parameter $E_{\mathbf{k}}$ is its eigenvalue. A simple evaluation shows that the matrix eigenvalues are simply $\pm E_{\mathbf{k}}$, where

$$E_{\mathbf{k}} = \sqrt{(\epsilon_{\mathbf{k}} - \mu)^2 + |\Delta|^2}. \tag{6.58}$$

We will see in the next section that its physical significance is the excitation energy for an extra electron or hole added to the BCS ground state. We can also evaluate the eigenvector $(u_{\mathbf{k}}, v_{\mathbf{k}})$. After some manipulation of the eigenvector equation Eq. 6.57, it is possible to show that

$$|u_{\mathbf{k}}|^2 = \frac{1}{2}\left(1 + \frac{\epsilon_{\mathbf{k}} - \mu}{E_{\mathbf{k}}}\right) \tag{6.59}$$

$$|v_{\mathbf{k}}|^2 = \frac{1}{2}\left(1 - \frac{\epsilon_{\mathbf{k}} - \mu}{E_{\mathbf{k}}}\right). \tag{6.60}$$

We now have a completely closed system of equations, which we can solve to find Δ. First we need to eliminate $u_{\mathbf{k}}$ and $v_{\mathbf{k}}$. From the matrix eigenvector

Eq. 6.57 one can show that

$$u_{\mathbf{k}} v_{\mathbf{k}}^* = \frac{\Delta}{2E_{\mathbf{k}}}. \tag{6.61}$$

Combining this result with Eq. 6.55 gives a very important result known as the **BCS gap equation**

$$\Delta = |g_{\text{eff}}|^2 \sum_{\mathbf{k}} \frac{\Delta}{2E_{\mathbf{k}}}. \tag{6.62}$$

Or, simplifying by cancelling the Δ, which appears on both sides, we find

$$1 = \frac{|g_{\text{eff}}|^2}{2} \sum_{\mathbf{k}} \frac{1}{\left((\epsilon_{\mathbf{k}} - \mu)^2 + |\Delta^2|\right)^{1/2}}. \tag{6.63}$$

This equation is central to the BCS theory since it determines the magnitude of the gap parameter $|\Delta|$ at zero temperature. In evaluating the **k** sum, it is important to not forget that we are always working within the thin shell of Bloch states around the Fermi energy with $|\epsilon_{\mathbf{k}} - \mu| < \hbar\omega_D$, where the BCS electron–phonon interaction is attractive. Within this thin shell we can replace the sum over **k** by an integration over energy using

$$\sum_{\mathbf{k}} \to g(\epsilon_F) \int d\epsilon,$$

where $g(\epsilon_F)$ is the density of states (per spin) at the Fermi energy. The gap equation then becomes

$$1 = \lambda \int_0^{\hbar\omega_D} \frac{1}{(\epsilon^2 + |\Delta|^2)^{1/2}} d\epsilon, \tag{6.64}$$

where we have defined the dimensionless **electron–phonon coupling constant**

$$\lambda = |g_{\text{eff}}|^2 g(\epsilon_F). \tag{6.65}$$

The energy integral is dominated by a near logarithmic divergence at the upper limit, and approximately the result is,

$$1 = \lambda \ln\left(\frac{2\hbar\omega_D}{|\Delta|}\right). \tag{6.66}$$

This finally gives us the famous BCS value of the gap parameter at zero temperature

$$|\Delta| = 2\hbar\omega_D e^{-1/\lambda}. \tag{6.67}$$

Interestingly this is quite similar to the binding energy of a single Cooper pair, which we found in Section 6.3. It is also clear that there is always a solution to the gap equation, however small the coupling constant λ. We can also see from this result that the typical energy scale relevant to superconductivity is very much less than the Debye energy $\hbar\omega_D$. From this it is immediately obvious why superconducting T_c values are typically so much lower than other relevant energy scales in metals, such as the Fermi energy or the phonon energies.

6.6 The BCS energy gap and quasiparticle states

The BCS wave function is a brilliant example of a correlated many-particle quantum ground state. There are only a handful of problems in many-body physics for which we have essentially exact ground states with highly nontrivial properties. In this case the BCS state is a **mean-field** ground state, as we shall see below. As such, it is not an exact solution to the many-body problem, but becomes nearly exact, to all intents and purposes, within a certain limit. In the BCS theory case the limit is the **weak coupling** limit, in which we assume that the dimensionless coupling parameter is much less than 1. If this is true then $|\Delta|$ is very much smaller than all other relevant energy scales in the problem, such as ϵ_F and $\hbar\omega_D$. It is possible to extend BCS theory to larger values of λ, giving **strong coupling** theory, but this would take us beyond the scope of this book.[5]

[5] See Alexandrov (2003) for an extensive introduction to strong coupling superconductivity.

Having found the variational ground state wave function $|\Psi_{\rm BCS}\rangle$, the next step is to explore its predictions for the physical properties of superconductors. An immediate question is how to extend the discussion of the previous section to deal with the case of finite temperature, and to discover the energies of the lowest energy excited states. The method used to find the excited states is similar to that which we used in the previous chapter to find the quasiparticle excitations of the weakly interacting Bose gas. The idea is to consider $|\Psi_{\rm BCS}\rangle$ as a reference state, and to consider small excitations (such as adding one extra particle or hole) relative to that state.

The method to find the excitation energies for adding a single particle relies on the approximation

$$c^+_{\mathbf{k}\uparrow}c^+_{-\mathbf{k}\downarrow}c_{-\mathbf{k}'\downarrow}c_{\mathbf{k}\uparrow} \approx \langle c^+_{\mathbf{k}\uparrow}c^+_{-\mathbf{k}\downarrow}\rangle c_{-\mathbf{k}'\downarrow}c_{\mathbf{k}'\uparrow} + c^+_{\mathbf{k}\uparrow}c^+_{-\mathbf{k}\downarrow}\langle c_{-\mathbf{k}'\downarrow}c_{\mathbf{k}'\uparrow}\rangle. \qquad (6.68)$$

This is derived from a result in many-body theory known as Wick's theorem, in which expectation values of products of four particle operators can be approximated by averages over pairs of operators. According to the theorem, all possible pairings are included in these averages, but here we have dropped the "uninteresting" averages such as $\langle c^+_{\mathbf{k}\uparrow}c_{\mathbf{k}\uparrow}\rangle$. These averages are essentially the same in the normal state as in the superconducting state, and so they can be absorbed into the definitions of the single particle energies $\epsilon_{\mathbf{k}}$ without any significant error.

Using this approximation based on Wick's theorem the BCS Hamiltonian Eq. 6.44 can be replaced by the following effective Hamiltonian,

$$\hat{H} = \sum_{\mathbf{k}\sigma}(\epsilon_{\mathbf{k}} - \mu)c^+_{\mathbf{k}\sigma}c_{\mathbf{k}\sigma} - |g_{\rm eff}|^2\sum_{\mathbf{k}\mathbf{k}'}\Big(\langle c^+_{\mathbf{k}\uparrow}c^+_{-\mathbf{k}\downarrow}\rangle c_{-\mathbf{k}'\downarrow}c_{\mathbf{k}'\uparrow} + c^+_{\mathbf{k}\uparrow}c^+_{-\mathbf{k}\downarrow}\langle c_{-\mathbf{k}'\downarrow}c_{\mathbf{k}'\uparrow}\rangle\Big). \qquad (6.69)$$

Given the fact that

$$\Delta = |g_{\rm eff}|^2\sum_{\mathbf{k}}\langle c_{-\mathbf{k}\downarrow}c_{\mathbf{k}\uparrow}\rangle \qquad (6.70)$$

this Hamiltonian can be rewritten in the form

$$\hat{H} = \sum_{\mathbf{k}\sigma}(\epsilon_{\mathbf{k}} - \mu)c^+_{\mathbf{k}\sigma}c_{\mathbf{k}\sigma} - \sum_{\mathbf{k}}\Big(\Delta^* c_{-\mathbf{k}\downarrow}c_{\mathbf{k}\uparrow} + \Delta c^+_{\mathbf{k}\uparrow}c^+_{-\mathbf{k}\downarrow}\Big). \qquad (6.71)$$

This effective Hamiltonian is quadratic in the particle field operators, unlike the original Hamiltonian, which was quadratic. Such quadratic Hamiltonians can be solved exactly by a **Bolgoliubov–Valatin transformation**. The method is very similar to that used in the previous chapter for the case of the weakly interacting Bose gas. The idea is to find new field operators, which make the Hamiltonian diagonal. The new operators should be defined to obey the same fermion (or boson) anti-commutation (or commutation) rules as the original ones.

In the case of the BCS theory the appropriate transformation can most easily be derived by rewriting the Hamiltonian Eq. 6.71 explicitly as a matrix quadratic form

$$\hat{H} = \sum_{\mathbf{k}} \begin{pmatrix} c^+_{\mathbf{k}\uparrow} & c_{-\mathbf{k}\downarrow} \end{pmatrix} \begin{pmatrix} \epsilon_{\mathbf{k}} - \mu & -\Delta \\ -\Delta^* & -(\epsilon_{\mathbf{k}} - \mu) \end{pmatrix} \begin{pmatrix} c_{\mathbf{k}\uparrow} \\ c^+_{-\mathbf{k}\downarrow} \end{pmatrix}. \tag{6.72}$$

The matrix

$$\begin{pmatrix} \epsilon_{\mathbf{k}} - \mu & -\Delta \\ -\Delta^* & -(\epsilon_{\mathbf{k}} - \mu) \end{pmatrix}$$

is essentially the same one we encountered in the previous section, except for the minus sign in front of Δ. It has two eigenvectors

$$\begin{pmatrix} u_{\mathbf{k}} \\ -v_{\mathbf{k}} \end{pmatrix} \qquad \begin{pmatrix} v^*_{\mathbf{k}} \\ u^*_{\mathbf{k}} \end{pmatrix}$$

with energies $E_{\mathbf{k}} = +\sqrt{(\epsilon_{\mathbf{k}} - \mu)^2 + |\Delta|^2}$ and $-E_{\mathbf{k}}$, respectively.

We can use this matrix representation to transform to a new "basis" in which the matrix is diagonalized. A well known piece of matrix algebra tells us that any Hermitian matrix can be transformed into diagonal form by multiplying on left and right by a unitary transformation U. Therefore we must have,

$$U^+ \begin{pmatrix} \epsilon_{\mathbf{k}} - \mu & \Delta \\ \Delta^* & -(\epsilon_{\mathbf{k}} - \mu) \end{pmatrix} U = \begin{pmatrix} E_{\mathbf{k}} & 0 \\ 0 & -E_{\mathbf{k}} \end{pmatrix}. \tag{6.73}$$

The transformation matrix U, which achieves this is a matrix where each column is an eigenvector of the original matrix. In our case this will be,

$$U = \begin{pmatrix} u_{\mathbf{k}} & v^*_{\mathbf{k}} \\ -v_{\mathbf{k}} & u^*_{\mathbf{k}} \end{pmatrix}. \tag{6.74}$$

One can easily check that this matrix is unitary ($UU^+ = I$) because of the identity $|u_{\mathbf{k}}|^2 + |v_{\mathbf{k}}|^2 = 1$. Making use of this unitary transformation the quadratic Hamiltonian becomes

$$\hat{H} = -\sum_{\mathbf{k}} \begin{pmatrix} c^+_{\mathbf{k}\uparrow} & c_{-\mathbf{k}\downarrow} \end{pmatrix} U \begin{pmatrix} E_{\mathbf{k}} & 0 \\ 0 & -E_{\mathbf{k}} \end{pmatrix} U^+ \begin{pmatrix} c_{\mathbf{k}\uparrow} \\ c^+_{-\mathbf{k}\downarrow} \end{pmatrix}. \tag{6.75}$$

We can now introduce a new pair of operators defined by,

$$\begin{pmatrix} b_{\mathbf{k}\uparrow} \\ b^+_{-\mathbf{k}\downarrow} \end{pmatrix} = U^+ \begin{pmatrix} c_{\mathbf{k}\uparrow} \\ c^+_{-\mathbf{k}\downarrow} \end{pmatrix} \tag{6.76}$$

or written out explicitly,

$$b_{\mathbf{k}\uparrow} = u^*_{\mathbf{k}} c_{\mathbf{k}\uparrow} - v^*_{\mathbf{k}} c^+_{-\mathbf{k}\downarrow} \tag{6.77}$$

$$b^+_{-\mathbf{k}\downarrow} = v_{\mathbf{k}} c_{\mathbf{k}\uparrow} + u_{\mathbf{k}} c^+_{-\mathbf{k}\downarrow}. \tag{6.78}$$

In terms of these new operators the Hamiltonian is diagonal,

$$\hat{H} = -\sum_{\mathbf{k}} \begin{pmatrix} b^+_{\mathbf{k}\uparrow} & b_{-\mathbf{k}\downarrow} \end{pmatrix} \begin{pmatrix} E_{\mathbf{k}} & 0 \\ 0 & -E_{\mathbf{k}} \end{pmatrix} \begin{pmatrix} b_{\mathbf{k}\uparrow} \\ b^+_{-\mathbf{k}\downarrow} \end{pmatrix}. \tag{6.79}$$

Writing it out explicitly,

$$\hat{H} = \sum_{\mathbf{k}} \left(E_{\mathbf{k}} b^+_{\mathbf{k}\uparrow} b_{\mathbf{k}\uparrow} - E_{\mathbf{k}} b_{-\mathbf{k}\downarrow} b^+_{-\mathbf{k}\downarrow} \right),$$

$$= \sum_{\mathbf{k}} E_{\mathbf{k}} \left(b^+_{\mathbf{k}\uparrow} b_{\mathbf{k}\uparrow} + b^+_{-\mathbf{k}\downarrow} b_{-\mathbf{k}\downarrow} \right), \tag{6.80}$$

where we have dropped some constant (non-operator) terms arising from the reordering of the final expression into normal order form.

What is the physical interpretation of these new operators? First, we should note that they are genuinely fermion particle operators, since they obey the standard anti-commutation laws

$$\{b_{\mathbf{k}\sigma}, b^+_{\mathbf{k}'\sigma'}\} = \delta_{\mathbf{k}\mathbf{k}'}\delta_{\sigma\sigma'} \tag{6.81}$$

$$\{b^+_{\mathbf{k}\sigma}, b^+_{\mathbf{k}'\sigma'}\} = 0 \tag{6.82}$$

$$\{b_{\mathbf{k}\sigma}, b_{\mathbf{k}'\sigma'}\} = 0. \tag{6.83}$$

These are simple to prove from the definitions given above. Second it is easy to prove that the new "particles" created and annihilated by these operators are not present in the variational BCS ground state, since

$$b_{\mathbf{k}\uparrow}|\Psi_{\text{BCS}}\rangle = 0 \tag{6.84}$$

$$b_{-\mathbf{k}\downarrow}|\Psi_{\text{BCS}}\rangle = 0. \tag{6.85}$$

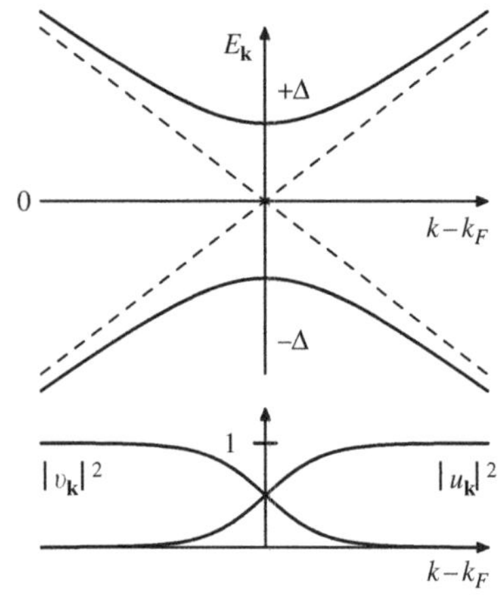

Fig. 6.8 Top: Energy eigenvalues $E_{\mathbf{k}}$ as a function of $\mathbf{k}$ near the Fermi wave vector $\mathbf{k}_F$. The dashed lines show the electron and hole energy levels $\epsilon_{\mathbf{k}} - \epsilon_F$ and $-\epsilon_{\mathbf{k}} + \epsilon_F$ in the normal metal. In the superconductor these states become hybridized, and the resulting eigenvalues are $\pm E_{\mathbf{k}}$ relative to ϵ_F. One can see that there are no states with energy less than $\pm\Delta$ near the Fermi energy. **Bottom:** The BCS wave function parameters $|u_{\mathbf{k}}|^2$ and $|v_{\mathbf{k}}|$ for $\mathbf{k}$ near to the Fermi surface. The state is predominantly electron like well below k_F ($|v_{\mathbf{k}}|^2 \approx 1$) and predominantly hole like far above the Fermi surface, ($|u_{\mathbf{k}}|^2 \approx 1$). But near to $\mathbf{k}_F$ the quasiparticle has mixed electron and hole character.

Therefore the BCS ground state is the "vacuum" for these particles, a state where no particles are present. The excited states then correspond to adding one, two or more of these new quasiparticles to this ground state. The excitation energy to do this is $E_{\mathbf{k}}$.

At finite temperature T, the quasiparticles will have occupations given by the Fermi–Dirac distribution. Therefore

$$\langle b^+_{\mathbf{k}\uparrow} b_{\mathbf{k}\uparrow}\rangle = f(E_{\mathbf{k}}), \tag{6.86}$$

$$\langle b_{-\mathbf{k}\downarrow} b^+_{\mathbf{k}\downarrow}\rangle = 1 - f(E_{\mathbf{k}}), \tag{6.87}$$

where $f(E) = 1/(e^{\beta E} + 1)$. From this one can show that the finite temperature value of the BCS gap parameter is given by

$$\Delta = |g_{\text{eff}}|^2 \langle c_{-\mathbf{k}\downarrow} c_{\mathbf{k}\uparrow}\rangle$$

$$= |g_{\text{eff}}|^2 u_{\mathbf{k}} v^*_{\mathbf{k}} (1 - 2f(E)) \tag{6.88}$$

generalizing the zero temperature result of Eq. 6.55.

Figure 6.8 shows energies of the excitations created by the $b^+_{\mathbf{k}}$ operators, $\pm E_{\mathbf{k}}$, as a function of $\mathbf{k}$. It gives the following physical picture. In the normal state $\Delta = 0$ and the excitation energies are $+\epsilon_{\mathbf{k}}$ for adding an electron to an empty state, or $-\epsilon_{\mathbf{k}}$ for removing an electron (adding a hole).

In the superconducting state these become modified to $+E_{\mathbf{k}}$ for adding a b particle, or $-E_{\mathbf{k}}$ for removing one. Because $+E_{\mathbf{k}}$ is greater than Δ and $-E_{\mathbf{k}}$ is less than $-\Delta$ the minimum energy to make an excitation is 2Δ. Thus this is the **energy gap** of the superconductor. The b particles are called quasiparticles.

The b^+, b operators are a mixture of the creation c^+ and annihilation c operators. This implies that the states they create or destroy are neither purely electron or purely hole excitations, instead they are a quantum superposition of electron and hole. In fact u and v have the physical interpretation that

$$|v_{\mathbf{k}}|^2$$

is the probabilities that the excitation is an electron if one measures its charge, and

$$|u_{\mathbf{k}}|^2$$

is the probability that it is a hole.[6]

Finally in order to find Δ it is again necessary to invoke self-consistency. Δ was defined by

$$\Delta = |g_{\mathrm{eff}}|^2 \sum_{\mathbf{k}} \langle c_{-\mathbf{k}\downarrow} c_{\mathbf{k}\uparrow} \rangle. \tag{6.89}$$

At temperature T the occupation of the quasiparticle state with energy $E_{\mathbf{k}}$ is given by the Fermi–Dirac distribution. Determining the expectation value from Eqs 6.88 and 6.57, Bardeen Cooper and Schrieffer obtained

$$\Delta = |g_{\mathrm{eff}}|^2 \sum_{\mathbf{k}} \frac{\Delta}{2E_{\mathbf{k}}} \tanh\left(\frac{E_{\mathbf{k}}}{2k_{\mathrm{B}}T}\right), \tag{6.90}$$

or converting the sum into an integral over energy we arrive at **the BCS gap equation**,

$$1 = \lambda \int_0^{\hbar\omega_D} d\epsilon \frac{1}{E} \tanh\left(\frac{E}{2k_{\mathrm{B}}T}\right), \tag{6.91}$$

where $E = \sqrt{\epsilon^2 + |\Delta|^2}$ and $\lambda = |g_{\mathrm{eff}}|^2 g(\epsilon_F)$ is the dimensionless electron–phonon coupling parameter.

The BCS gap equation implicitly determines the gap $\Delta(T)$ at any temperature T. It is the central equation of the theory, since it predicts both the transition temperature T_c and the value of the energy gap at zero temperature $\Delta(0)$. The temperature dependence of $\Delta(T)$ is shown in Fig. 6.9.

From the BCS gap equation, taking the limit $\Delta \to 0$ one can obtain an equation for T_c

$$k_{\mathrm{B}}T_c = 1.13\hbar\omega_D \exp(-1/\lambda), \tag{6.92}$$

which has almost exactly the same form as the formula for the binding energy in the Cooper problem. Also at $T = 0$ one can also do the integral and determine $\Delta(0)$. The famous BCS result

$$2\Delta(0) = 3.52k_{\mathrm{B}}T_c \tag{6.93}$$

is obeyed very accurately in a wide range of different superconductors.

[6] Again there are nice analogies to particle physics. The neutral K-meson, K_0, has an antiparticle, $\bar{K}_0$. Neither of them are eigenstates of total energy (mass), and so when the particle propagates it oscillates between these two states. If it is measured at any point there is a certain probability that it will be found to be K_0 and another for it to be $\bar{K}_0$. Here, the BCS quasiparticles are the energy (mass) eigenstates, and they are quantum superpositions of electron and hole states.

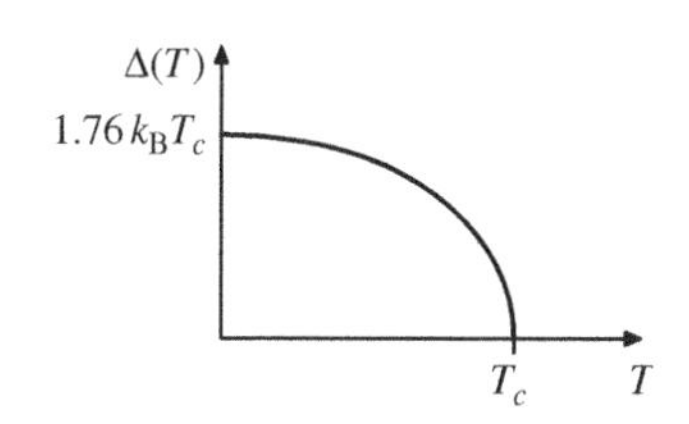

Fig. 6.9 Δ as a function of temperature in the BCS theory.

6.7 Predictions of the BCS theory

The BCS theory went on to predict many other physical properties of the superconducting state. For most simple metallic superconductors, such as Al, Hg, etc., these agreed very well with experimental facts, providing strong evidence in support of the theory. For example, two key predictions where the behavior of the nuclear magnetic resonance (NMR) relaxation rate, $1/T_1$ below the critical temperature T_c, and the temperature dependence of the attenuation coefficient for ultrasound. These are both sensitive to the electronic density of states in the superconductor, Fig. 6.1, but also depend on **coherence factors**, which are certain combinations of the BCS parameters $u_{\mathbf{k}}$ and $v_{\mathbf{k}}$. It turns out (Schrieffer 1964), that the good agreement between theory and experiment depends in detail on the values of these parameters. In particular, the NMR relaxation rate, $1/T_1$, has a characteristic peak just below T_c. The existence of this Hebel Slichter peak was predicted by BCS and depends strongly on the coherence factors. Therefore one can say that the BCS theory has been tested not just at the level of the quasiparticle energies, $E_{\mathbf{k}}$, but also at a more fundamental level. Therefore one can say that not only the existence of Cooper pairs, but also their actual wave functions, $u_{\mathbf{k}}$ and $v_{\mathbf{k}}$, have been confirmed experimentally.

A further confirmation of both the existence of Cooper pairs, and the BCS energy gap is provided by **Andreev scattering**. Consider an interface between a normal metal and a superconductor, as shown in Fig. 6.10. Consider an electron moving in the metal in a Bloch state $\mathbf{k}$ with energy $\epsilon_{\mathbf{k}}$. If its energy is below the superconductor energy gap,

$$\epsilon_{\mathbf{k}} - \epsilon_F < \Delta, \tag{6.94}$$

then the electron cannot propagate into the superconductor, and so it is perfectly reflected at the interface. This is normal particle reflection. But Andreev noticed that another process is possible. The electron can combine with another electron and form a Cooper pair, which will pass freely into the superconductor. By conservation of charge, a hole must be left behind. By conservation of momentum this hole will have to have momentum exactly opposite to the original electron, $-\mathbf{k}$. For the same reason it will also have opposite spin, therefore we have the situation shown in Fig. 6.10. The incoming electron can be reflected either as an electron, with a specularly reflected $\mathbf{k}$ vector, or it can be reflected as a hole of opposite spin and momentum, which travels back exactly along the original electron's path! Direct evidence for such scattering events can be found by injecting electrons into such an interface, say by electron tunnelling. Since the returning hole carries a positive charge and is moving in the opposite direction to the injected electron the tunnel current is actually twice what it would have been if $\Delta = 0$, or if the tunnelling electron is injected with at a voltage above the energy gap, $V > \Delta$.

An interesting feature of Andreev reflection is that the electron and hole are exactly **time reversed** quantum states with the correspondences,

$$\begin{aligned} -e &\to e, \\ \mathbf{k} &\to -\mathbf{k}, \\ \sigma &\to -\sigma, \end{aligned} \tag{6.95}$$

in charge, momentum, and spin. Fundamentally this arises because the Cooper pairs in the BCS wave function are pairs of time reversed single particle states. A very surprising implication of this fact was pointed out by P.W. Anderson. He noted that if the crystal lattice is disordered due to impurities, then Bloch's theorem no longer applies and the crystal momentum $\mathbf{k}$ is not a good quantum number. But, even in a strongly disordered system the single particle wave functions still come in **time reversed pairs**

$$\psi_{i\uparrow}(\mathbf{r}) \qquad \psi^*_{i\downarrow}(\mathbf{r}). \tag{6.96}$$

The single particle Hamiltonian operator $\hat{H} = -\hbar^2\nabla^2/2m + V(\mathbf{r})$ is real even if the potential $V(\mathbf{r})$ is not periodic, and it turns out that this implies that $\psi_{i\uparrow}(\mathbf{r})$ and

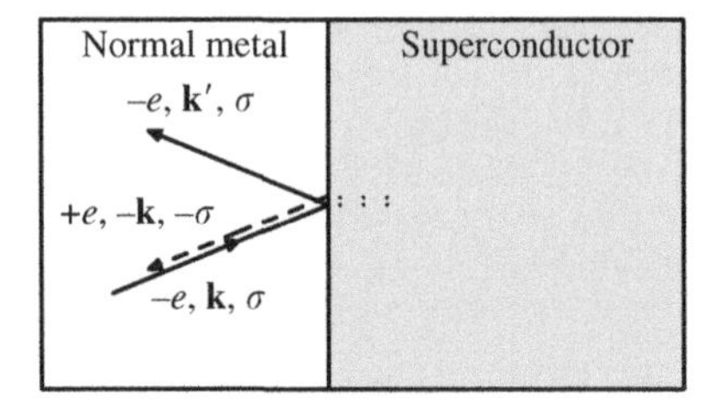

Fig. 6.10 Andreev scattering of electrons in a normal metal. The electron incident on the superconductor can either be reflected normally, remaining an electron of the same spin, or it can be Andreev reflected, becoming a hole of opposite momentum and spin. In the Andreev scattering process a net charge of $-2e$ is passed into the superconducting condensate. The conductivity of the junction is two times that for electrons with energies V above the gap Δ.

$\psi^*_{i\downarrow}(\mathbf{r})$ must be both eigenstates and be **exactly degenerate**. Anderson argued that that one could reformulate BCS theory entirely in terms of these new states of the disordered crystal lattice, and that to a first approximation quantities like T_c (which depend only on $g(\epsilon_F)$ and λ) will be essentially unchanged. This explains why the BCS theory works well even in highly disordered systems, such as metallic alloys. If the mean free path l for the electrons in the solid is greater than the coherence length,

$$l > \xi_0$$

then the alloy is said to be in the **clean limit**, but if it is shorter,

$$l < \xi_0$$

the alloy is said to be in the **dirty limit**. On the other hand, Anderson's argument does not apply if the crystal impurities themselves break the time-reversal symmetry, such as magnetic impurities.[7] Therefore superconductivity is heavily influenced (and rapidly destroyed) by magnetic impurities. They are said to be **pair breaking** since they break up the Cooper pairs.[8]

Finally, in some superconductors one has to extend the original BCS theory to allow for **strong coupling**. The assumptions made by BCS are essentially exact in the weak-coupling limit, namely when $\lambda \ll 1$. But when the coupling parameter λ becomes larger, say $0.2-0.5$, then one has to self-consistently take into account both the effects of the phonons on the electrons, and the effects of the electrons on the phonons. For example, the phonon frequencies are affected by the coupling to the electrons. All of these effects can be included consistently, systematically keeping all effects which are of order m/M, where M is the mass of the crystal lattice ions. In terms of the Migdal theorem stated above, that each electron–phonon vertex is of order $\sqrt{m/M}$, it is only necessary to systematically include all Feynman diagrams which have two electron–phonon vertices. When all such effects are included it is also necessary to fully include the phonon denstity of states, and the electron–phonon coupling matrix elements. The theory developed by Eliashberg characterizes both of these with a single function $\alpha^2(\omega)F(\omega)$, where $F(\omega)$ is the phonon density of states and $\alpha(\omega)$ is an effective electron–phonon matrix element. In terms of these quantities the electron–phonon coupling constant becomes

$$\lambda = 2\int_0^\infty \frac{\alpha^2(\omega)F(\omega)}{\omega}\,d\omega. \tag{6.97}$$

An approximate expression for the critical temperature was found by McMillan,

$$k_B T_c = \frac{\hbar\omega_D}{1.45}\exp\left(\frac{1.04(1+\lambda)}{\lambda - \mu^*(1+0.62\lambda)}\right). \tag{6.98}$$

Here the parameter μ^* is the **Coulomb pseudopotential**, which takes into account the direct (screened) Coulomb repulsion between the electrons. This formula works well in superconductors such as Pb and Nb, where there are significant deviations from BCS theory. For example, it explains the reduced isotope effect in these materials, as shown in Table 6.1. See Alexandrov (2003) for a more complete discussion.

[7] Under time reversal spin becomes reversed, so magnetic impurity atoms break time reversal symmetry. An external magnetic field would also break the symmetry.

[8] Interestingly, for superconductors with magnetic impurities states start to fill in the energy gap, Δ. As more impurities are added the transition temperature T_c decreases as more and more states fill the gap. It turns out that there is a small regime of **gapless superconductivity** in which the energy gap has completely disappeared, even though the system is still superconducting and below T_c. Therefore the presence of the energy gap is not completely essential to the existence of superconductivity.

Further reading

There are many excellent text books on the BCS theory. These include Schrieffer (1964), de Gennes (1966), Tinkham (1996), Ketterson and Song (1999) and

Waldram (1996), Alexandrov (2003) as well as many others. These include many more details which we have not had space to include here. The description of the BCS state given here is similar to those given in most of these books.

At a more advanced level one should first learn many-body theory formally. Books at this level include Fetter and Walecka (1971), Abrikosov Gorkov Dzyaloshinski (1963), and Rickayzen (1980). Schrieffer (1964) also introduces these methods as part of his discussion of the BCS theory.

Exercises

(6.1) (a) Show that the pair operators $\hat{P}^+_{\mathbf{k}}$ and $\hat{P}^+_{\mathbf{k}'}$ commute, as given in Eqs 6.32.

Show that they do not obey boson commutator equations, that is,

$$\left[\hat{P}_{\mathbf{k}}, \hat{P}^+_{\mathbf{k}'}\right] \neq \delta_{\mathbf{k},\mathbf{k}'}.$$

(6.2) Dervie Eq. 6.49 using the method of permuting operators into normal order, as described below Eq. 6.46.

(6.3) Confirm that the quasiparticle operators $b^+_{\mathbf{k}_\sigma}$ and $b_{\mathbf{k}_\sigma}$ obey fermionic anti-commutation rules

$$\left\{b_{\mathbf{k}\sigma}, b^+_{\mathbf{k}'\sigma'}\right\} = \delta_{\mathbf{k},\mathbf{k}'} \cdot \delta_{\sigma,\sigma'}$$

(6.4) (a) Show that the BCS coherent state, Eq. 6.38, has the property that $b_{\mathbf{k}\sigma}|\Psi_{\rm BCS}\rangle = 0$

(b) Show that

$$\langle \Psi_{\rm BCS}|c^+_{\mathbf{k}\uparrow}c^+_{-\mathbf{k}\downarrow}|\Psi^*_{\rm BCS}\rangle = u^*_{\mathbf{k}}v_{\mathbf{k}}.$$

(6.5) (a) The BCS gap equation becomes

$$1 = \lambda \int_0^{\hbar\omega_D} \frac{1}{\epsilon} \tanh\left(\frac{\epsilon}{2k_{\rm B}T_c}\right) d\epsilon$$

at the critical temperature T_c. Show that the integrand can be reasonably well approximated by

$$\frac{1}{\epsilon}\tanh\left(\frac{\epsilon}{2k_{\rm B}T_c}\right) \approx \begin{array}{ll} 1/\epsilon, & \epsilon > 2k_{\rm B}T_c, \\ 0, & \text{otherwise.} \end{array}$$

Hence, write down a simple analytical estimate of T_c. How close is your estimate to the exact BCS value?

(b) Show that the gap equation becomes

$$1 = \lambda \int_0^{\hbar\omega_D} \frac{1}{(\epsilon^2 + |\Delta|^2)^{1/2}} d\epsilon$$

at $T = 0$. Making the approximation

$$\frac{1}{(\epsilon^2 + |\Delta|^2)^{1/2}} \approx \begin{array}{ll} 1/\epsilon, & \epsilon > |\Delta|, \\ 0, & \text{otherwise.} \end{array}$$

find a simple analytical estimate of $|\Delta|$. Compare your results with the famous BCS result $|\Delta| = 1.76 k_{\rm B}T_c$.

Superfluid ^{3}He and unconventional superconductivity

7.1 Introduction

By about 1970 the Bardeen Cooper and Schrieffer (BCS) theory of superconductivity was well established, and had led to major new discoveries, such as the Josephson effect. The theory of superfluidity in ^{4}He was also clear in the outline, although the numerical problems of calculating observable quantities for a dense and strongly interacting Bose liquid were considerable with the computational tools available at that time. So, for many scientists the field of low temperature physics seemed to be nearing its completion.

This complacency was shattered in 1972 by the surprising discovery of superfluid phases of ^{3}He.[1] The superfluid transition temperature was about three orders of magnitude lower than for ^{4}He, occurring at around 2.7 mK. Another surprise was that two transitions were observed, corresponding to two distinct phases called A and B. Figure 7.1 shows part of the $P-T$ phase diagram of liquid ^{3}He in the region 2–3 mK (Wheatley 1975).

[1] The first observation of two novel phase transitions in ^{3}He, was made at Cornell, by Lee, Oscheroff, and Richardson. Originally they had thought that the specific heat anomalies they observed were associated with a solid ^{3}He phase, possibly a magnetic state. But it was soon established that this was in fact a liquid phase, and the two transitions (called A and B) corresponded to two different superfluid phases. Leggett shared the 2003 Nobel Prize in Physics for identifying these states.

The fact that the ^{3}He atoms are fermions immediately suggested that these new superfluid states were analogs of BCS superconductivity. However, in ^{3}He there was no obvious analog of the phonons to provide an attractive potential binding for the Cooper pairs. The direct van der Waals forces between ^{3}He atoms, as shown in Fig. 2.1 are dominated by the strong interatomic repulsion, and the attractive part of the interaction is probably too weak to even account for the 2 mK transition temperature.

In fact, the relevant theoretical models had already been explored in the 1960s, by Pitaevskii, Emery and Sessler, Anderson and Morel, and Balian and Werthamer. These theories had explored the idea of BCS pairing in systems where the Cooper pair bound state is not in the usual, $l = 0$, zero angular momentum *s*-**wave** state, but in an $l = 1$ or $l = 2$ bound state. For these pairing states the Cooper pair wave functions $\varphi(\mathbf{r} - \mathbf{r}')$ vanish when $\mathbf{r} = \mathbf{r}'$. Therefore the paired particles are never at the same position. In a system with strongly repulsive interaction, such as ^{3}He, this is obviously advantageous. In the case of superfluid ^{3}He it turns out that the Cooper pairs become bound in an $l = 1$, or *p*-**wave** pairing state.

The possibility of similar **unconventional** Cooper pairs also arises in superconductors. Currently, many superconductors are generally believed to be examples of this, although often there is still quite some controversy. The longest established of these cases are in the so called "heavy fermion" materials, such as UPt_3 and UBe_{13}, known since the 1980s. The case of the high temperature superconductors, such as $YBa_2Cu_3O_7$ is more controversial. But

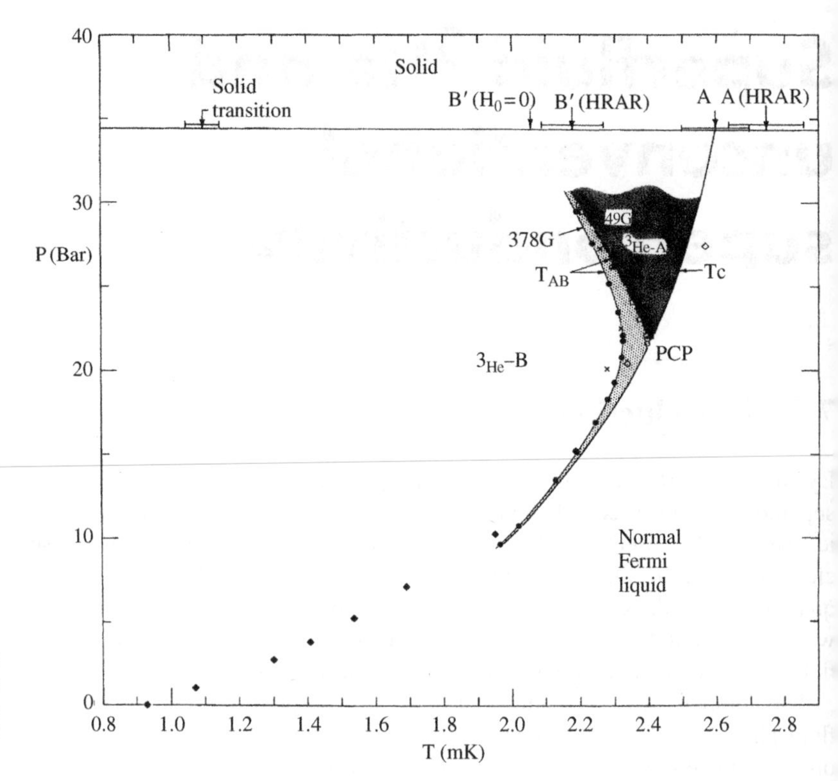

Fig. 7.1 (a) $P-T$ phase diagram of ^{3}He at small magnetic field. (b) $H-T$ phase diagram of ^{3}He at low pressure. Reprinted figure with permission from Wheatley (1975). Copyright (1975) by the American Physical Society.

now there is extremely strong experimental evidence that these materials have Cooper pairs with d-**wave** Cooper pairs corresponding to an $l = 2$ spin singlet bound state.

In this chapter we shall present a very brief introduction to this subject. The field is vast, and the exotic properties of superfluid ^{3}He alone could fill a whole book (in fact there are several!). The discussion of different exotic superconductors could also go on in great depth. Here the discussion has been kept as brief as possible. We will only discuss the most well known examples of exotic superconductors, especially concentrating on those which are currently subject of wide scientific interest.

7.2 The Fermi liquid normal state of ^{3}He

The pairwise van der Waals interactions between ^{3}He atoms, $V(\mathbf{r}_i - \mathbf{r}_j)$, are essentially identical to those in ^{4}He, as shown in Fig. 2.1. Therefore exactly the same many-particle Hamiltonian applies,

$$\hat{H} = \sum_{i=1,N} -\frac{\hbar^2}{2m}\nabla_i^2 + \frac{1}{2}\sum_{i\neq j} V(\mathbf{r}_i - \mathbf{r}_j). \tag{7.1}$$

However, there is a fundamental difference, in that the ^{3}He atoms are spin 1/2 fermions, rather than bosons. Therefore the many-particle Schrödinger equation

$$\hat{H}\Psi_n(\mathbf{r}_1, \sigma_1, \ldots, \mathbf{r}_N, \sigma_N) = E_n\Psi_n(\mathbf{r}_1\sigma_1, \ldots, \mathbf{r}_N\sigma_N) \tag{7.2}$$

has eigenfunctions which must be odd under exchange of any pair of particle spatial and spin coordinates, $\{\mathbf{r}_i, \sigma_i\}$ with $\{\mathbf{r}_j, \sigma_j\}$. Here $\sigma_i = \uparrow$ or $\downarrow$, denoting to two spin states per particle.

This difference between Fermi and Bose statistics is the fundamental source of the very different properties of ^{3}He and ^{4}He. If we neglect the interactions between particles, then we can write down a complete set of such antisymmetric functions. The appropriate functions are **Slater determinants**

$$\Psi_n^{(0)}(\mathbf{r}_1\sigma_1, \ldots, \mathbf{r}_N\sigma_N) = \frac{1}{\sqrt{N!}} \begin{vmatrix} \psi_1(\mathbf{r}_1\sigma_1) & \psi_1(\mathbf{r}_2\sigma_2) & \ldots & \psi_1(\mathbf{r}_N\sigma_N) \\ \psi_2(\mathbf{r}_1\sigma_1) & \psi_2(\mathbf{r}_2\sigma_2) & \ldots & \psi_2(\mathbf{r}_N\sigma_N) \\ \ldots & & & \\ \psi_N(\mathbf{r}_1\sigma_1) & \ldots & \ldots & \psi_N(\mathbf{r}_N\sigma_N) \end{vmatrix}. \tag{7.3}$$

The single particle states $\psi_i(\mathbf{r}\sigma)$ are simply plane wave states

$$\psi(\mathbf{r}\sigma) = \frac{1}{\sqrt{V}} e^{i\mathbf{k}\cdot\mathbf{r}} \chi(\sigma), \tag{7.4}$$

where V is the total system volume. Here $\chi(\sigma)$ is the spin part of the wave function,

$$\chi(\uparrow) = \begin{pmatrix} 1 \\ 0 \end{pmatrix} \quad \chi(\downarrow) = \begin{pmatrix} 0 \\ 1 \end{pmatrix} \tag{7.5}$$

corresponding to up and down spin states, with respect to the (arbitrarily chosen) z-axis.

Assuming that we could neglect the particle interactions, $V(\mathbf{r}_i - \mathbf{r}_j)$, the zero temperature ground state, $\Psi_0^{(0)}$, is simply found by occupying all of the plane wave states up to a Fermi surface. Because ^{3}He is isotropic (unlike a crystal), the Fermi surface must be just a sphere of radius k_F, where

$$k_F = (3\pi^2 n)^{1/3}, \tag{7.6}$$

exactly as in the Sommerfeld theory of electrons in metals (Singelton 2001). Given the mass density of ^{3}He of about $\rho \approx 81$ kgm^{-3}, this corresponds to $k_F = 0.78$ Å^{-1}, and a Fermi energy, $\epsilon_F = (\hbar k_F)^2/2m$, of about 0.49 meV or 4.9 K. Therefore the temperatures where the superfluidity occurs are at least a factor of 1000 smaller than the energy scale set by ϵ_F, implying we have a **degenerate Fermi gas**.

However, the particle–particle interactions, $V(\mathbf{r}_i - \mathbf{r}_j)$, are very strong in ^{3}He and cannot be neglected. It is a very dense fluid of nearly perfectly hard spheres. Instead of being an ideal noninteracting quantum gas, we say that it is a dense **Fermi liquid**. This concept of a Fermi liquid was developed by Landau. The key idea is that even though the system is strongly interacting and has an extremely complicated ground state wave function, the excitations relative to this ground state act as weakly interacting particles. These weakly interacting excitations are called **quasiparticles**.

Let us consider first the ground state wave function in an interacting system of fermions. We can consider a system in which the interaction can be turned on or off continuously, with the Hamiltonian,

$$\hat{H}^{(\lambda)} = \sum_{i=1,N} -\frac{\hbar^2}{2m}\nabla_i^2 + \frac{\lambda}{2}\sum_{i\neq j} V(\mathbf{r}_i - \mathbf{r}_j). \tag{7.7}$$

The case $\lambda = 0$ corresponds to the ideal noninteracting Fermi gas, for which we know the exact ground state $\Psi_0^{(0)}$ is the Slater determinant state given above.

On increasing the value of the parameter λ, we will obtain a family of ground state wave functions obeying,

$$\hat{H}^{(\lambda)}\Psi_0^{(\lambda)}(\mathbf{r}_1\sigma_1,\ldots,\mathbf{r}_N\sigma_N) = E_0^{(\lambda)}\Psi_0^{(\lambda)}(\mathbf{r}_1\sigma_1,\ldots,\mathbf{r}_N\sigma_N). \tag{7.8}$$

When $\lambda = 1$ this becomes the ground state wave function of the physical system, $\Psi_0^{(1)}$.

The underlying assumption of Landau's Fermi-liquid theory is that we can assume that there is **adiabatic continuity**. This means that the wave functions change continuously as we slowly increase λ from 0 to 1,[2]

$$\Psi_n^{(0)} \to \Psi_n^{(\lambda)} \to \Psi_n^{(1)}. \tag{7.9}$$

[2]This assumption of continuity will not be true if there is a sudden change in the character of the wave function as a function of λ. For example, it will not hold if there is a phase transition from the liquid state to a solid phase or from normal liquid to superfluid.

Under this assumption, we can consider how physical properties of the system evolve as the interaction λ is slowly increased. For example, consider the **momentum distribution**, the average particle occupation of a single $\mathbf{k}, \sigma$ quantum state,

$$\langle \hat{n}_{\mathbf{k}\sigma} \rangle = \langle \Psi^{(\lambda)} | c^+_{\mathbf{k}\sigma} c_{\mathbf{k}\sigma} | \Psi^{(\lambda)} \rangle. \tag{7.10}$$

For $\lambda = 0$ this is simply the zero temperature Fermi–Dirac distribution

$$n_{\mathbf{k}\sigma} = \begin{cases} 1 & k < k_F, \\ 0 & k > k_F. \end{cases} \tag{7.11}$$

Now, because the wave function evolves continuously with increasing λ, so must this momentum distribution. It follows that the momentum distribution of the interacting system continues to have a discontinuity at k_F. The height of the discontinuity need no longer be unity, and becomes reduced by a factor $Z < 1$, but a discontinuity must remain. This is illustrated in Fig. 7.2. The position of this discontinuity defines the Fermi wave vector, k_F even in the interacting Fermi liquid. Furthermore, for a spherically symmetric system such as ^{3}He, this implies that the value of k_F is given by Eq. 7.6 and is unchanged even in the strongly interacting Fermi liquid system (a result known as Luttinger's theorem).

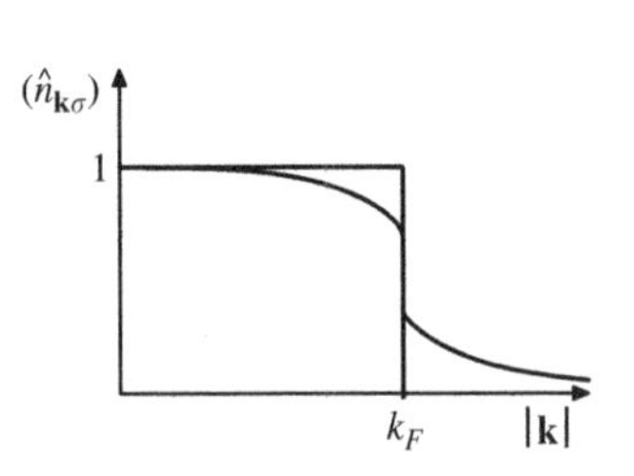

Fig. 7.2 Momentum distribution of the ideal (noninteracting) Fermi gas, and an interacting Fermi liquid. The discontinuity at k_F remains, although the height of the discontinuity is reduced from 1 to a smaller number, Z.

We can apply the same adiabatic continuity argument to the low energy excited states of the system. It implies that we can expect a similar continuity of the excited state energy eigenvalues and wave functions, at least for the low energy states. In the noninteracting $\lambda = 0$ limit the excitations are easy to find. They are simply given by adding or removing electrons from the occupied states in the Slater determinant Eq. 7.3.

For example, the simplest excited state of the noninteracting system is to simply add one extra particle to an empty state above k_F. The corresponding many-body wave function is[3]

[3]Of course one could equally well create a hole below k_F by acting with the annihilation operator $\hat{c}_{\mathbf{k}\sigma}$.

$$|\Psi_{\mathbf{k}\sigma}^{(0)}\rangle = \hat{c}^+_{\mathbf{k}\sigma}|\Psi_0^{(0)}\rangle, \tag{7.12}$$

In the noninteracting system the energy of this excited state is (relative to the noninteracting system Fermi energy, $\hbar^2 k_F^2/2m$)

$$\epsilon_{\mathbf{k}}^{(0)} = \frac{\hbar^2(k^2 - k_F^2)}{2m}. \tag{7.13}$$

Applying the principle of adiabatic continuity, we expect that the particle energy will have a similar functional form even in the interacting system. It must still go to zero as $k \to k_F$, and so we expect

$$\epsilon_{\mathbf{k}} \approx \frac{\hbar^2(k^2 - k_F^2)}{2m^*} \tag{7.14}$$

for the fully interacting system. This defines the **effective mass** m^* for the system. Near k_F this is approximately linear,

$$\epsilon_{\mathbf{k}} \approx \hbar(|\mathbf{k}| - k_F)v_F \tag{7.15}$$

where v_F is the **Fermi velocity** of the interacting system. For ^{3}He, the effective mass is of order $m^* \sim 3m$ at low pressure, rising to about $m^* \sim 6m$ at pressures near the solid–liquid phase boundary in Fig. 7.1. The interacting system Fermi energy, $\epsilon_F = \hbar^2 k_F^2/2m^*$, is therefore of order 1 K.

Following the same logic, we can also imagine creating a pair of particles (or a pair of holes, or one particle and one hole), as shown in Fig. 7.3. Creating one particle at $\mathbf{k}\sigma$ and an other at $\mathbf{k}'\sigma'$ gives a wave function of the noninteracting system

$$|\Psi^{(0)}_{\mathbf{k}\sigma\mathbf{k}'\sigma'}\rangle = \hat{c}^{+}_{\mathbf{k}\sigma}\hat{c}^{+}_{\mathbf{k}'\sigma'}|\psi^{(0)}_0\rangle. \tag{7.16}$$

The corresponding noninteracting particle energy is just a sum of the individual particle energies

$$\epsilon^{(0)}_{\mathbf{k}} + \epsilon^{(0)}_{\mathbf{k}'}.$$

However, in the interacting system the corresponding two particle state energy will **not** be simply a sum of the noninteracting particle energies. We can write the two particle energy as

$$\epsilon_{\mathbf{k}} + \epsilon_{\mathbf{k}'} + f(\mathbf{k}\sigma, \mathbf{k}'\sigma'),$$

where the term $f(\mathbf{k}\sigma, \mathbf{k}'\sigma')$ is an **effective interaction** between the two quasiparticles. Extending this to a system with many quasiparticles, we can write the total energy,

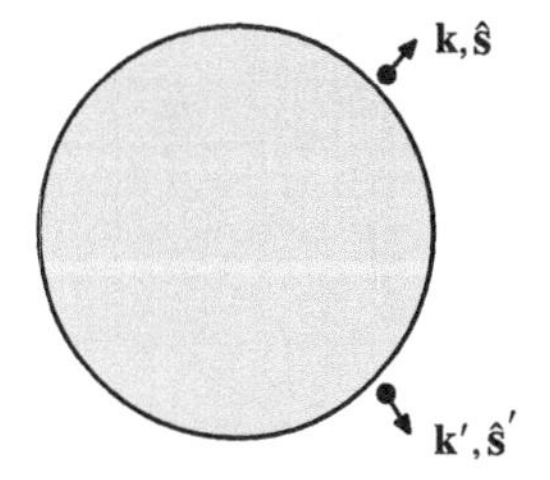

Fig. 7.3 Interactions between quasiparticles near the Fermi surface in Landau Fermi liquid. The interaction depends on two contributions: one which does not depend on the relative spin orientations $\hat{\mathbf{s}}$, $\hat{\mathbf{s}}'$, and one which does. Both interactions are functions of $\mathbf{k}$ and $\mathbf{k}'$.

$$E = E_0 + \sum_{\mathbf{k}\sigma} \epsilon_{\mathbf{k}}\delta n_{\mathbf{k}\sigma} + \frac{1}{2}\sum_{\mathbf{k}\sigma,\mathbf{k}'\sigma'} f(\mathbf{k}\sigma, \mathbf{k}'\sigma')\delta n_{\mathbf{k}\sigma}\delta n_{\mathbf{k}'\sigma'}, \tag{7.17}$$

where E_0 is the ground state energy (no quasiparticles) and $\delta n_{\mathbf{k}\sigma}$ is the change in occupation of the quasiparticle state $\mathbf{k}\sigma$ (equal to $+1$ for an added particle at $\mathbf{k}$, -1 for a hole).

We can use symmetry to describe the effective interaction $f(\mathbf{k}\sigma, \mathbf{k}'\sigma')$ with just a small number of numerical parameters. First it is convenient to represent the quasiparticle more generally by its spin direction in three dimensional space. We can define a quasiparticle with a spin in an orbitrary direction with a 2×2 matrix,

$$\delta n_{\mathbf{k}\alpha\beta} = \delta n_{\mathbf{k}}\left(\tfrac{1}{2}\delta_{\alpha\beta} + \hat{\mathbf{s}}.\boldsymbol{\sigma}_{\alpha\beta}\right), \tag{7.18}$$

where $\boldsymbol{\sigma} = \{\sigma_x, \sigma_y, \sigma_z\}$ is the vector of the Pauli matrices,

$$\sigma_x = \begin{pmatrix} 0 & 1 \\ 1 & 0 \end{pmatrix},$$

$$\sigma_y = \begin{pmatrix} 0 & -i \\ i & 0 \end{pmatrix},$$

$$\sigma_z = \begin{pmatrix} 1 & 0 \\ 0 & -1 \end{pmatrix},$$

and $\hat{\mathbf{s}}$ is a vector of length 1/2 in the direction of the quasiparticle spin. Then, the Fermi liquid total energy can be written in the form

$$E = E_0 + \sum_{\mathbf{k}\sigma} \epsilon_{\mathbf{k}} \delta n_{\mathbf{k}\sigma} + \frac{1}{2} \sum_{\mathbf{k},\mathbf{k}'} \delta n_{\mathbf{k}} \delta n_{\mathbf{k}'} \left(f_1(\mathbf{k},\mathbf{k}') + f_2(\mathbf{k},\mathbf{k}') \hat{\mathbf{s}}_{\mathbf{k}} \cdot \hat{\mathbf{s}}_{\mathbf{k}'} \right). \quad (7.19)$$

The spin directions can only enter via the scalar product $\hat{\mathbf{s}}_{\mathbf{k}} \cdot \hat{\mathbf{s}}_{\mathbf{k}'}$ because of overall rotational invariance in spin.[4]

[4]Equivalently we can see that the four possible spin combinations of the two quasiparticles (↑↑, ↑↓, ↓↑, and ↓↓) can be divided into one singlet and three triplet spin contributions. For two spin 1/2 particles $\hat{\mathbf{s}}_1 \cdot \hat{\mathbf{s}}_2$ equals $-3/4$ for the spin singlet combination, and equals $+1/4$ for the three spin triplet states. Singlet and triplet pairs of quasiparticles have interactions f_1 and f_2, respectively.

The quasiparticle wave vectors $\mathbf{k}$, $\mathbf{k}'$ must lie very close to the Fermi surface (since $k_B T \ll \epsilon_F$), and so we can take these as unit vectors lying on the Fermi sphere. Also by rotational invariance, the quasiparticle energy can depend only on $\mathbf{k} \cdot \mathbf{k}' = k_F^2 \cos\theta$. Therefore we can expand the functions $f_1(\mathbf{k},\mathbf{k}')$ and $f_2(\mathbf{k},\mathbf{k}')$ in series of Legendre polynomials $P_l(\cos\theta)$. We define expansion parameters F_l and Z_l by,

$$g(\epsilon_F) f_1(\mathbf{k},\mathbf{k}') = \sum_l F_l P_l(\cos\theta),$$

$$g(\epsilon_F) f_2(\mathbf{k},\mathbf{k}') = \sum_l Z_l P_l(\cos\theta).$$

The factor of $g(\epsilon_F)$, the density of states, is included to make the parameters F_l and Z_l dimensionless.

The success of the Landau Fermi liquid theory for ^{3}He arises from the fact that just a handful of numerical parameters, m^*, F_0, F_1, Z_0, and Z_1 need be obtained from fitting experiments. All higher F_l and Z_l coefficients can be neglected. Once this fitting is done, then the theory contains an essentially complete description of the low temperature thermodynamics and excitations of the normal state fluid. According to Leggett (1975) F_0 and F_1 are large and positive, Z_0 is about -3 and Z_1 is small.

It is the negative parameter Z_0 which shows that the normal liquid state of ^{3}He is close to ferromagnetism. A calculation of the spin susceptibility shows that

$$\chi = \frac{\chi_0}{1 + (Z_0/4)}, \quad (7.20)$$

where χ_0 is the Pauli spin susceptibility of the noninteracting Fermi gas (Blundell 2001). Clearly, from this expression, with $Z_0 = -3$ the susceptibility χ is about four times its noninteracting value. This is indicative of a tendency towards ferromagnetism, and suggests that there are strong **ferromagnetic spin fluctuations** present in the normal fluid state (Leggett 1975). It is these spin-fluctuations which are believed to drive the transition to superfluidity in ^{3}He.

7.3 The pairing interaction in liquid ^{3}He

The superfluidity in ^{3}He cannot arise from a Bose–Einstein condensation (BEC) process, such as in ^{4}He, but instead must be more similar to BCS superconductivity. We can expect that there will be some sort of Cooper pairs, and a coherent quantum state with off-diagonal long ranged order, as discussed in Chapters 5 and 6. But, since the ^{3}He atoms are neutral, there will not be any

Meissner effect. However, we should still obtain persistent currents and hence superfluidity.

To describe a Cooper pairing instability in ^{3}He, we need to consider the effective pairing interaction between the particles. But, unlike BCS superconductivity there is (at first sight) no analog of the electron–phonon interaction in ^{3}He. Therefore, if BCS-like Cooper pairing is to occur, then the attractive binding force between the particles must arise directly from the ^{3}He particles themselves. Naively one would expect that the appropriate interparticle interaction would be just the pairwise van der Waals potential $V(\mathbf{r}_i - \mathbf{r}_j)$ of Fig. 2.1. The corresponding pairing Hamiltonian will have the k-space interaction,

$$V(\mathbf{k}, \mathbf{k}') = \int e^{-i(\mathbf{k}-\mathbf{k}')\cdot\mathbf{r}} V(\mathbf{r})\, d^3r. \tag{7.21}$$

Since we are interested in low temperatures, much less than ϵ_F/k_B, we can restrict ourselves to values of $\mathbf{k}$ and $\mathbf{k}'$ lying on the Fermi sphere. Because of spherical symmetry the interaction can only depend on $\mathbf{k}\cdot\mathbf{k}' = k_F^2 \cos\theta$, or the angle θ. Therefore we can express the effective pairing interaction in terms of Legendre polynomials in $\cos\theta$

$$V(\mathbf{k}, \mathbf{k}') = \sum_l \frac{(2l+1)}{2} V_l P_l(\cos\theta), \tag{7.22}$$

where

$$V_l = \frac{1}{4\pi} \int V(\hat{\mathbf{n}}(\theta,\phi), \hat{e}_z) P_l(\cos\theta) \sin\theta\, d\theta\, d\phi, \tag{7.23}$$

and $\hat{\mathbf{n}}(\theta,\phi)$ is a unit vector in the θ,ϕ direction on the Fermi sphere. The factor of $(2l+1)/2$ arises from the normalization of Legendge polynomials

$$\int P_l^2(\cos\theta) \sin\theta\, d\theta = \frac{2}{2l+1}. \tag{7.24}$$

If we were to use the bare van der Waals interaction between two helium atoms in this expression, then we could directly calculate the parameters V_l. The result is that V_0 is strongly repulsive, because of the strong hard core repulsion between the atoms, while V_2 and V_3 are weakly attractive (Leggett 1975). But these are too small to account for the observed superfluid transition temperature in ^{3}He.

However, as we have seen, at low temperatures we should instead use an **effective interaction** between quasiparticles, rather than this "bare" interaction. The effective interaction will include the effects of the direct interaction, but also indirect effects, such as those arising from spin fluctuations. A quasiparticle with spin $\sigma = \uparrow$ causes a local polarization of the neighboring quasiparticles, via the Pauli exchange interaction.[5] This tendency for neighboring quasiparticles to align ferromagnetically leads to an attractive potential, which can lead to a Cooper instability.

It is not directly possible to calculate the effective pairing interaction $V(\mathbf{k}, \mathbf{k}')$ using the Landau fermi liquid parameters. However, for small momentum transfers, $|\mathbf{k} - \mathbf{k}'| \ll k_F$, it is possible to estimate this interaction using Fermi liquid theory. The result is that (Leggett 1975)

$$V(\mathbf{k}, \mathbf{k}') \approx \frac{1}{g(\epsilon_F)} \frac{Z_0}{1 + \frac{1}{4}Z_0} \hat{\mathbf{s}}_{\mathbf{k}} \cdot \hat{\mathbf{s}}_{\mathbf{k}'} \tag{7.25}$$

in this region. The interaction is spin dependent, as expected from its physical origin from spin fluctuation exchange. Since $Z_0 \sim -3$ this pairing potential

[5] The direct spin–spin interaction between ^{3}He atoms is the direct dipole–dipole force between the $s = 1/2$ helium nucleii. But this direct interaction is very small. The Pauli exclusion principle requires that parallel spin particles remain spatially separated, leading to **exchange energy**. This exchange energy is a much stronger effect, and is the dominant spin–spin interaction in ^{3}He. It tends to favor parallel spin, ferromagnetic, alignment of the spins. The mechanism is very similar to the mechanism which produces ferromagnetism in iron and nickel.

is negative when the spins are parallel, that is, for spin triplet pairs (for which $\hat{\mathbf{s}}_\mathbf{k} \cdot \hat{\mathbf{s}}_{\mathbf{k}'} = +1/4$). In contrast, for spin singlet quasiparticle pairs, $\hat{\mathbf{s}}_\mathbf{k} \cdot \hat{\mathbf{s}}_{\mathbf{k}'} = -3/4$ and there is a net repulsion. Furthermore, the fact that Z_0 is close to -3 also means that this interaction is quite strongly enhanced by the factor in the denominator.

7.4 Superfluid phases of ^{3}He

If we wish to model BCS pairing with a more general **k** and spin dependent interaction potential, we can consider the Hamiltonian

$$\hat{H} = \sum_{\mathbf{k},\sigma}(\epsilon_\mathbf{k} - \mu)c^+_{\mathbf{k}\sigma}c_{\mathbf{k}\sigma} + \hat{H}_{\text{int}}, \tag{7.26}$$

where

$$\hat{H}_{\text{int}} = \sum_{\mathbf{k}\mathbf{k}'\alpha\beta\gamma\delta} V_{\alpha\beta\gamma\delta}(\mathbf{k},\mathbf{k}')c^+_{\mathbf{k}'\alpha}c^+_{-\mathbf{k}'\beta}c_{-\mathbf{k}\gamma}c_{\mathbf{k}\delta}. \tag{7.27}$$

Here we have allowed any possible dependence of the interaction V on the four particle spin indices, $\alpha, \beta, \gamma, \delta$, but limited the k-dependence to the scattering of a Cooper pair from $(\mathbf{k}, -\mathbf{k})$ to $(\mathbf{k}', -\mathbf{k}')$. These are the terms which give the usual Cooper instability in metals, corresponding to Cooper pairs with zero center of mass total momentum.

Following through the usual mean-field argument of the BCS theory, we must replace the four fermion interaction term, H_{int} with its average over all pairs of fermions.

$$\hat{H}_{\text{int}} \approx \sum_{\mathbf{k}\mathbf{k}'\alpha\beta\gamma\delta} V_{\alpha\beta\gamma\delta}(\mathbf{k},\mathbf{k}')\Big(\langle c^+_{\mathbf{k}'\alpha}c^+_{-\mathbf{k}'\beta}\rangle c_{-\mathbf{k}\gamma}c_{\mathbf{k}\delta} + c^+_{\mathbf{k}'\alpha}c^+_{-\mathbf{k}'\beta}\langle c_{-\mathbf{k}\gamma}c_{\mathbf{k}\delta}\rangle\Big) + \cdots, \tag{7.28}$$

where the neglected terms are not important for BCS pairing. Therefore we see that the analog of the BCS order parameter is the expectation value

$$F_{\alpha\beta}(\mathbf{k}) = \langle c_{-\mathbf{k}\alpha}c_{\mathbf{k}\beta}\rangle. \tag{7.29}$$

It is a matrix of four complex pairing amplitudes

$$F(\mathbf{k}) = \begin{pmatrix} \langle c_{-\mathbf{k}\uparrow}c_{\mathbf{k}\uparrow}\rangle & \langle c_{-\mathbf{k}\uparrow}c_{\mathbf{k}\downarrow}\rangle \\ \langle c_{-\mathbf{k}\downarrow}c_{\mathbf{k}\uparrow}\rangle & \langle c_{-\mathbf{k}\downarrow}c_{\mathbf{k}\downarrow}\rangle \end{pmatrix}. \tag{7.30}$$

The analog of the BCS gap parameter, Δ is also spin and **k** dependent,

$$\Delta_{\alpha\beta}(\mathbf{k}) = \sum_{\mathbf{k}'\gamma\delta} V_{\alpha\beta\gamma\delta}(\mathbf{k},\mathbf{k}')\langle c_{-\mathbf{k}'\gamma}c_{\mathbf{k}'\delta}\rangle, \tag{7.31}$$

so that the effective mean-field Hamiltonian is of the form

$$\hat{H} = \sum_{\mathbf{k},\sigma}(\epsilon_\mathbf{k} - \mu)c^+_{\mathbf{k}\sigma}c_{\mathbf{k}\sigma} + \sum_{\mathbf{k},\alpha,\beta} c^+_{\mathbf{k}\alpha}c^+_{-\mathbf{k}\beta}\Delta_{\alpha\beta}(\mathbf{k}) + \Delta^*_{\alpha\beta}(\mathbf{k})c_{-\mathbf{k}\alpha}c_{\mathbf{k}\beta} \tag{7.32}$$

which can be compared with the usual BCS expression Eq. 6.70.

This mean field Hamiltonian can be diagonalized by a spin dependent generalization of the Bogoliubov–Valatin transformation used in the BCS theory of Chapter 6. The BCS gap equation given in the previous chapter becomes a 4×4 matrix equation

$$\begin{pmatrix} \epsilon_{\mathbf{k}} - \mu & 0 & \Delta_{\uparrow\uparrow}(\mathbf{k}) & \Delta_{\uparrow\downarrow}(\mathbf{k}) \\ 0 & \epsilon_{\mathbf{k}} - \mu & \Delta_{\downarrow\uparrow}(\mathbf{k}) & \Delta_{\downarrow\downarrow}(\mathbf{k}) \\ \Delta^*_{\uparrow\uparrow}(\mathbf{k}) & \Delta^*_{\downarrow\uparrow}(\mathbf{k}) & -\epsilon_{\mathbf{k}} + \mu & 0 \\ \Delta^*_{\uparrow\downarrow}(\mathbf{k}) & \Delta^*_{\downarrow\downarrow}(\mathbf{k}) & 0 & -\epsilon_{\mathbf{k}} + \mu \end{pmatrix} \begin{pmatrix} u_{\mathbf{k}\uparrow n} \\ u_{\mathbf{k}\downarrow n} \\ v_{\mathbf{k}\uparrow n} \\ v_{\mathbf{k}\downarrow n} \end{pmatrix} = E_{\mathbf{k}n} \begin{pmatrix} u_{\mathbf{k}\uparrow n} \\ u_{\mathbf{k}\downarrow n} \\ v_{\mathbf{k}\uparrow n} \\ v_{\mathbf{k}\downarrow n} \end{pmatrix}, \tag{7.33}$$

where there are now two positive energy eigenvalues, $E_{\mathbf{k}n}$, labeled $n = 1, 2$, and two negative eigenvalues $-E_{\mathbf{k}n}$. The corresponding Bogoliubov transformation of operators is

$$b_{\mathbf{k}n} = \sum_{\sigma} \left(u^*_{\mathbf{k}\sigma n} c_{\mathbf{k}\sigma} - v^*_{\mathbf{k}\sigma n} c^+_{-\mathbf{k}\sigma} \right), \tag{7.34}$$

extending the usual expression from Eq. 6.76.

This general gap equation allows both **singlet** or **triplet** spin pairing states. But, these can be distinguished by symmetry. First, we can see that the anti-commutation properties of the fermion operators imply that the pairing amplitude has the following symmetry property,

$$\begin{aligned} F_{\alpha\beta}(\mathbf{k}) &= \langle c_{-\mathbf{k}\alpha} c_{\mathbf{k}\beta} \rangle \\ &= -\langle c_{\mathbf{k}\beta} c_{-\mathbf{k}\alpha} \rangle \\ &= -F_{\beta\alpha}(-\mathbf{k}). \end{aligned} \tag{7.35}$$

Similarly it must follow that the gap parameters have the same symmetry property,

$$\Delta_{\alpha\beta}(\mathbf{k}) = -\Delta_{\beta\alpha}(-\mathbf{k}). \tag{7.36}$$

It is now convenient to rewrite the four gap components $\Delta_{\alpha\beta}(\mathbf{k})$ in terms of a scalar $\Delta_{\mathbf{k}}$ and a vector $\mathbf{d}(\mathbf{k})$, using

$$\begin{pmatrix} \Delta_{\uparrow\uparrow}(\mathbf{k}) & \Delta_{\uparrow\downarrow}(\mathbf{k}) \\ \Delta_{\downarrow\uparrow}(\mathbf{k}) & \Delta_{\downarrow\downarrow}(\mathbf{k}) \end{pmatrix} = i(\Delta_{\mathbf{k}} I + \mathbf{d}(\mathbf{k}) \cdot \boldsymbol{\sigma})\sigma_y, \tag{7.37}$$

where $\boldsymbol{\sigma} = (\sigma_x, \sigma_y, \sigma_z)$ is a vector of the Pauli matrices, and I is the 2×2 unit matrix. Written out explicitly,

$$\begin{pmatrix} \Delta_{\uparrow\uparrow}(\mathbf{k}) & \Delta_{\uparrow\downarrow}(\mathbf{k}) \\ \Delta_{\downarrow\uparrow}(\mathbf{k}) & \Delta_{\downarrow\downarrow}(\mathbf{k}) \end{pmatrix} = \begin{pmatrix} -d_x(\mathbf{k}) + i d_y(\mathbf{k}) & \Delta_{\mathbf{k}} + d_z(\mathbf{k}) \\ -\Delta_{\mathbf{k}} + d_z(\mathbf{k}) & d_x(\mathbf{k}) + i d_y(\mathbf{k}) \end{pmatrix}. \tag{7.38}$$

The parity symmetry property above now implies that

$$\Delta_{\mathbf{k}} = \Delta_{-\mathbf{k}}, \tag{7.39}$$

$$\mathbf{d}(\mathbf{k}) = -\mathbf{d}(-\mathbf{k}). \tag{7.40}$$

The scalar component is thus even under the parity operation $\mathbf{k} \to -\mathbf{k}$, while the vector component is odd. It follows that in general we will find that the gap equation has solutions which are either even or odd, but never a mixture.[6] Therefore we can distinguish the two possibilities, spin singlet pairing, where

[6] This is always guaranteed to be the case near to T_c, since there the gap equation is linear and solutions will have a definite parity. Below T_c the gap equation is nonlinear, and in principle may allow mixed parity solutions. However, in practise this never occurs, since the two different parity solutions will have very different critical temperatures, and any given pairing mechanism will strongly favor one solution over the other.

the gap is

$$\begin{pmatrix} \Delta_{\uparrow\uparrow}(\mathbf{k}) & \Delta_{\uparrow\downarrow}(\mathbf{k}) \\ \Delta_{\downarrow\uparrow}(\mathbf{k}) & \Delta_{\downarrow\downarrow}(\mathbf{k}) \end{pmatrix} = \begin{pmatrix} 0 & \Delta_{\mathbf{k}} \\ -\Delta_{\mathbf{k}} & 0 \end{pmatrix}, \tag{7.41}$$

and spin triplet pairing, where the gap parameters are

$$\begin{pmatrix} \Delta_{\uparrow\uparrow}(\mathbf{k}) & \Delta_{\uparrow\downarrow}(\mathbf{k}) \\ \Delta_{\downarrow\uparrow}(\mathbf{k}) & \Delta_{\downarrow\downarrow}(\mathbf{k}) \end{pmatrix} = \begin{pmatrix} -d_x(\mathbf{k}) + id_y(\mathbf{k}) & d_z(\mathbf{k}) \\ d_z(\mathbf{k}) & d_x(\mathbf{k}) + id_y(\mathbf{k}) \end{pmatrix}. \tag{7.42}$$

In superfluid ^{3}He there is strong experimental evidence (Leggett 1975) that the pairing is in the spin triplet state. This contrasts strongly with the original BCS pairing theory of low T_c superconductors, in which (as we saw in the previous chapter) the pairing is always spin singlet. The difference arises because of the very different effective particle–particle interaction in ^{3}He compared with the BCS electron–phonon pairing mechanism. The BCS pairing potential is strongly attractive for all **k** vectors near the Fermi surface. In contrast, as we have seen in the Fermi liquid theory of the normal state of liquid ^{3}He, the quasiparticle interactions at the Fermi level are both **k** and spin dependent.

If we restrict ourselves to **k** vectors lying on, or very near to, the Fermi surface of liquid ^{3}He, then we can use spherical polar coordinates to expand the components of $\mathbf{d}(\mathbf{k})$ in terms of spherical Harmonic functions,

$$d_\nu(\mathbf{k}) = \sum \eta_{\nu lm} Y_{lm}(\theta_{\mathbf{k}}, \phi_{\mathbf{k}}), \tag{7.43}$$

where the subscript $\nu = x, y, z$ counts the three vector components. The symmetry property Eq. 7.40 implies that only **odd values** of the angular momentum quantum number, l, are allowed for spin triplet states. Conversely, only even values of l occur for spin singlet.

In principle the pairing state can include all possible angular momenta in the sum, Eq. 7.43. But in practise only a single l value will be relevant. Examining the BCS gap equation near to T_c, one finds that each different l value corresponds to a different T_c. One of these will inevitably be much higher than the others, because the effective pairing interactions will be quite different in each angular momentum channel. Recalling how rapidly the BCS expression for T_c decreases with coupling constant λ, going as $T_c \sim e^{-1/\lambda}$, it is clear that only the pairing state with the largest λ value will matter. Therefore we shall assume that only one l value is relevant. But which l value is expected for superfluid ^{3}He? Initially it was proposed that superfluid ^{3}He could be in an $l = 3$ state (called an f-wave state by the analogy with atomic spherical harmonic functions). But it was soon realized that the most attractive pairing potential would arise from spin fluctuations, and that this would correspond to $l = 1$, or ***p*-wave** pairing.

Assuming p-wave pairing to be the case, we can rewrite Eq. 7.43 in the form

$$d_\nu(\mathbf{k}) = \sum \eta_{\nu i} f_i(\theta_{\mathbf{k}}, \phi_{\mathbf{k}}), \tag{7.44}$$

where the functions $f_i(\theta_{\mathbf{k}}, \phi_{\mathbf{k}})$ can be taken as the three $l = 1$ p_x, p_y, and p_z functions which are familiar from atomic physics

$$f_x(\theta_{\mathbf{k}}, \phi_{\mathbf{k}}) = \left(\frac{3}{4\pi}\right)^{1/2} \cos\phi_{\mathbf{k}} \sin\theta_{\mathbf{k}}, \tag{7.45}$$

$$f_x(\theta_{\mathbf{k}}, \phi_{\mathbf{k}}) = \left(\frac{3}{4\pi}\right)^{1/2} \sin\phi_{\mathbf{k}} \sin\theta_{\mathbf{k}}, \tag{7.46}$$

$$f_z(\theta_{\mathbf{k}}, \phi_{\mathbf{k}}) = \left(\frac{3}{4\pi}\right)^{1/2} \cos\theta_{\mathbf{k}}. \tag{7.47}$$

The gap parameter $\mathbf{d}(\mathbf{k})$ on the Fermi surface is thus dependent on a total of **nine complex coefficients**,

$$[\eta_{\nu i}] = \begin{pmatrix} \eta_{xx} & \eta_{xy} & \eta_{xz} \\ \eta_{yx} & \eta_{yy} & \eta_{yz} \\ \eta_{zx} & \eta_{zy} & \eta_{zz} \end{pmatrix}.$$

Given the number of subscripts, it is useful to distinguish between those related to the vector direction of the $\mathbf{d}$ vector (spin orientation) by using a Greek letter for the subscript , and to use Roman letters (such as i) for the orbital or spatial directions $(\theta_{\mathbf{k}}, \phi_{\mathbf{k}})$ on the Fermi surface.

To determine these coefficients it is necessary to solve the analog of the BCS gap equation. It turns out that a number of different possible states can occur. The most important ones are the **Anderson–Brinkman–Morrel** (ABM) state,

$$[\eta_{\nu i}] = \eta \begin{pmatrix} 1 & i & 0 \\ 0 & 0 & 0 \\ 0 & 0 & 0 \end{pmatrix} \tag{7.48}$$

(implying $\mathbf{d}(\mathbf{k}) = (f_x + if_y, 0, 0)$), and the **Balain–Werthamer** (BW) state,

$$[\eta_{\mu\nu}] = \eta \begin{pmatrix} 1 & 0 & 0 \\ 0 & 1 & 0 \\ 0 & 0 & 1 \end{pmatrix} \tag{7.49}$$

(implying $\mathbf{d}(\mathbf{k}) = (f_x, f_y, f_z)$). These are illustrated in Fig. 7.4. In the ABM state the $\mathbf{d}(\mathbf{k})$ vector has a constant direction in space, and vanishes at the two points $\mathbf{k} = (0, 0, \pm k_f)$. In the BW case the vector $\mathbf{d}(\mathbf{k})$ always points outward at every point on the Fermi sphere, and has a constant magnitude. The analog of the BCS quasiparticle excitation energy in a triplet superconductor is

$$E^n_{\mathbf{k}} = \sqrt{(\epsilon_{\mathbf{k}} - \mu)^2 + |\mathbf{d}(\mathbf{k})|^2 \pm |\mathbf{d}(\mathbf{k}) \times \mathbf{d}^*(\mathbf{k})|}. \tag{7.50}$$

For both the ABM and BW solutions the cross product is zero $\mathbf{d}(\mathbf{k}) \times \mathbf{d}^*(\mathbf{k}) = 0$ (states obeying this are termed **unitary states**), and so the quasiparticle energies are just

$$E^n_{\mathbf{k}} = \pm\sqrt{(\epsilon_{\mathbf{k}} - \mu)^2 + |\mathbf{d}(\mathbf{k})|^2}. \tag{7.51}$$

It is clear from this that the magnitude of the $\mathbf{d}(\mathbf{k})$ vector, $|\mathbf{d}(\mathbf{k})|$, has the same role as the BCS energy gap $|\Delta|$.

Comparing with Fig. 7.4 one can see that the BW state has a gap, which has a constant magnitude over all the Fermi surface, rather like a BCS superconductor, while the ABM state has a gap which vanishes at the two points on the Fermi surface, $\mathbf{k} = (0, 0, \pm k_f)$. This difference leads to quite different physical properties in each phase, which led (Leggett 1975) to the identification of the

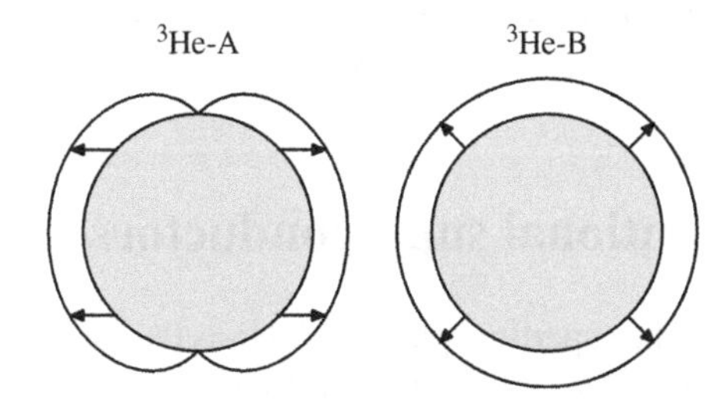

Fig. 7.4 The main two superfluid phases of ^{3}He. The A phase has a $\mathbf{d}$ vector in a constant direction and has two gap nodes at the "north" and "south" poles of the spherical Fermi surface. The B phase has a constant magnitude $\mathbf{d}$ vector everywhere on the Fermi surface, and hence a constant gap value.

ABM state with the ^{3}He A-phase in Fig. 7.1, while the BW state corresponds to the ^{3}He B-phase in Fig. 7.1. As Fig. 7.1 shows, the B-phase is generally the more stable one, except in the high pressure region near to T_c.

There are many interesting consequences of the complex gap structure in superfluid ^{3}He. The books by Vollhardt and Wölfle (1990) and Volovik (1992) give a comprehensive modern introduction. Perhaps the most interesting aspects of the theory, compared with the usual BCS theory of superconductivity, is the role of **topological defects**. In BCS theory, or indeed in superfluid ^{4}He, the gap parameter, or macroscopic wave function or Ginzburg–Landau (GL) order parameter correspond to a single complex function $\psi(\mathbf{r})$. In Fig. 2.9 we saw that a nontrivial topological property of such a single complex wave function is that it has quantized winding numbers around a ring. This led us to quantization of vorticity in superfluid ^{4}He and to flux quantization in superconductors. But in the case of superfluid ^{3}He we have seen that we need nine complex numbers to describe the gap function. Therefore the analog of the GL order parameter is a three by three matrix of complex numbers

$$[\psi_{\nu i}] = \begin{pmatrix} \psi_{xx} & \psi_{xy} & \psi_{xz} \\ \psi_{yx} & \psi_{yy} & \psi_{yz} \\ \psi_{zx} & \psi_{zy} & \psi_{zz} \end{pmatrix}.$$

The corresponding GL free energy density is also much more complicated than usual

$$\begin{aligned} f_s^{\text{bulk}} - f_n = {} & a\psi_{\nu i}^*\psi_{\nu i} + \frac{b_1}{2}\psi_{\nu i}^*\psi_{\nu i}^*\psi_{\mu j}\psi_{\mu j} + \frac{b_2}{2}\psi_{\nu i}^*\psi_{\nu i}\psi_{\mu j}^*\psi_{\mu j} \\ & + \frac{b_3}{2}\psi_{\nu i}^*\psi_{\mu i}^*\psi_{\nu j}\psi_{\mu j} + \frac{b_4}{2}\psi_{\nu i}^*\psi_{\mu i}^*\psi_{\mu j}^*\psi_{\nu j} \\ & + \frac{b_5}{2}\psi_{\nu i}^*\psi_{\mu i}\psi_{\mu j}\psi_{\nu j}^* \end{aligned} \tag{7.52}$$

in the bulk, where we have used Einstein summation convention to handle the sums over all the possible subscripts. Each of the five quartic terms corresponds to a different way of contracting the subscript indices or a different positioning of the complex congugates. Similarly there are three different gradient terms, corresponding to different contractions of the subscripts

$$f_s = f_s^{\text{bulk}} + \frac{\hbar^2}{2m_1}\nabla_i\psi_{\nu j}^*\nabla_i\psi_{\nu j} + \frac{\hbar^2}{2m_2}\nabla_i\psi_{\nu i}^*\nabla_j\psi_{\nu j} + \frac{\hbar^2}{2m_3}\nabla_i\psi_{\nu j}^*\nabla_j\psi_{\nu i}\psi_{\mu\nu}. \tag{7.53}$$

Such complicated expressions have some similarities to the corresponding theories of liquid crystals (Jones 2002). As in nematic liquid crystals, the nontrivial structure of the order parameter leads to a variety of novel topological defects, which have no analogs in ordinary superconductors. Superfluid ^{3}He has a rich structure of such topological defects. These include **disclinations** (similar to those occurring in liquid crystals), vortices with half-integer circulation, and an exotic defect occurring on surfaces, which is known as the Boojum![7]

[7] The Boojum will be well known to those who have read Lewis Carroll's nonsense poem *The Hunting of the Snark*. In the poem the Snark is a strange elusive creature which resists all attempts to hunt it. The final stanza reveals the reason: *"He had softly and suddenly vanished away – -, For the Snark was a Boojum, you see."*

7.5 Unconventional superconductors

The identification of the superfluid phases of ^{3}He as the spin-triplet p-wave generalization of BCS Cooper pairs, led naturally to the question of whether any other superconductors also have such exotic symmetry pairing states. It is clear that the original BCS model of electron–phonon coupling, Eq. 6.11, does not allow any solution other than the usual BCS one. But one can question whether other types of interactions might be important in some superconductors. In the case of ^{3}He it is clear that there are two main reasons for the p-wave pairing state. First the bare particle–particle interaction is strongly repulsive at short range, which is due to the strong energy cost for having two hard-sphere like helium atoms too close together. Second the quasiparticle–quasiparticle interactions near the Fermi surface are spin dependent, because of the strong Stoner enhanced ferromagnetic spin susceptibility. Basically the quasiparticles are close to a ferromagnetic instability, which leads to effective attractive interactions for spin triplet Cooper pairs. By choosing an $l = 1$ spatial wave function the Cooper pairs have no probability amplitude for finding both particles at the same point in space because the Cooper pair wave function $\phi(\mathbf{r}_1, \mathbf{r}_2)$ is zero when $\mathbf{r}_1 = \mathbf{r}_2$. On the other hand, the pair wave function in k-space has plenty of amplitude on the Fermi surface, and so it can take advantage of the attractive interaction.

There is no reason at all why a similar pairing mechanism could not also occur in a superconductor. The conduction electrons in metals have strong Coulomb repulsions, which are somewhat similar to the repulsive hard-sphere potentials in helium. To minimize this strong repulsive energy it is favorable to form Cooper pairs in a pair wave function which is zero when $\mathbf{r}_1 = \mathbf{r}_2$. This can be accomplished by adopting pairing states with $l \neq 0$. On the other hand, the essential feature for pairing to occur is that the effective quasiparticle–quasiparticle interactions are attractive near to the Fermi surface. It is quite reasonable to expect that this occurs in some systems, even when the electron–phonon coupling is not strong. Again one could look for systems where the normal state Landau fermi liquid is near to an instability (e.g. ferromagnetism or antiferromagnetism).

There are now a number of good candidates for examples of **unconventional superconductivity**. To be precise, this is defined as superconductivity where the ground state has a different **symmetry** from the usual BCS ground state. Since we are considering electrons in metals we should in general consider the symmetry of the first Brillouin zone or of the Fermi surface as our reference. In a crystal, the Fermi surface will be unchanged under all of the **point group** symmetry operations of the crystal (e.g. in a cubic crystal point group symmetries include rotation by 90° about crystal axes, and reflection in mirror planes). In the original BCS theory, as described in Chapter 6, there is just a single parameter Δ, but as we have seen in this chapter this can be generalized to possibly $\mathbf{k}$ dependent function, $\Delta_{\mathbf{k}}$. We shall define the superconductivity as **conventional** if

$$\Delta_{\hat{R}\mathbf{k}} = \Delta_{\mathbf{k}}, \tag{7.54}$$

where $\hat{R}$ is any symmetry operation in the point group. For example, in a cubic crystal the symmetry operation of rotation by 90° about the z-axis corresponds to

$$\hat{R}\begin{pmatrix} k_x \\ k_y \\ k_z \end{pmatrix} = \begin{pmatrix} -k_y \\ k_x \\ k_z \end{pmatrix}. \tag{7.55}$$

Conversely superconductivity will be defined to be unconventional if $\Delta_{\hat{R}\mathbf{k}} \neq \Delta_{\mathbf{k}}$ for at least one symmetry operation $\hat{R}$.

Interestingly, the Fermi surfaces of all non-magnetic metals have inversion symmetry, even if the real-space crystal does not, therefore

$$\epsilon_{\mathbf{k}} = \epsilon_{-\mathbf{k}}. \tag{7.56}$$

Conventional superconductivity must also respect this symmetry, implying that

$$\Delta_{\mathbf{k}} = \Delta_{-\mathbf{k}}. \tag{7.57}$$

In fact, we have already seen that this property arises in Eq. 7.39 for spin-singlet pairing, and is a consequence of the fermion anti-commutation laws. On the other hand, spin triplet superconductivity always has $\mathbf{d}(\mathbf{k}) = -\mathbf{d}(\mathbf{k})$, from Eq. 7.40, and so triplet pairing is always by definition unconventional.

One further symmetry in non-magnetic solids is the freedom to rotate the quantization axis for spin (at least ignoring spin–orbit coupling). The BCS theory for spin singlet pairing is invariant under such rotations of axis (although this was not explicitly obvious in our formulation in Chapter 6). Spin singlet pairing is always invariant under rotations of spin. But, on the other hand, spin triplet pairing necessarily breaks spin–rotation symmetry, since this corresponds to rotating the vector $\mathbf{d}(\mathbf{k})$.

The full theory of unconventional superconductivity proceeds by using methods of group theory to analyze the possible point group symmetries of a given system. One first looks up the irreducible representations and character table for a given crystal structure, and then classifies the distinct pairing states that can be identified (Annett 1990; Mineev and Samokhin 1999). The spherical Harmonic expansion of the gap function for superfluid ^{3}He is replaced by an expansion in terms of a set of functions classified in term of the **irreducible representations** of the point group symmetry. For single pairing superconductors this becomes[8]

$$\Delta_{\mathbf{k}} = \sum_{\Gamma m} \eta_{\Gamma m} f_{\Gamma m}(\mathbf{k}), \tag{7.58}$$

where Γ is the irreducible representation and the functions $f_{\Gamma m}(\mathbf{k})$ are a complete set of basis functions with the given symmetry representation Γ. If the irreducible representation has dimension d, then it will be necessary to include at least a set of functions for $m = 1, \ldots, d$. For a one-dimensional representation Γ only one basis function is necessary, and so the gap function can be described by a single complex GL order parameter, just as in conventional superconductors. On the other hand, two or three dimensional representations Γ are possible in cubic, tetragonal, or hexagonal crystals, and for these the order parameter will have correspondingly two or three expansion coefficients $\eta_{\Gamma m}$. For triplet superconductors the expansion also includes the three components of the spin, or the vector components of $\mathbf{d}(\mathbf{k})$,

$$d_\nu(\mathbf{k}) = \sum_{\Gamma m} \eta_{\Gamma m \nu} f_{\Gamma \nu}(\mathbf{k}) \tag{7.59}$$

and therefore there are at least $3 \times d$ expansion coefficients for a d dimensional representation Γ.

The superconductor UPt_3 is probably the clearest example of a superconductor with multiple components of the gap. Figure 7.5 shows the specific heat of UPt_3 near to its transition temperature. Clearly there are actually two phase transitions, T_{c+}, and T_{c-}. It is believed that these correspond to two different gap components $\eta_{\Gamma m}$ becoming nonzero. At the higher T_c first one of them

[8] In practice it is usual to simply refer to the pairing states as "*s*-wave", "*p*-wave", "*d*-wave" or "*f*-wave", depending on the minimum number of spherical harmonic functions, which are necessary to describe the variation of $\Delta_{\mathbf{k}}$ or $\mathbf{d}(\mathbf{k})$ over the Fermi surface. This naming convention is strictly incorrect, since the true expansion functions are the basis functions $f_{\Gamma m}(\mathbf{k})$ not the spherical harmonics, but in fact no ambiguity arises. Note that *s*-wave and *d*-wave pairing states must be spin singlet, while *p*-wave and *f*-wave pairing states must be spin triplet.

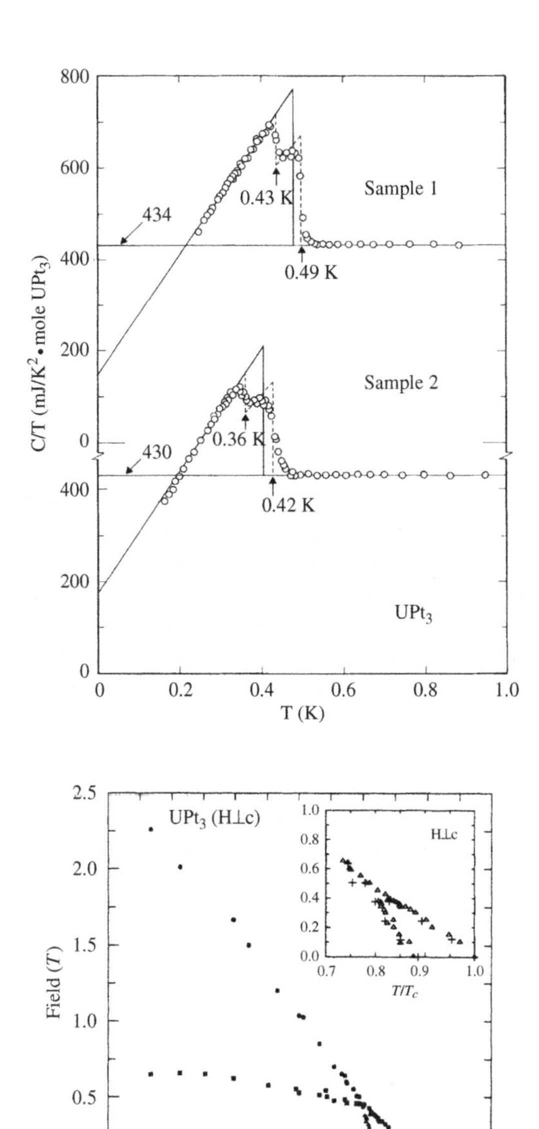

Fig. 7.5 The heat capacity of UPt_3 near to T_c. The two jumps indicate that there are two phase transitions, and hence the GL order parameter has multiple components. Reprinted figure with permission from Fisher *et al.* (1989). Copyright (1989) by the American Physical Society.

Fig. 7.6 The H–T phase diagram of UPt_3. There are three distinct superconducting phases, corresponding to different order parameter symmetries. These most probably indicate an unconventional superconducting phase with a symmetry degenerate order parameter. Reprinted figure with permission from Adenwalla *et al.* (1990). Copyright (1990) by the American Physical Society.

becomes nonzero, and then at a slightly lower T_c the second follows. The fact that the two jumps are so close together is generally believed to be because the two gap components belong to the same irreducible representation Γ, and so are essentially degenerate. The double transition is seen because the degeneracy is slightly perturbed, possibly by a charge density or spin density wave, which occurs in the normal state above T_c. It is possible to map out the two phase transitions as a function of both temperature and magnetic field (UPt_3 is a type II superconductor), and this is shown in Fig. 7.6. In fact, as the figure shows, there are actually three distinct superconducting phases, separated by two transition lines. Unfortunately it has proved difficult to conclusively determine experimentally, which irreducible representation Γ is responsible for this remarkable phase diagram. Models, which have been proposed include a spin singlet (d-wave) state, a spin triplet (p-wave or f-wave) state, and even a mixture of two different irreducible representations Γ and Γ'. See Sauls (1994) and Joynt and Taillefer (2002) for a more detailed review of experiments and theoretical models of UPt_3. A number of other exotic superconductors also occur in the class of **heavy fermion metals**. Like UPt_3 these are usually compounds containing uranium, or other members of the actinide or lanthanide series in the periodic table (e.g. Ce). These materials have a number of unusual properties in both the normal state above T_c and in their superconducting states. These unusual effects (including the extremely heavy electron effective masses $m^* \gg m_e$) arise because of the very strong electron–electron interactions associated with the partially full atomic f-shell. Superconductivity appears to be often associated with the proximity to magnetic (anti-ferromagnetic or ferromagnetic) phase transitions. There is strong evidence for unconventional superconductivity in several members of the heavy fermion materials, such as UBe_{13}. Another especially notable example is UGe_2, which is simultaneously ferromagnetic and superconducting. Coexistence of ferromagnetism and superconductivity is surprising because the magnetic exchange splitting can easily destroy spin singlet pairing. For this reason it is likely that UGe_2 is a spin triplet superconductor, but at present this has not been been established conclusively.

The case of the high temperature superconductors, such as $La_{2-x}Sr_xCuO_4$ and $YBa_2Cu_3O_7$ has also been controversial, but now the experimental evidence strongly points toward an unconventional pairing state. These compounds are very complex in structure, but the common feature of all of them is that they contain two-dimensional layers of copper and oxygen. These layers give rise to the Fermi surface, which is also highly two-dimensional. Therefore in examining the gap equation we can safely restrict ourselves to a two-dimensional square (or nearly square) Brillouin zone. Early experiments on nuclear magnetic resonance (NMR) indicated that the pairing state is probably spin singlet. In which case it is natural to consider the possible pairing states described by

$$\Delta_{\mathbf{k}} = \Delta \qquad (s), \tag{7.60}$$

$$\Delta_{\mathbf{k}} = \Delta(\cos(k_x a) + \cos(k_y a))/2 \qquad (s^-), \tag{7.61}$$

$$\Delta_{\mathbf{k}} = \Delta(\cos(k_x a) - \cos(k_y a))/2 \qquad (d_{x^2-y^2}), \tag{7.62}$$

$$\Delta_{\mathbf{k}} = \Delta \sin(k_x a)\sin(k_y a), \qquad (d_{xy}), \tag{7.63}$$

where a is the crystal lattice constant of the square copper-oxide plane. In the first Brillouin zone, $(-\pi/a \leq k_x \leq \pi/a, -\pi/a \leq k_y \leq \pi/a)$ these all have a maximum possible value of Δ, but they vary considerably as functions of $\mathbf{k}$, as shown in Fig. 7.7.

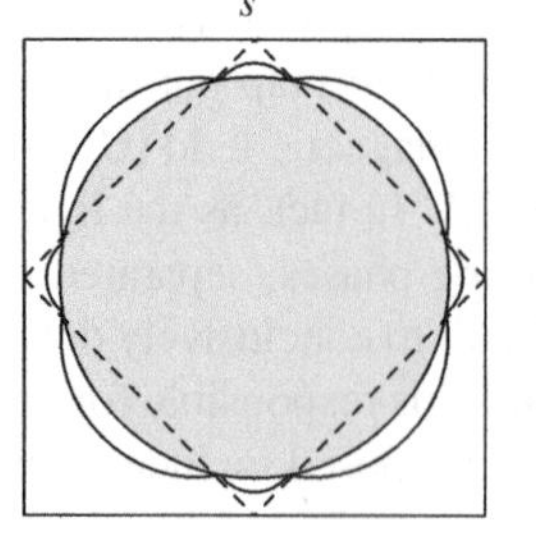

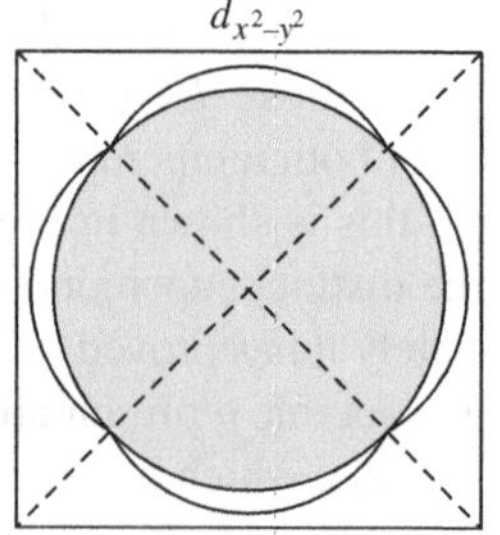

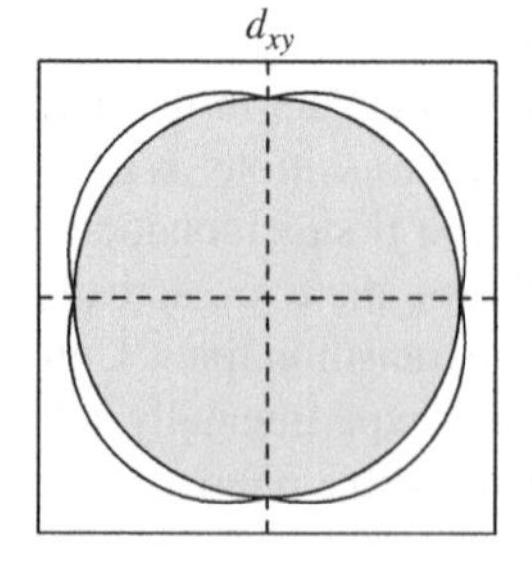

Fig. 7.7 Three possible superconducting phases of a tetragonal superconductor, such as the high T_c cuprates. The first, extended-s, has the full lattice symmetry. As shown it can have eight nodes of the gap on the Fermi surface. The remaining two functions d-$x^2$$-$$y^2$ and d-xy both have four gap nodes on the Fermi surface. Much experimental evidence now points to the d-$x^2$$-$$y^2$ state as the one which is present in the high T_c cuprate superconductors.

The first of the possible gap functions, Eq. 7.60, is simply a constant, and so this is exactly as one would expect in the standard BCS theory as derived in Chapter 6. This obviously is constant under all possible symmetry operations of the square Brilloiuin zone (parity, rotation by 90°, or mirror reflections), therefore it is conventional or s-wave pairing. But now contrast this with the second possible gap function, Eq. 7.61, to the left-hand side in Fig. 7.7. This function is also constant under all possible symmetries, therefore it is also an example of conventional (or s-wave) pairing. But as shown in the figure, the magnitude of this gap function becomes zero at eight points on the Fermi surface. These are the points, where the Fermi surface intersects the dotted lines shown in the figure. It is easy to confirm that the function $\cos k_x + \cos k_y$ is zero on these lines (where $k_y = \pm\pi/a \pm k_x$). Therefore this pairing state is one for which the quasiparticle energy gap $2|\Delta_{\mathbf{k}}|$ becomes zero at several points on the Fermi surface, but which is nevertheless a conventional s-wave symmetry solution. Usually this is termed "extended-s".

In contrast, the two gap functions given by Eqs 7.62 and 7.63, shown to the center and right in Fig. 7.7, both have unconventional symmetry. They both change sign under rotation of the square by 90° ($k_x \to k_y$, $k_y \to -k_x$). The $d_{x^2-y^2}$ gap is odd under mirror reflections in the diagonals of the square, while the d_{xy} gap is odd under mirror reflections about the k_x and k_y axes ($k_x \to -k_x$ and $k_y \to -k_y$). These are normally termed "d-wave" gap functions, as indicated, because the gap functions have the same symmetry as atomic d spherical harmonic functions, $d_{x^2-y^2}$ and d_{xy}. In both cases the quasiparticle gap $2|\Delta_{\mathbf{k}}|$ vanishes at four points on the Fermi surface, as can be seen in Fig. 7.7.

How can experiments distinguish between these various gap functions? First one can try to determine whether the quasiparticle gap $2|\Delta_{\mathbf{k}}|$ is finite everywhere on the Fermi surface, or has zeros. If it is finite everywhere the density of states in the superconducting state will be qualitatively similar to the usual BCS density of states, shown in Fig. 6.1, with a gap region of energies where no quasiparticle excitations are present. On the other hand, if $2|\Delta_{\mathbf{k}}|$ goes to zero at one or more points on the Fermi surface, then there will be a nonzero density of states for quasiparticle excitations at any energy. In a two-dimensional d-wave superconductor the density of states will be linear near to the Fermi energy, as shown in Fig. 7.8. If this occurs, then it will be always possible to excite some number of quasiparticles, even at very low temperatures, and this will have observable experimental consequences. For example the specific heat capacity will have a contribution from these "nodal" quasiparticles, typically giving a term $C_V \sim T^2$ at low temperatures. By carefully subtracting the phonon contribution ($C_V \sim T^3$ in the Debye model) one can determine whether or not a T^2 term is present, and hence detect the presence of the gap zeros (or nodes). Another experiment which more directly detects the gap nodes is a measurement of the London penetration depth as a function of temperature, $\lambda(T)$. Converting the London penetration depth into the superfluid density n_s (as defined in Chapter 3), one can see that this experiment measures $n_s(T)$, the superfluid density as a function of temperature. Writing

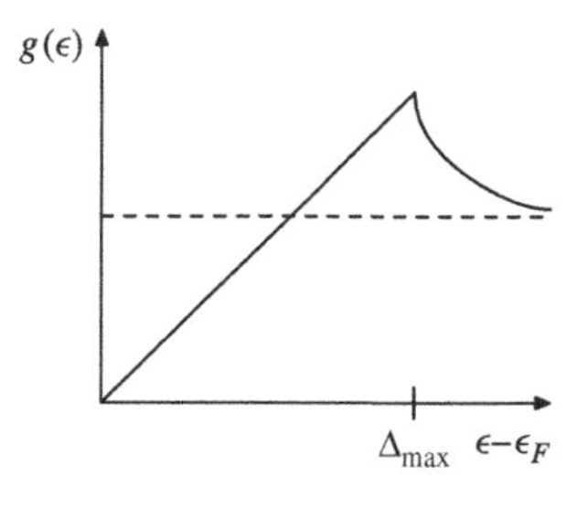

Fig. 7.8 The density of states in a d-wave superconductor. There is no true gap, since there are states present at all energies. The linear increase in density of states near to $\epsilon = 0$ leads to a linear change in the London penetration depth, λ, with temperature near to $T = 0$.

$$n_s(T) = n_s(0) - n_n(T), \qquad (7.64)$$

where $n_s(0)$ is the superfluid density at absolute zero, the experiment effectively measures the normal fluid density $n_n(T)$ as a function of temperature. In a BCS superconductor, with a finite gap $2|\Delta|$ everywhere on the Fermi surface,

$n_n(T) \sim e^{-2|\Delta|/k_B T}$, and the probability of exciting quasiparticles is very low for $k_B T \ll 2|\Delta|$. But for a gap with nodes, it is always possible to excite quasiparticles even at very low temperatures. The result is that

$$n_n(T) \sim T \tag{7.65}$$

for the three states shown in Fig. 7.7. By measuring the penetration depth at low temperatures one should therefore expect to observe a linear dependence on temperature

$$\lambda(T) = \lambda(0) + cT, \tag{7.66}$$

where c is a constant which can be calculated for each given $\Delta_{\mathbf{k}}$ function. Figure 7.9 shows the experimental results for the superconductor $YBa_2Cu_3O_7$ at low temperatures. Clearly there is a strong linear behavior, which is strong evidence that the gap does indeed have zeros. By studying the effect of impurities on this linear behavior, one can further argue that the gap function is more likely to be one of the d-wave solutions, Eq. 7.62 or 7.63 than the extended s-wave state of Eq. 7.61.

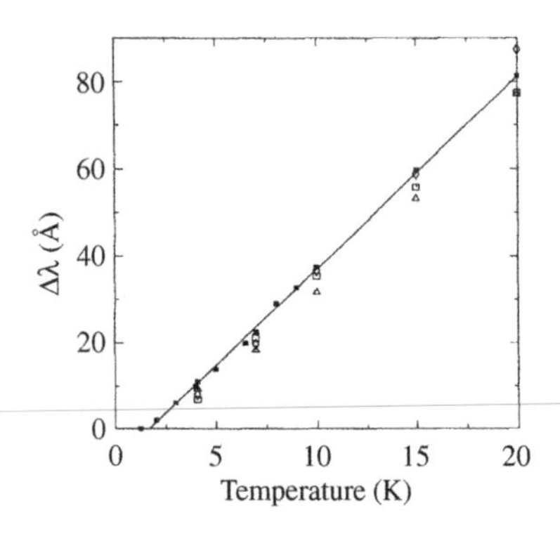

Fig. 7.9 The temperature dependence of the penetration depth of $YBa_2Cu_3O_7$ near to $T=0$. The linear dependence indicates point nodes of the gap function, as present in all of the possible phases in Fig. 7.7. The dependence of this linear T effect on disorder implies that the state is d wave not extended s. Reprinted figure with permission from Hardy *et al.* (1993). Copyright (1993) by the American Physical Society.

How can one distinguish between the two possible d-wave states, Eqs 7.62 and 7.63? Basically both states are rather similar. They only differ in the location of the gap nodes relative to the Brillouin zone axes. In the case of the high temperature superconductors two independent types of experiments have now shown very convincingly that the gap nodes are on the Brillouin zone diagonal, as in the $d_{x^2-y^2}$ state shown in the central panel of Fig. 7.7. The first of these experiments is Angle Resolved Photoemission (ARPES). An X-ray photon of known energy and wave vector excites an electron out of the surface of the superconductor. By measuring the energy E and parallel components of the wave vector (k_x, k_y) of the emitted electron one can deduce its initial energy and crystal momentum. By carefully comparing the spectra as a function of temperature, one can essentially directly map out the energy gap $2|\Delta_{\mathbf{k}}|$ at any point on the Fermi surface. The results of these experiments show that the gap is at its maximum value near to the points $(k_x, k_y) = (\pi/a, 0)$ and $(k_x, k_y) = (0, \pi/a)$, and zero (or as near to zero as can be measured) near the Brillouin zone diagonals $(k_x = \pm k_y)$. Only the $d_{x^2-y^2}$ state has this pattern of gap nodes.

The second type of experiment which can distinguish between the different d-wave states operates on completely different principles, namely making use of the Josephson effect. We saw in Chapter 5 that the Josephson effect arises from a coherent tunnelling of Cooper pairs between two superconductors. The current in the junction depends on the phase difference between the BCS coherent states on either side of the junction $I = I_c \sin(\theta_1 - \theta_2)$. For d-wave superconductors it turns out that the relevant phase θ often depends on the gap function $\Delta_{\mathbf{k}}$ at one particular point on the Fermi surface, usually the point in the direction of a vector normal to the Josephson junction. Therefore by making junctions with different orientations relative to the crystal axes one can be sensitive to the values of $\Delta_{\mathbf{k}}$ at different points on the Fermi surface. A single junction experiment is possible, but so far, these have not proved conclusive. On the other hand experiments with two or more junctions, arranged in SQUID ring type geometries (as described in Chapter 5), have shown that the gap function $\Delta_{\mathbf{k}}$ **must change sign** exactly as given by Eq. 7.62. First Wollman *et al.* (1993) (see also Van Harlingen 1995) made a SQUID consisting of two different crystal faces of a $YBa_2Cu_3O_7$ crystal, connected into a superconducting ring with niobium metal. They observed that the usual SQUID interference pattern (Fig. 5.10) became phase shifted by exactly one half of the interference fringe width ($\Phi_0/2$). This indicated a sign change in $\Delta_{\mathbf{k}}$ between the k_x and k_y orientations in the Brillouin zone, consistent with Eq. 7.62. A secondly, somewhat different Josephson coupling experiment was performed by Tsuei and Kirtley. They constructed supercondcucting rings, as shown in Fig. 7.10. Each ring is fabricated by growing a thin superconducting layer on an insulating substrate, and then using lithographic techniques to cut away material to leave an isolated superconducting ring. The substrate chosen was a tricrystal, meaning that it has regions with three different crystal orientations, separated by crystal grain boundaries. Knowing that the superconductor grows in the same directions as the substrate, it turns out that the superconducting rings will contain either, three, two or no grain boundaries, as shown in Fig. 7.10. The rings containing two or no grain boundaries were found to have the usual flux quantum $\Phi = n\Phi_0$, as expected from the discussion in Chapter 4. However, the special ring in the center has three grain boundaries, and was observed to have a quite different flux quantization

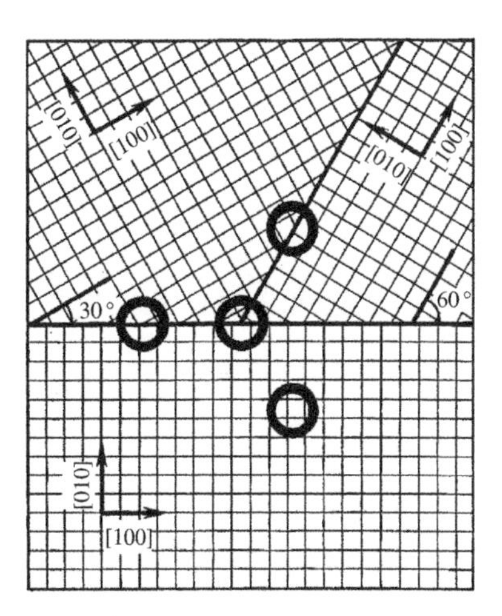

Fig. 7.10 A superconducting SQUID ring made with three weak links. The magnetic flux through the ring has quanta $(n+1/2)\Phi_0$, not the usual $n\Phi_0$. This indicates a sign change in the order parameter associated with the ring geometry. For this ring geometry the sign change indicates d-x^2-y^2 pairing. Reprinted figure with permission from Tsuei and Kirtley (2000). Copyright (2000) by the American Physical Society.

$$\Phi = \left(n + \tfrac{1}{2}\right)\Phi_0 \tag{7.67}$$

(Tsuei and Kirtley 2000). The explanation for this surprising behavior is a sign change arising from sign changes between $\Delta_{\mathbf{k}}$ for a $d_{x^2-y^2}$ pairing state in each of the three sections of the central superconducting ring. This half integer flux quantization is especially interesting because it is reflecting a **topological property**, namely winding number (recall Fig. 2.9). In superfluid ^{4}He, or in the usual GL theory of superconductivity flux (or vorticity) quantization arises because of the requirement that the macroscopic wave function be single valued. The fact that half-integer quantization occurs in these tricrystal rings shows that the macroscopic wave function itself has internal structure, namely the sign changes associated with the $d_{x^2-y^2}$ gap function $\Delta_{\mathbf{k}}$. The topological nature of the effect means that it is independent of disorder or other minor perturbations. Indeed the half-integer flux quantum has now been observed in a variety of different superconducting materials, and in several different ring geometries (Tsuei and Kirtley 2000).

To what extent does this $d_{x^2-y^2}$ gap function help to determine the **pairing mechanism** of high temperature superconductivity? Unfortunately we cannot immediately say that it uniquely tells us the pairing mechanism, since a number of different mechanisms are still are consistent with $d_{x^2-y^2}$ Cooper pairs. However this unusual symmetry does at least suggest that the usual BCS electron–phonon pairing mechanism does not apply, since that normally leads to s-wave pairing. Further evidence for this is the apparent lack of an isotope effect in optimally doped $YBa_2Cu_3O_7$. The class of pairing mechanisms that most naturally lead to $d_{x^2-y^2}$ pairing are those based on the principle of **strong electron–electron repulsion**. The cuprate superconductors are all close to being anti-ferromagnetic Mott insulators because this interaction. For example, the compound La_2CuO_4 is an anti-ferromagnetic insulator, which becomes metallic and superconducting when "doped" with barium or strontium, such as $La_{2-x}Sr_xCuO_4$ with $x \sim 5-15\%$. It is widely believed that $d_{x^2-y^2}$ pairing is possible in the metallic state close to the transition to antiferromagnetism. In some sense the mechanism is similar to the spin-fluctuation exchange model of superfluid ^{3}He, but here the dominant spin fluctuations are anti-ferromagnetic in character. Anti-ferromagnetic spin fluctuations naturally favor $d_{x^2-y^2}$ type pairing. Unfortunately, at the time of writing, this issue is still not settled conclusively, and so the pairing mechanism of high temperature superconductivity remains a very much open question.

Finally, there are many other exciting classes of novel superconductors where the pairing mechanism or pairing state may be quite different from BCS. New superconductors are being discovered every year, and so the field is continuing to grow at a rapid pace. Over the past few years just some of the many interesting and novel superconducting compounds that have been discovered include: doped fulerences (such as K_3C_{60}) and carbon nanotubes, organic superconductors, magnesium boride (MgB_2), and borocarbides (such as YNi_2B_2C), ferromagnetic superconductors (such as $ZrZn_2$ and UGe_2), and strotium ruthenate (Sr_2RuO_4 a possible spin-triplet p-wave superconductor) among many others. Understanding all of these will keep experimentalists and theorists busy for many years, even if room temperature superconductvity remains elusive!

Further reading

There are several good books and review articles describing the experimental properties and theories of superfluid ^{3}He. Tilley and Tilley (1990) is a good introduction, while the books by Volovik (1992) and by Vollhardt and Wölfle (1990) go into greater depth. The classic review papers by Wheatley (1975) and Leggett (1975) are also excellent. The Nobel lectures by Lee (1997) and Osheroff (1997) give an overview and also some interesting details about the history of the original discovery of superfluidity.

The ideas of topological defects arising on condensed matter systems, such as liquid crystals are introduced in books such as Jones (2002) and Chakin and Lubesky (1995). Detail about the exotic vortices occurring in superfluid ^{3}He are described in Salomaa and Volovik (1987). For an amusing anecdote about the origin of the term "boojum" and how it can to be accepted as a technical term in superfluid ^{3}He, see the book of essays on science and science communication by Mermin (1990).

Unconventional superconductivity is described in several books and articles. Recent books by Ketterson and Song (1999) and Mineev and Samhokin (1999) give a detailed account. For a short introduction to the underlying symmetry principles and group theoretic methods see Annett (1990).

A description of the superconducting phases of of UPt_3 is given in the reviews by Joynt and Taillefer (2002) and Sauls (1994). Descriptions of the possible superconducting states of the high temperature superconductors and analysis of experiments is given in Annett Goldenfeld, and Renn (1990) and Annett, Goldenfeld, and Leggett (1996). Reviews of Josephson based experiments to determine the order parameter are given by Van Harlingen (1995) and Tsuei and Kirtley (2000). A (very incomplete) anaylsis of possible unconventional superconductivity in a wide variety of compounds is given in Annett (1999). Mackenzie and Maeno (2003) discuss the possible spin-triplet pairing in Sr_2RuO_4.

Solutions and hints to selected exercises

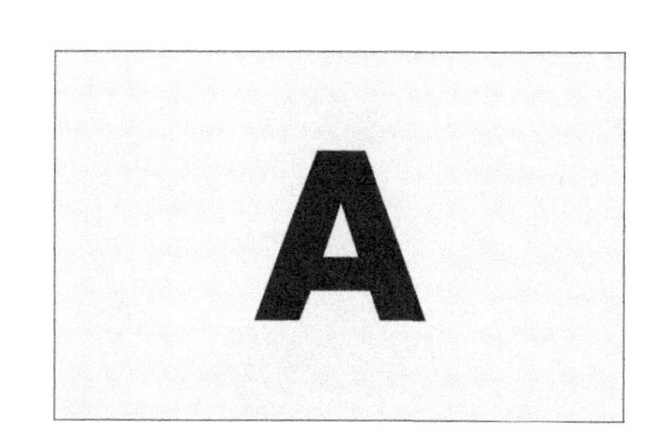

A.1 Chapter 1

(1.1) This is the same as the classic problem of how many ways to choose r identical objects out of a total of n,

$$ {}_nC_r = \binom{n}{r} = \frac{n!}{r!(n-r)!}. $$

Here, considering Fig. 1.1, we see that we have a total of M_s boxes and we must choose N_s out of them to be occupied with a fermion particle (one per box). Thus there are ${}_{M_s}C_{N_s}$ possible microstates.

(1.3) The sum over all possible states of the system are specified by the summing over the set of occupation numbers $\{n_{\mathbf{k}}\}$, of each plane wave quantum state, $\mathbf{k}$. Therefore the grand partition function is given by

$$ \begin{aligned} \mathcal{Z} &= \sum_{\{n_{\mathbf{k}}\}} \exp\left(\beta \sum_{\mathbf{k}} (\epsilon_{\mathbf{k}} - \mu) n_{\mathbf{k}}\right), \\ &= \prod_{\mathbf{k}} \left[\sum_{n=0,1,\ldots} \exp\left(\beta(\epsilon_{\mathbf{k}} - \mu)n\right) \right], \\ &= \prod_{\mathbf{k}} \mathcal{Z}_{\mathbf{k}}, \end{aligned} $$

where

$$ \mathcal{Z}_{\mathbf{k}} = \frac{1}{1 - e^{-\beta(\epsilon_{\mathbf{k}} - \mu)}}, $$

as defined by the sum of the geometric progression in the square brackets.

The average occupation of state $\mathbf{k}$ is

$$ \begin{aligned} \langle n_{\mathbf{k}} \rangle &= \frac{1}{\mathcal{Z}} \sum_{\{n_{\mathbf{k}}\}} n_{\mathbf{k}} \exp\left(-\beta \sum_{\mathbf{k}} (\epsilon_{\mathbf{k}} - \mu) n_{\mathbf{k}}\right), \\ &= -\frac{1}{\beta \mathcal{Z}} \frac{\partial}{\partial \epsilon_{\mathbf{k}}} \sum_{\{n_{\mathbf{k}}\}} \exp\left(-\beta \sum_{\mathbf{k}} (\epsilon_{\mathbf{k}} - \mu) n_{\mathbf{k}}\right), \\ &= -k_{\mathrm{B}} T \frac{\partial \ln \mathcal{Z}}{\partial \epsilon_{\mathbf{k}}}. \end{aligned} $$

It is easy to see that only $\mathcal{Z}_{\mathbf{k}}$ contributes to the derivative, and that we obtain the Bose–Einstein distribution $f_{\mathrm{BE}}(\epsilon)$ by differentiating the expression found above for $\mathcal{Z}_{\mathbf{k}}$.

(1.5) The integral substitution $y = ze^{-x}$ yields the result

$$\int_0^\infty \frac{ze^{-x}}{1 - ze^{-x}}\, dx = -\ln(1 - z),$$

for $0 < z < 1$. Therefore using $z = e^{\beta\mu}$ the chemical potential is given explicitly by

$$\mu = k_{\rm B}T \ln\left(1 - e^{-2\pi\hbar^2 n/(mk_{\rm B}T)}\right).$$

In the limit where $n/k_{\rm B}T$ is small this is

$$\mu \approx k_{\rm B}T \ln\left(\frac{2\pi\hbar^2 n}{mk_{\rm B}T}\right),$$

which is negative. Similarly, expanding in the limit where $n/k_{\rm B}T$ is large we obtain

$$\mu \approx -k_{\rm B}Te^{-2\pi\hbar^2 n/(mk_{\rm B}T)},$$

which is also negative. Therefore μ never becomes equal to zero, and this is why there is no Bose–Einstein condensation (BEC) in two-dimensional systems.

(1.6) Using the specific heat below T_c given in Eq. 1.52 and integrating we obtain the total entropy per particle at temperature T

$$s(T) = \frac{5}{2}\frac{g_{5/2}(1)}{g_{3/2}(1)}\frac{T^{3/2}}{T_c^{3/2}}k_{\rm B}.$$

But the total number of particles in the normal fluid, N_n, is also proportional to $T^{3/2}/T_c^{3/2}$, and so to total entropy of the gas is

$$S(T) = N_n s(T_c).$$

We can interpret this as a statistical mixture of two "fluids" the condensate and the normal fluid since

$$S(T) = N_0 s(0) + N_n s(T_c),$$

where the entropy at absolute zero is (by the third law of thermodynamics) $s(0) = 0$.

(1.7) This is just a matter of working out $\langle \hat{S}_{2z} \rangle$ for each of the states given in Eqs 1.55 and 1.56, for example, the $S = 2$, $M_s = +1$ state

$$|S = 2, M_S = 1\rangle = \tfrac{1}{2}\left(\sqrt{3}\left|\tfrac{1}{2}, \tfrac{1}{2}\right\rangle + \left|\tfrac{3}{2}, -\tfrac{1}{2}\right\rangle\right),$$

has the expectation value

$$\langle S = 2, M_S = 1|\hat{S}_{2z}|S = 2, M_S = 1\rangle = \tfrac{3}{4}\left(+\tfrac{1}{2}\right) + \tfrac{1}{4}\left(-\tfrac{1}{2}\right) = +\tfrac{1}{4}.$$

On the other hand, the $S = 1$ $M_S = +1$ state

$$|S = 1, M_S = 1\rangle = \tfrac{1}{2}\left(\left|\tfrac{1}{2}, \tfrac{1}{2}\right\rangle + \sqrt{3}\left|\tfrac{3}{2}, -\tfrac{1}{2}\right\rangle\right),$$

has an expectation value of the opposite sign,

$$\langle S = 1, M_S = 1|\hat{S}_{2z}|S = 1, M_S = 1\rangle = \tfrac{1}{4}\left(+\tfrac{1}{2}\right) + \tfrac{3}{4}\left(-\tfrac{1}{2}\right) = -\tfrac{1}{4}.$$

This explains the sign reversal between the field dependencies of the $S = 2$ and $S = 1$ energy levels, as shown in Fig. 1.8.

(1.8)

(a) This is a straightforward application of separation of variables. The three-dimensional wave function $\psi(x, y, z)$ can be represented as a product, $X(x)Y(y)Z(z)$ and each of these functions obeys a one-dimensional simple harmonic oscillator equation. The total energy

$$\epsilon_{n_x n_y n_z} = \hbar\omega\left(n_x + n_y + n_z + \tfrac{3}{2}\right).$$

is just a sum of three contributions, $\hbar\omega(n_x + 1/2)$ *etc.*

(b) Requires counting the total number of states with energy less than ϵ. It is similar to the counting argument used in obtaining Eq. 1.4, but here the state energies depend on n_x, n_y, and n_z, instead of k_x, k_y, and k_z. Drawing a three-dimensional set of axes, in the n_x, n_y, and n_z directions, Fig. A.1, it is clear that there is one quantum state per unit volume in this space. But all of the states with energy less than ϵ must lie in the region

$$n_x + n_y + n_z < \frac{\epsilon}{\hbar\omega},$$

which is a tetrahedron of side $\epsilon/\hbar\omega$ as shown in Fig. A.1. The volume of the tetrahedron is

$$N(\epsilon) = \frac{1}{6}\left(\frac{\epsilon}{\hbar\omega}\right)^3,$$

giving the total number of states. Differentiating, we find that the density of states at energy ϵ is

$$g(\epsilon) = \frac{\epsilon^2}{2(\hbar\omega)^3}.$$

(c) The chemical potential equals zero when

$$N = \int_0^\infty g(\epsilon)\frac{1}{e^{\beta\epsilon} - 1}\,d\epsilon.$$

Using the density of states from (b) this is when

$$N = (k_B T_c)^3 \frac{1}{2(\hbar\omega)^3}\int_0^\infty \frac{x^2}{e^x - 1}\,dx.$$

Combining the numerical constants which are of order unity, we thus obtain $k_B T_c \sim N^{1/3}\hbar\omega$.

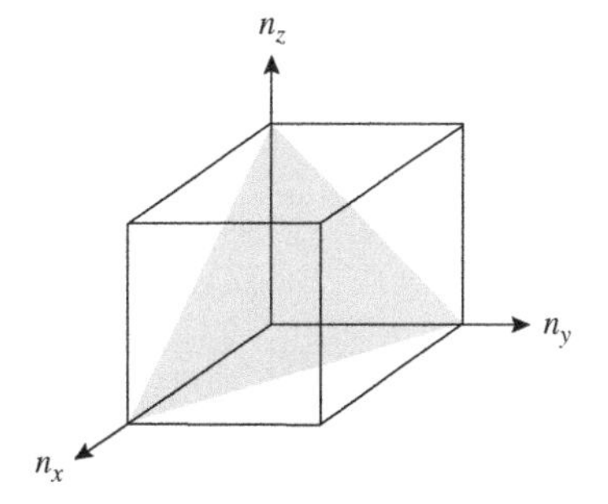

Fig. A.1 Exercise 1.8: counting the number of states with energy less than ϵ for a three-dimensional harmonic atom trap. In the three-dimensional space $\{n_x, n_y, n_z\}$ there is one quantum state per unit volume. In this space all the states with energy below ϵ lie in the tetrahedron shown. The volume of the tetrahedron is $1/6$ the total volume of the cube, where the cube has sides of length, $\epsilon/(\hbar\omega)$.

A.2 Chapter 2

(2.3) If we make the association $\mathbf{B} \leftrightarrow \mathbf{v}_s$, then clearly we must identify the circulation κ with the $\mu_0 I$ in Ampère's law. The two energy densities will be equal if we identify ρ_s with $1/\mu_0$.

(2.4) Using the results from 2.3 the force per unit length between two vortices is

$$\begin{aligned} F &= \frac{1}{\rho_s}\frac{(\rho_s\kappa_1)(\rho_s\kappa_2)}{2\pi R}, \\ &= \frac{\rho_s\kappa_1\kappa_2}{2\pi R}. \end{aligned}$$

You can also use dimensional analysis to check that this result is correct. The interaction energy between vortices is found by integrating the work done FdR

to move the wires apart from separation R_0 to R. The sign is such that two parallel wires or vortices repel, and so the total energy decreases when R is increased.

(2.6)

(a) Using spherical polar coordinates we find

$$\rho_1(r) = \frac{1}{(2\pi)^3} \int n_{\mathbf{k}} e^{-ikr\cos\theta} k^2 \sin\theta \, d\theta d\phi dk,$$

where the z-axis has been chosen to be in the direction of the vector $\mathbf{r}$. (Obviously the result does not, in the end, depend on the direction of $\mathbf{r}$.) Using the substitution $x = -\cos\theta$ the integral from $\theta = 0$ to π becomes $x = -1$ to $x = 1$, and gives the desired result since,

$$\int_{-1}^{1} e^{ikrx} dx = \frac{1}{ikr}\left(e^{ikr} - e^{-ikr}\right) = 2\frac{\sin(kr)}{(kr)}.$$

The ϕ integral just gives a 2π factor in the result.

(b) Near $k = 0$ we can expand

$$\frac{1}{e^{\beta(\epsilon_k - \mu)} - 1} \approx \frac{1}{z^{-1} - 1 + \hbar^2 k^2/(2mk_B T) + \cdots},$$

where $z = e^{\beta\mu}$ is less than 1, since μ is negative. Therefore $a = z^{-1} - 1$, $b = \hbar^2/(2mk_B T)$.

Using this expansion in the integral from (a) we find

$$\rho_1(r) = \frac{4\pi}{(2\pi)^3} \int_0^\infty \frac{1}{a + bk^2} k^2 \frac{\sin(kr)}{(kr)} dk.$$

Changing variables to $x = kr$ and taking the constants outside the integrand leads to

$$\rho_1(r) = \frac{4\pi}{(2\pi)^3} \frac{1}{br} \int_0^\infty \frac{x}{c^2 + x^2} \sin(x) \, dx,$$

where $c^2 = ar^2/b$. Therefore we obtain a density matrix of the form

$$\rho_1(r) \sim \frac{1}{r} e^{-r/d}.$$

(c) In part (b) we find that the distance d is given by

$$d = \left(\frac{b}{a}\right)^{1/2} = \left(\frac{\hbar^2}{2mk_B T(z^{-1} - 1)}\right)^{1/2}.$$

As $T \to T_c$, the chemical potential μ approaches zero, and z approaches 1. Therefore the denominator goes to zero and the distance d diverges. Infinitesimally above T_c, μ will be very small and so $e^{\beta\mu} \sim 1 + \beta\mu$. Therefore $z^{-1} - 1 \sim -\beta\mu = \beta|\mu|$. Hence we find,

$$d \sim \left(\frac{\hbar^2}{2m|\mu|}\right)^{1/2}$$

just above T_c, and $d \to \infty$ as $T \to T_c$.

(d) This is similar to parts (b) and (c), except that now we expand

$$\frac{1}{e^{\beta\epsilon_{\mathbf{k}}}-1} \sim \frac{2mk_{\mathrm{B}}T}{\hbar^2 k^2},$$

since $\mu = 0$ below T_c. The corresponding density matrix is

$$\rho_1(r) = n_0 + \frac{4\pi}{(2\pi)^3}\frac{2mk_{\mathrm{B}}T}{\hbar^2}\frac{1}{r}\int_0^{\infty}\frac{\sin(x)}{x}\,dx.$$

Hence there is a constant plus $1/r$ behavior at large r. Obviously this function can only be valid asymptotically for large r and must break down at smaller r values, since we know that at $r = 0$ the density matrix is finite and given by the total particle density, $\rho_1(0) = n$.

A.3 Chapter 3

(3.1)

(a) Just use $\nabla \times \mathbf{B} = \mu_0 \mathbf{j}$ to eliminate $\mathbf{j}$ from the London equation. The numerical constants n_s, m_e, and μ_0 can be combined into a single constant, λ, which must have dimensions of length.

(b) Since we are told that $\mathbf{B}$ is only in the z direction, and it is only a function of x, $\mathbf{B} = (0, 0, B_z(x))$, we have

$$\nabla \times \mathbf{B} = \begin{vmatrix} \mathbf{i} & \mathbf{j} & \mathbf{k} \\ \dfrac{\partial}{\partial x} & \dfrac{\partial}{\partial y} & \dfrac{\partial}{\partial z} \\ 0 & 0 & B_z(x) \end{vmatrix} = -\mathbf{j}\frac{dB_z}{dx}.$$

This is also just a function of x, and so taking the curl again we obtain,

$$\nabla \times (\nabla \times \mathbf{B}) = \begin{vmatrix} \mathbf{i} & \mathbf{j} & \mathbf{k} \\ \dfrac{\partial}{\partial x} & \dfrac{\partial}{\partial y} & \dfrac{\partial}{\partial z} \\ 0 & -\dfrac{d}{dx}B_z(x) & 0 \end{vmatrix} = \mathbf{k}\frac{d^2 B_z}{dx^2}.$$

(c) This is a standard second order differential equation. The constant coefficient is positive and so the solutions are exponentials not sines and cosines. The general solution is

$$B_z(x) = Ce^{x/\lambda} + De^{-x/\lambda}.$$

But we know that the field is equal to B_0 at $x = 0$ and the field cannot increase to infinity far from the surface, and so $C = 0$, $D = B_0$, giving the desired result.

(3.2) This is similar to part (c) of 3.1, except now the superconductor is finite, so we must obey the conditions $B_z(L) = B_z(-L) = B_0$. Therefore in the general solution

$$B_0 = Ce^{L/\lambda} + De^{-L/\lambda} \qquad B_0 = Ce^{-L/\lambda} + De^{+L/\lambda}.$$

Solving the simultaneous equations to find C and D gives the required result. Interestingly the same result can be obtained with less algebra by noticing that instead of exponentials we are equally well entitled to write the general solution in terms of hyperbolic sines and cosines,

$$B_z(x) = C\cosh(x/\lambda) + D\sinh(-x/\lambda).$$

In this case symmetry tells us immediately that $D = 0$.

(3.3)

(a) We have to again take two curls, as in Exercise (3.1), but here we must use the cylindrical polar form of curl. Since $\mathbf{B}$ is in the z direction and depends only on r (by the cylindrical symmetry of the vortex) we have

$$\nabla \times \mathbf{B} = \frac{1}{r}\begin{vmatrix} \mathbf{e}_r & r\mathbf{e}_\phi & \mathbf{e}_z \\ \dfrac{\partial}{\partial r} & \dfrac{\partial}{\partial \phi} & \dfrac{\partial}{\partial z} \\ 0 & 0 & B_z(r) \end{vmatrix} = -\mathbf{e}_\phi \frac{dB_z}{dr}.$$

Taking curl again gives

$$\nabla \times (\nabla \times \mathbf{B}) = \frac{1}{r}\begin{vmatrix} \mathbf{e}_r & r\mathbf{e}_\phi & \mathbf{e}_z \\ \dfrac{\partial}{\partial r} & \dfrac{\partial}{\partial \phi} & \dfrac{\partial}{\partial z} \\ 0 & -r\dfrac{d}{dr}B_z(r) & 0 \end{vmatrix} = -\mathbf{e}_z \frac{1}{r}\frac{d}{dr}\left(r\frac{dB_z}{dr}\right).$$

(b) Using the approximation given, we must solve

$$\frac{1}{r}\frac{d}{dr}\left(r\frac{dB_z}{dr}\right) = 0.$$

Clearly, the term in brackets must be constant,

$$r\frac{dB_z}{dr} = a,$$

and we can integrate this one variable ordinary equation to obtain the given result.

(c) The current is simply found from $\mu_0\mathbf{j} = \nabla \times \mathbf{B}$, giving

$$\mathbf{j} = -\frac{a}{\mu_0 r}\mathbf{e}_\phi.$$

Using the London equation this gives the vector potential,

$$\mathbf{A} = -\frac{a\lambda^2}{r}\mathbf{e}_\phi.$$

To find the flux we use Stokes theorem and $\mathbf{B} = \nabla \times \mathbf{A}$ to find

$$\Phi = \int \mathbf{B}\cdot d\mathbf{S}\, d^2r = \oint \mathbf{A}\cdot d\mathbf{r} = -2\pi a\lambda^2$$

hence the constant a is given by

$$a = -\frac{\Phi}{2\pi\lambda^2}.$$

(d) This is exactly the same second order differential equation as in Exercise (3.1(c)), and again only the decreasing exponential $e^{-r/\lambda}$ is possible.

(e) Taking $B_z = r^p e^{-r/\lambda}$ we find

$$\frac{1}{r}\frac{d}{dr}\left(r\frac{dB_z}{dr}\right) = \left(p^2 r^{p-2} - \frac{(2p+1)r^{p-1}}{\lambda} + \frac{r^p}{\lambda^2}\right)e^{-r/\lambda}.$$

Equating this to B/λ^2 we find that the term r^p/λ^2 cancels and we have

$$p^2 r^{p-2} - \frac{(2p+1)r^{p-1}}{\lambda} = 0.$$

Obviously this can never be true for all r, but we can make it true in the large r limit. We must make the coefficient of the largest term vanish, implying that $2p + 1 = 0$ or $p = -1/2$. This is the correct asymptotic limit of the Bessel function.

(3.4) Using $\mathbf{j} = -(a/r)\mathbf{e}_\phi$ and $j = -en_s v$ gives the kinetic energy density

$$\frac{1}{2}m_e n_s v^2 d^3r = \frac{1}{2}m_e \frac{a^2}{e^2 n_s r^2} d^3r.$$

Integrating over the two-dimensional cross section of the vortex, the energy per unit length is

$$E = \frac{1}{2}m_e \frac{a^2}{e^2 n_s} \int \frac{1}{r^2}\, d^2r.$$

The integration region extends roughly from the vortex core, ξ_0, to the end of the $1/r$ region, that is, until $r \sim \lambda$. Therefore

$$E = \frac{1}{2}m_e \frac{a^2}{e^2 n_s} \int_{\xi_0}^{\lambda} \frac{1}{r^2} 2\pi r\, dr.$$

This is equivalent to the result given in the question, once we use the definition of λ and $a = -\Phi/(2\pi\lambda^2)$ from Exercise (3.3).

(3.5)

(a) We just insert $\mathrm{Re}[\sigma(\omega')] = \pi e^2 n_s \delta(\omega')/m_e$ into

$$\mathrm{Im}[\sigma(\omega)] = -\frac{1}{\pi}\mathcal{P}\int_{-\infty}^{\infty} \frac{\mathrm{Re}[\sigma(\omega')]}{\omega' - \omega} d\omega'$$

to obtain

$$\mathrm{Im}[\sigma(\omega)] = +\frac{e^2 n_s}{m_e \omega},$$

which is consistent with the London theory.

(b) The Drude expression for $\sigma(\omega)$ Eq. 3.38 has a pole where

$$\omega = -i\tau^{-1}$$

(shown as a dot on the negative y-axis in Fig. 3.14). This always lies below the x-axis, and is the only pole, therefore there are no poles in the upper half-plane in Fig. 3.14.

(c) The $1/(\omega' - \omega)$ part of the integrand has a pole at $\omega' = \omega$, lying on the $x - axis$ in (shown as a dot in Fig. 3.14). The contour shown avoids this point, and so neither of these poles lies inside the closed contour. Hence using Cauchy's theorem we can say that $I = 0$.

The contour integral has three contributions. First there is the contribution from the large semicircle. If the radius of this semicircle is R, then one can see that the contribution to the integral from the semicircle is of order $1/R$ since $\sigma(\omega') = O(1/R)$. Therefore we can neglect this in the $R \to \infty$ limit. The contribution from the small semicircle is easily calculated as

$$\int_{\pi}^{0} \frac{\sigma(\omega)}{re^{i\theta}} ire^{i\theta}\, d\theta = -i\pi\sigma(\omega),$$

where $\omega' = \omega + re^{i\theta}$. The remaining contribution to the contour integral is the integral along the x-axis, which becomes a principal part integral in the limit $r \to 0$ (because of the singularity at ω). Therefore

$$I = 0 - i\pi\sigma(\omega) + \mathcal{P}\int_{-\infty}^{\infty} \frac{\sigma(\omega')}{\omega' - \omega}\, d\omega'.$$

Using $I = 0$ gives

$$\sigma(\omega) = -\frac{i}{\pi}\mathcal{P}\int_{-\infty}^{\infty} \frac{\sigma(\omega')}{\omega' - \omega}\, d\omega'.$$

Taking the real and imaginary parts of this expression finally gives us the Kramers–Kronig relations.

A.4 Chapter 4

(4.1) Equilibrium between phases ensures that $G_s = G_n$ at every point on the phase boundary $H_c(T)$. Therefore

$$G_s(H + \delta H, T + \delta T) = G_n(H + \delta H, T + \delta T)$$

and $G_s(H, T) = G_n(H, T)$. Assuming that δH and δT are infinitesimal,

$$\frac{\partial G_s}{\partial H}\delta H + \frac{\partial G_s}{\partial T}\delta T = \frac{\partial G_n}{\partial H}\delta H + \frac{\partial G_n}{\partial T}\delta T,$$

and so

$$-\mu_0 V M_s \delta H - S_s \delta T = \mu_0 V M_n \delta H - S_n \delta T.$$

Rearranging gives the analog of the Clausius–Claperyon equation

$$\frac{dH}{dT} = -\frac{1}{\mu_0 V}\frac{S_s - S_n}{M_s - M_n} = -\frac{L}{\mu_0 H T},$$

where we use $M_s = -H$ for the type I superconductor and $M_n \approx 0$ for the normal phase.

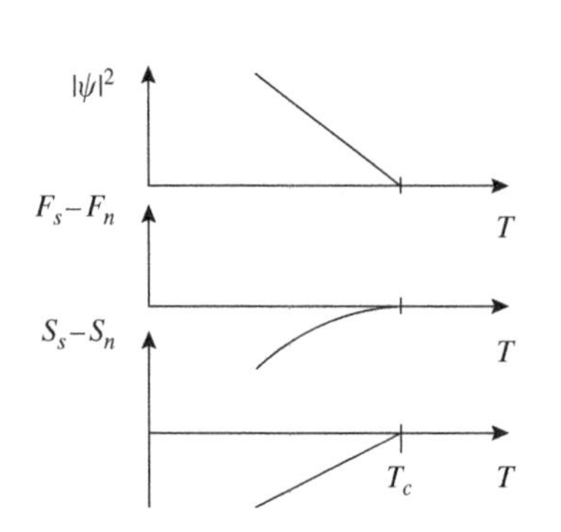

Fig. A.2 Exercise 4.2. The order parameter squared, $|\psi|^2$, free energy $F_s - F_n$ and entropy $S_s - S_n$ of a superconductor near to T_c. The fact that the entropy is continuous, but with a change of slope, implies that T_c is a second-order phase transition.

(4.2) It is easy to see that for $T < T_c$,

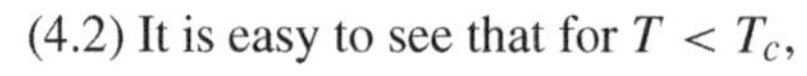

$$|\psi|^2 = \frac{\dot{a}(T_c - T)}{b},$$

$$F_s - F_n = -V\frac{\dot{a}^2(T_c - T)^2}{2b},$$

$$S_s - S_n = -V\frac{\dot{a}^2(T_c - T)}{b},$$

and they are all zero for $T > T_c$. Therefore $|\psi|^2$ and $S_s - S_n$ have discontinuities in derivative at T_c, while $F_s - F_n$ has a discontinuity in the second derivative, as shown in Fig. A.2.

Fig. A.3 Exercise 4.3. The order parameter near the surface of a superconductor: Top, the case $\psi(0) = 0$, middle panel the case $0 < \psi(0) < \psi_0$, and bottom graph the order parameter in a normal metal in proximity to a superconductor.

(4.3)
(a) Substituting $\psi(x) = |\psi_0| f(y)$ into Eq. 4.39 and assuming $f(y)$ is real leads directly to the cubic differential equation given in the question.
(b) Substituting $f(y) = \tanh(cy)$ into this cubic equation gives

$$-\ddot{f} - f + f^3 = 2c^2 t(1 - t^2) - t + t^3 = 0$$

which is obeyed when $2c^2 = 1$, or $c = 1/\sqrt{2}$. The resulting solution is shown in Fig. A.3.

(c) If the boundary condition is $\psi(0) \neq 0$, and $\psi(0) < \psi_0$, then we can just translate the solution found in (b) sideways,

$$\psi(x) = \psi_0 \tanh\left(\frac{(x + x_0)}{\sqrt{2}\xi(T)}\right),$$

where x_0 is chosen so that

$$\psi(0) = \psi_0 \tanh\left(\frac{x_0}{\sqrt{2}\xi(T)}\right),$$

as shown in Fig. A.3.

(d) In the normal metal we can assume that ψ is small, and so we can approximate Eq. 4.39 by a linear equation,

$$-\frac{\hbar^2}{2m^*}\frac{d^2\psi}{dx^2} + a\psi = 0,$$

resulting in the exponential solution,

$$\psi(x) = \psi(0)e^{-x/\xi(T)},$$

as shown in Fig. A.3.

(4.4)

(a) In Eq. 4.120 calculate the internal energy and specific heat in the gaussian model using

$$U = -\frac{\partial \ln Z}{\partial \beta} \qquad C_v = \frac{dU}{dT}.$$

All of the steps are independent of the dimension of space, until we replace the sum over **k** points by a three-dimensional integral. Therefore in the two dimensional case we use

$$\sum_{\mathbf{k}} \rightarrow \frac{1}{(2\pi)^2}\int d^2k$$

to find

$$\begin{aligned} C_V &\sim k_B T^2 \left(\frac{\dot{a}}{a}\right)^2 \frac{1}{(2\pi)^2}\int \frac{1}{(1+\xi(T)^2k^2)^2} d^2k \\ &= k_B T^2 \left(\frac{\dot{a}}{a}\right)^2 \frac{1}{(2\pi)^2} 2\pi \int_0^\infty \frac{1}{(1+\xi(T)^2k^2)^2} k\,dk \\ &= k_B T^2 \left(\frac{\dot{a}}{a}\right)^2 \frac{1}{(2\pi)^2} 2\pi \frac{1}{\xi(T)^2}\int_0^\infty \frac{1}{(1+y^2)^2} y\,dy. \end{aligned}$$

Using $a = \dot{a}(T - T_c)$ and $\xi(T)^2 = \hbar^2/(2m^*|a|)$ and ignoring the numerical prefactors we find,

$$C_V \sim \frac{1}{|T - T_c|}$$

or $\alpha = 1$ in two dimensions.

(b) The argument in (a) is easy to generalize to d-dimensions, by writing

$$\sum_{\mathbf{k}} \to \frac{1}{(2\pi)^d} \int d^d k \to \frac{\Omega_d}{(2\pi)^d} \int_0^\infty k^{d-1}\, dk,$$

where Ω_d is the surface area of a "sphere" in d-dimensions (2π in $d = 2$, 4π in $d = 3$, etc.). The integral over k is always of the form

$$\frac{\Omega_d}{(2\pi)^d} \int_0^\infty \frac{1}{(1+\xi(T)^2 k^2)^2} k^{d-1}\, dk \sim \frac{1}{\xi(T)^d}.$$

Therefore the specific heat singularity is proportional to

$$\left(\frac{\dot{a}}{a}\right)^2 \frac{1}{\xi(T)^d} \sim \frac{1}{|T-T_c|^{2-(d/2)}},$$

so that $\alpha = 2 - d/2$ in the gaussian model.

A.5 Chapter 5

(5.2) Using the definition of the coherent state $|\alpha\rangle$ from Eq. 5.13 gives

$$\begin{aligned}
\langle\alpha|\beta\rangle &= e^{-|\alpha|^2/2} e^{-|\beta|^2/2} \int \left(\sum_n \frac{\alpha^{*n}}{n!^{1/2}} \psi_n^*(x)\right)\left(\sum_m \frac{\beta^m}{m!^{1/2}} \psi_m(x)\right) dx \\
&= e^{-|\alpha|^2/2} e^{-|\beta|^2/2} \sum_n \frac{(\alpha^*\beta)^n}{n!} \\
&= e^{-|\alpha|^2/2} e^{-|\beta|^2/2} e^{\alpha^*\beta}.
\end{aligned}$$

Finally, this can be put into the form of Eq. 5.33 because

$$|\langle\alpha|\beta\rangle|^2 = e^{-|\alpha|^2 - |\beta|^2 + \alpha^*\beta + \alpha\beta^*}.$$

(5.4)

(a) Using the definition of the coherent state $|\alpha\rangle$ from Eq. 5.13, we see that

$$|\alpha\rangle\langle\alpha| = e^{-|\alpha^2|} \left(\sum_m \frac{\alpha^m}{m!^{1/2}} |\psi_m\rangle\right)\left(\sum_n \frac{\alpha^{*n}}{n!^{1/2}} \langle\psi_n|\right),$$

which gives the desired expression.

(b) Transforming to polar coordinates in the complex plane, $\alpha = re^{i\theta}$, the integration unit area in the complex plane is

$$d^2\alpha = r dr d\theta.$$

Therefore

$$\begin{aligned}
\frac{1}{4\pi} \int d^2\alpha |\alpha\rangle\langle\alpha| &= \frac{1}{4\pi} \sum_{mn} \left(\int r dr d\theta \frac{r^{m+n} e^{i(m-n)\theta}}{(n!m!)^{1/2}} e^{-r^2}\right) |\psi_m\rangle\langle\psi_n|, \\
&= \frac{1}{2} \sum_n \frac{1}{n!} \left(\int r dr r^{2n} e^{-r^2}\right) |\psi_n\rangle\langle\psi_n|, \\
&= \sum_n |\psi_n\rangle\langle\psi_n|.
\end{aligned}$$

where we used the integral

$$\int_0^\infty x^n e^{-x}\, dx = n!$$

with $x = r^2$.

(c) Starting with an arbitrary state $|\psi\rangle$ expanded in the basis of the harmonic oscillator eignestates

$$|\psi\rangle = \sum_m c_m |\psi_m\rangle$$

then it is easy to see that

$$\begin{aligned}\left(\sum_n |\psi_n\rangle\langle\psi_n|\right)|\psi\rangle &= \sum_{nm} |\psi_n\rangle\langle\psi_n|\psi_m\rangle c_m \\ &= \sum_n c_n|\psi_n\rangle \\ &= |\psi\rangle.\end{aligned}$$

(5.5)

(a) This is simply a question of inverting the matrix

$$\begin{pmatrix} u_{\mathbf{k}} & v_{\mathbf{k}} \\ v_{\mathbf{k}} & u_{\mathbf{k}} \end{pmatrix}$$

giving

$$\begin{pmatrix} u_{\mathbf{k}} & v_{\mathbf{k}} \\ v_{\mathbf{k}} & u_{\mathbf{k}} \end{pmatrix}^{-1} = \begin{pmatrix} u_{\mathbf{k}} & -v_{\mathbf{k}} \\ -v_{\mathbf{k}} & u_{\mathbf{k}} \end{pmatrix}$$

because $u^2 - v^2 = 1$.

(d) Multiplying the product of three matrices in (c) gives

$$M_{12} = M_{21} = \frac{2u_{\mathbf{k}}v_{\mathbf{k}}}{u_{\mathbf{k}}^2 + v_{\mathbf{k}}^2} - \frac{n_0 g}{\epsilon_{\mathbf{k}} + n_0 g}$$

and so the matrix is diagonal if u and v are chosen as specified.

(e) Again, multiplying the matrices gives

$$M_{11} = (\epsilon_{\mathbf{k}} + n_0 g)u_{\mathbf{k}}^2 - n_0 g u_{\mathbf{k}} v_{\mathbf{k}}$$

and

$$M_{22} = (\epsilon_{\mathbf{k}} + n_0 g)v_{\mathbf{k}}^2 - n_0 g u_{\mathbf{k}} v_{\mathbf{k}}$$

giving the sum $M_{11} + M_{22}$ specified. The trace of the matrix can be interpreted as the excitation energy, E, because we have

$$\begin{aligned}\hat{H} &= \sum_{\mathbf{k}} M_{11} b_{\mathbf{k}}^{+} b_{\mathbf{k}} + M_{22} b_{-\mathbf{k}} b_{-\mathbf{k}}^{+} \\ &= \sum_{\mathbf{k}} M_{11} b_{\mathbf{k}}^{+} b_{\mathbf{k}} + M_{22}(1 + b_{-\mathbf{k}}^{+} b_{-\mathbf{k}}).\end{aligned}$$

In the sum over all possible **k** values the number operator $b_{\mathbf{k}}^{+} b_{\mathbf{k}}$ for any given **k** point occurs twice in the sum, once from **k** and once from $-\mathbf{k}$. Therefore

$$\hat{H} = \sum_{\mathbf{k}} (M_{11} + M_{22}) b_{\mathbf{k}}^{+} b_{\mathbf{k}} + \text{const.}$$

Since $u^2 - v^2 = 1$ we can use the representation $u_{\mathbf{k}} = \cosh\theta$, $v_{\mathbf{k}} = \sinh\theta$. In this representation $2uv = \sinh(2\theta)$ and $u^2 + v^2 = \cosh(2\theta)$. Therefore the equation from (d) becomes

$$\tanh(2\theta) = \frac{n_0 g}{\epsilon_{\mathbf{k}} + n_0 g}.$$

The expression for the excitation energy becomes

$$E = (\epsilon_{\mathbf{k}} + n_0 g) \cosh(2\theta) - n_0 g \sinh(2\theta),$$

which can be shown to be equivalent to Eq. 5.95 after a few steps of algebra.

A.6 Chapter 6

(6.1) The commutator $[P^+_{\mathbf{k}}, P^+_{\mathbf{k}'}]$ is obviously zero, as is $[P_{\mathbf{k}}, P^+_{\mathbf{k}'}]$ for $\mathbf{k} \neq \mathbf{k}'$. The only difficult case is when $\mathbf{k} = \mathbf{k}'$. Then we have

$$\begin{aligned}
[\hat{P}_{\mathbf{k}}, \hat{P}^+_{\mathbf{k}}] &= c_{-\mathbf{k}\downarrow} c_{\mathbf{k}\uparrow} c^+_{\mathbf{k}\uparrow} c^+_{-\mathbf{k}\downarrow} - c^+_{\mathbf{k}\uparrow} c^+_{-\mathbf{k}\downarrow} c_{-\mathbf{k}\downarrow} c_{\mathbf{k}\uparrow} \\
&= c_{-\mathbf{k}\downarrow} \left(1 - c^+_{\mathbf{k}\uparrow} c_{\mathbf{k}\uparrow}\right) c^+_{-\mathbf{k}\downarrow} - c^+_{\mathbf{k}\uparrow} c^+_{-\mathbf{k}\downarrow} c_{-\mathbf{k}\downarrow} c_{\mathbf{k}\uparrow} \\
&= \left(1 - c^+_{\mathbf{k}\uparrow} c_{\mathbf{k}\uparrow}\right)\left(1 - c^+_{-\mathbf{k}\downarrow} c_{-\mathbf{k}\downarrow}\right) - c^+_{\mathbf{k}\uparrow} c_{\mathbf{k}\uparrow} c^+_{-\mathbf{k}\downarrow} c_{-\mathbf{k}\downarrow} \\
&= 1 - \hat{n}_{\mathbf{k}\uparrow} - \hat{n}_{-\mathbf{k}\downarrow}.
\end{aligned}$$

Here, as usual, $\hat{n}_{\mathbf{k}\sigma} = c^+_{\mathbf{k}\uparrow} c_{\mathbf{k}\sigma}$ is the number operator. Therefore

$$[\hat{P}_{\mathbf{k}}, \hat{P}^+_{\mathbf{k}'}] = \delta_{\mathbf{k}\mathbf{k}'} \left(1 - \hat{n}_{\mathbf{k}\uparrow} - \hat{n}_{-\mathbf{k}\downarrow}\right),$$

which is different than the boson commutator. Therefore a Cooper pair is not simply a boson particle.

(6.2) The algebra is simplified if we write the interaction in terms of the pair operators as

$$c^+_{\mathbf{k}\uparrow} c^+_{-\mathbf{k}\downarrow} c_{-\mathbf{k}'\downarrow} c_{\mathbf{k}'\uparrow} = \hat{P}^+_{\mathbf{k}} \hat{P}_{\mathbf{k}'}.$$

Using the definitions of the BCS pairing state we need to evaluate

$$\langle \Psi_{\mathrm{BCS}} | \hat{P}^+_{\mathbf{k}} \hat{P}_{\mathbf{k}'} | \Psi_{\mathrm{BCS}} \rangle = \langle 0 | \Pi_{\mathbf{q}} (u_{\mathbf{q}} + v_{\mathbf{q}} \hat{P}_{\mathbf{q}}) \hat{P}^+_{\mathbf{k}} \hat{P}_{\mathbf{k}'} \Pi_{\mathbf{q}'} (u^*_{\mathbf{q}'} + v^*_{\mathbf{q}'} \hat{P}^+_{\mathbf{q}'}) | 0 \rangle.$$

In the product over all possible states $\mathbf{q}'$ there will be one term where $\mathbf{q}' = \mathbf{k}'$. This term gives a contribution, which acts on the vacuum state on the right like

$$\hat{P}_{\mathbf{k}'} (u^*_{\mathbf{k}'} + v^*_{\mathbf{k}'} \hat{P}^+_{\mathbf{k}'}) | 0 \rangle = v^*_{\mathbf{k}'} | 0 \rangle$$

because all other terms are of the form $\hat{P}_{\mathbf{k}'} | 0 \rangle = 0$. Similarly we can apply the same principle to the "bra" state on the left when $\mathbf{q} = \mathbf{k}$

$$\langle 0 | (u_{\mathbf{k}} + v_{\mathbf{k}} \hat{P}_{\mathbf{k}}) \hat{P}^+_{\mathbf{k}} = \langle 0 | v_{\mathbf{k}}.$$

The remaining terms include the products over all the other values of $\mathbf{q}$ and $\mathbf{q}'$,

$$\langle \Psi_{\mathrm{BCS}} | \hat{P}^+_{\mathbf{k}} \hat{P}_{\mathbf{k}'} | \Psi_{\mathrm{BCS}} \rangle = v_{\mathbf{k}} v^*_{\mathbf{k}'} \langle 0 | \Pi_{\mathbf{q} \neq \mathbf{k}} (u_{\mathbf{q}} + v_{\mathbf{q}} \hat{P}_{\mathbf{q}}) \Pi_{\mathbf{q}' \neq \mathbf{k}'} (u^*_{\mathbf{q}'} + v^*_{\mathbf{q}'} \hat{P}^+_{\mathbf{q}'}) | 0 \rangle.$$

In this product typical values of $\mathbf{q}$ and $\mathbf{q}'$ will just give a factor of 1, because of the normalization of the BCS wave function coefficients $|u_{\mathbf{k}}|^2 + |v_{\mathbf{k}}|^2 = 1$. However, in the product there are two "extra" terms, where the creation and annihilation operators do not match. These are $\mathbf{q}' = \mathbf{k}$ and $\mathbf{q} = \mathbf{k}'$. These terms contribute as follows

$$\begin{aligned}\langle\Psi_{\mathrm{BCS}}|\hat{P}^+_{\mathbf{k}}, \hat{P}_{\mathbf{k}'}|\Psi_{\mathrm{BCS}}\rangle &= v_{\mathbf{k}}v^*_{\mathbf{k}'}\langle 0|(u_{\mathbf{k}'} + v_{\mathbf{k}'}\hat{P}_{\mathbf{k}'})(u^*_{\mathbf{k}} + v^*_{\mathbf{k}}\hat{P}^+_{\mathbf{k}})|0\rangle \\ &= v_{\mathbf{k}}v^*_{\mathbf{k}'}u^*_{\mathbf{k}}u_{\mathbf{k}'},\end{aligned}$$

hence giving the desired result.

(6.3) According to Eqs 6.77 and 6.78 the $b_{\mathbf{k}\uparrow}$ and $b^+_{-\mathbf{k}\downarrow}$ operators are defined by

$$\begin{aligned}b_{\mathbf{k}\uparrow} &= u^*_{\mathbf{k}}c_{\mathbf{k}\uparrow} - v^*_{\mathbf{k}}c^+_{-\mathbf{k}\downarrow} \\ b^+_{-\mathbf{k}\downarrow} &= v_{\mathbf{k}}c_{\mathbf{k}\uparrow} + u_{\mathbf{k}}c^+_{-\mathbf{k}\downarrow}.\end{aligned}$$

Clearly these obey $\{b_{\mathbf{k}\uparrow}, b^+_{-\mathbf{k}'\downarrow}\} = 0$, because when $\mathbf{k} = \mathbf{k}'$

$$\begin{aligned}\{b_{\mathbf{k}\uparrow}, b^+_{-\mathbf{k}\downarrow}\} &= \{u^*_{\mathbf{k}}c_{\mathbf{k}\uparrow} - v^*_{\mathbf{k}}c^+_{-\mathbf{k}\downarrow}, v_{\mathbf{k}}c_{\mathbf{k}\uparrow} + u_{\mathbf{k}}c^+_{-\mathbf{k}\downarrow}\} \\ &= u^*_{\mathbf{k}}v_{\mathbf{k}}\{c_{\mathbf{k}\uparrow}, c_{\mathbf{k}\uparrow}\} + u^*_{\mathbf{k}}u_{\mathbf{k}}\{c_{\mathbf{k}\uparrow}, c^+_{-\mathbf{k}\downarrow}\} - v^*_{\mathbf{k}}v_{\mathbf{k}}\{c^+_{-\mathbf{k}\downarrow}, c_{\mathbf{k}\uparrow}\} \\ &\quad - v^*_{\mathbf{k}}u_{\mathbf{k}}\{c^+_{-\mathbf{k}\downarrow}, c^+_{-\mathbf{k}\downarrow}\} \\ &= 0.\end{aligned}$$

The other commutators to be checked involve the operator Hermitian conjugates of $b_{\mathbf{k}\uparrow}$ and $b^+_{-\mathbf{k}\downarrow}$. For example $b^+_{\mathbf{k}\uparrow}$ is defined by the conjugate equation to 6.77

$$b^+_{\mathbf{k}\uparrow} = u_{\mathbf{k}}c^{\dagger}_{\mathbf{k}\uparrow} - v_{\mathbf{k}}c_{-\mathbf{k}\downarrow}.$$

Then

$$\begin{aligned}\{b_{\mathbf{k}\uparrow}, b^+_{\mathbf{k}'\uparrow}\} &= \{u^*_{\mathbf{k}}c_{\mathbf{k}\uparrow} - v^*_{\mathbf{k}}c^+_{-\mathbf{k}\downarrow}, u_{\mathbf{k}'}c^+_{\mathbf{k}'\uparrow} - v_{\mathbf{k}'}c_{-\mathbf{k}'\downarrow}\} \\ &= u^*_{\mathbf{k}}u_{\mathbf{k}'}\{c_{\mathbf{k}\uparrow}, c^+_{\mathbf{k}'\uparrow}\} + v^*_{\mathbf{k}}v_{\mathbf{k}'}\{c^+_{-\mathbf{k}\downarrow}, c_{-\mathbf{k}'\downarrow}\} \\ &= (u^*_{\mathbf{k}}u_{\mathbf{k}'} + v^*_{\mathbf{k}}v_{\mathbf{k}'})\delta_{\mathbf{k}\mathbf{k}'} \\ &= \delta_{\mathbf{k}\mathbf{k}'}.\end{aligned}$$

Similarly, the Hermitian conjugate of $b^+_{-\mathbf{k}\downarrow}$ is defined by the Hermitian conjugate of Eq. 6.78,

$$b_{-\mathbf{k}\downarrow} = v^*_{\mathbf{k}}c^+_{\mathbf{k}\uparrow} + u^*_{\mathbf{k}}c_{-\mathbf{k}\downarrow}.$$

One can easily check that $\{b_{-\mathbf{k}\downarrow}, b^+_{-\mathbf{k}'\downarrow}\} = \delta_{\mathbf{k}\mathbf{k}'}$. Finally one should also check that $\{b_{-\mathbf{k}\downarrow}, b_{\mathbf{k}'\uparrow}\} = 0$. This follows from our definitions because

$$\begin{aligned}\{b_{-\mathbf{k}\downarrow}, b_{\mathbf{k}'\uparrow}\} &= \{v^*_{\mathbf{k}}c^+_{\mathbf{k}\uparrow} + u^*_{\mathbf{k}}c_{-\mathbf{k}\downarrow}, u^*_{\mathbf{k}'}c_{\mathbf{k}'\uparrow} - v^*_{\mathbf{k}'}c^+_{-\mathbf{k}'\downarrow}\} \\ &= v^*_{\mathbf{k}}u^*_{\mathbf{k}'}\{c^+_{\mathbf{k}\uparrow}, c_{\mathbf{k}'\uparrow}\} - u^*_{\mathbf{k}}v^*_{\mathbf{k}'}\{c_{-\mathbf{k}\downarrow}, c^+_{-\mathbf{k}'\downarrow}\} \\ &= (v^*_{\mathbf{k}}u^*_{\mathbf{k}'} - u^*_{\mathbf{k}}v^*_{\mathbf{k}'})\delta_{\mathbf{k}\mathbf{k}'} \\ &= 0.\end{aligned}$$

(6.4) Using the definition of $b_{\mathbf{k}\uparrow}$ from Eq. 6.77, when we evaluate, $b_{\mathbf{k}\uparrow}|\Psi_{\mathrm{BCS}}\rangle$ there will be a contribution of the form,

$$\begin{aligned}b_{\mathbf{k}\uparrow}(u^*_{\mathbf{k}} + v^*_{\mathbf{k}}c^+_{\mathbf{k}\uparrow}c^+_{-\mathbf{k}\downarrow})|0\rangle &= (u^*_{\mathbf{k}}c_{\mathbf{k}\uparrow} - v^*_{\mathbf{k}}c^+_{-\mathbf{k}\downarrow})(u^*_{\mathbf{k}} + v^*_{\mathbf{k}}c^+_{\mathbf{k}\uparrow}c^+_{-\mathbf{k}\downarrow})|0\rangle \\ &= (u^*_{\mathbf{k}}v^*_{\mathbf{k}}c^+_{-\mathbf{k}\downarrow} - v^*_{\mathbf{k}}c^+_{-\mathbf{k}\downarrow}u^*_{\mathbf{k}} - v^{*2}_{\mathbf{k}}c^+_{-\mathbf{k}\downarrow}c^+_{\mathbf{k}\uparrow}c^+_{-\mathbf{k}\downarrow})|0\rangle \\ &= 0.\end{aligned}$$

Here the last step follows because $c^+_{-\mathbf{k}\downarrow} c^+_{-\mathbf{k}\downarrow} |0\rangle = 0$.

Similarly, using the Hermitian conjugate of Eq. 6.78 to define $b_{-\mathbf{k}\downarrow}$, we have a contribution to $b_{-\mathbf{k}\downarrow}|\Psi_{\mathrm{BCS}}\rangle$ of the form,

$$\begin{aligned} b_{-\mathbf{k}\downarrow}(u^*_{\mathbf{k}} + v^*_{\mathbf{k}} c^+_{\mathbf{k}\uparrow} c^+_{-\mathbf{k}\downarrow})|0\rangle &= (v^*_{\mathbf{k}} c^+_{\mathbf{k}\uparrow} + u^*_{\mathbf{k}} c_{-\mathbf{k}\downarrow})(u^*_{\mathbf{k}} + v^*_{\mathbf{k}} c^+_{\mathbf{k}\uparrow} c^+_{-\mathbf{k}\downarrow})|0\rangle \\ &= (v^*_{\mathbf{k}} u^*_{\mathbf{k}} c^+_{\mathbf{k}\uparrow} + v^{*2}_{\mathbf{k}} c^+_{\mathbf{k}\uparrow} c^+_{\mathbf{k}\uparrow} c^+_{-\mathbf{k}\downarrow} - u^*_{\mathbf{k}} v^*_{\mathbf{k}} c^+_{\mathbf{k}\uparrow})|0\rangle \\ &= 0. \end{aligned}$$

Bibliography

Abo-Shaeer, J.R., Raman, C., Vogels, J.M., and Ketterle, W. (2001). *Science*, **292**, 476–479.

Abramowitz, M., and Stegun, I.A. (1965). *Handbook of Mathematical Functions*. Dover, New York.

Abrikosov, A.A., Gorkov, L.P., and Dzyaloshinski, I.E. (1963). *Methods of Quantum Field Theory in Statistical Physics*. Dover, New York.

Adenwalla *et al.* (1990). Phys. Rev. Lett. **65**, 2298.

Alexandrov, A.S. (2003). *Theory of Superconductivity from Weak to Strong Coupling*. Institute of Physics Publishing, Bristol.

Amit, D.J. (1984). *Field Theory Renormalization Group, and Critical Phenomena*. World Scientific, Singapore.

Anderson, P.W. (1984). *Basic Notions of Condensed Matter Physics*. Benjamin/Cummings, Melno Park.

Annett, J.F. (1990). *Advances in Physics*, **39**, 83–126.

Annett, J.F. (1999). *Physica C*, **317–318**, 1–8.

Annett, J.F., Goldenfeld, N.D., and Renn, S.R. (1990). In *Physical Properties of High Temperature Superconductors II*, D.M. Ginsberg (ed.), World Scientific, Singapore.

Annett, J.F., Goldenfeld, N.D., and Leggett, A.J. (1996). In *Physical Properties of High Temperature Superconductors V*, D.M. Ginsberg (ed.), World Scientific.

Annett, J.F., Gyorffy, B.L., and Spiller T.P. (2002). In *Exotic States in Quantum Nanostructures*, S. Sarkar (ed.), Kluwer.

Ashcroft, N. and Mermin, N.D. (1976). *Solid State Physics*, W.B. Saunders.

Blatter, G., Feigel'man, M.V., Geshkenbein, V.B., Larkin A.I., and Vinokur, V.M. (1994). *Rev. Mod. Phys.*, **66**, 1125–1388.

Blundell, S. (2001). *Magnetism in Condensed Matter*. Oxford University Press, Oxford.

Boas, M.L. (1983). *Mathematical Methods in the Physical Sciences* (2nd edn). John Wiley & Sons, New York.

Bogoliubov, N. (1947). *Journal of Physics*, **11**, 23.

Callen, H.B. (1960).*Thermodynamics*. John Wiley and Sons, New York.

Ceperley, D.M. (1995). *Rev. Mod. Phys.*, **67**, 279–355.

Chakin, P.M. and Lubensky, T.C. (1995). *Principles of Condensed Matter Physics*. Cambridge University Press.

Chiorescu *et al.* (2003). *Science*, **299**, 1869.

Chu, C.W., Gao, L., Chen, F., Huang, L.J., Meng, R.L., and Xue, Y.Y. (1993). *Nature*, **365**, 323–325.

Dalfovo, F., Giorgni, S., Pitaevskii, L.P., and Stringari, S. (1999). *Rev. Mod. Phys.*, **71**, 463–513.

Fetter, A.L. and Walecka, J.D. (1971). *Quantum Many-particle Theory*. McGraw Hill, New York.

Feynman, R.P. (1972). *Statistical Mechanics*. Addison Wesley, Redwood City.

Feynman R.P., Leighton, R.B., and Sands M. (1964). *The Feynman Lectures on Physics, Vol. II*. Addison Wesley, Reading MA.

Fisher, R.A. *et al.* (1989). Phys. Rev. Lett. **62**, 1411.

Friedman J.R. *et al.* (2000). *Nature*, **406**, 43.

de Gennes, P.-G., (1966). *Superconductivity of Metals and Alloys*, Addison Wesley Advanced Book Programme, Redwood City.

Goldendeld, N.D. (1992). *Lectures on Phase Transitions and the Renormalization Group*, Addison-Wesley, Reading, MA.

Hardy, W.N. *et al.* (1993). Phys. Rev. Lett. **70**, 3999.

Home, D. and Gribbin, J. (1994). *New Scientist* 9 January, 26.

Huang, K. (1987). *Statistical Physics* (2nd edn). John Wiley & Sons, New York.

Jones, R.A.L. (2002). *Soft Condensed Matter*. Oxford University Press, Oxford.

Joynt, R. and Taillefer, L. (2002). *Rev. Mod. Phys.* **74**, 235–294.

Ketterle, W. (2002). *Rev. Mod. Phys.*, **74**, 1131–1151.

Ketterle, W. (2003). http://cua.mit.edu/ketterle_group/Nice_pics.htm.

Ketterson, J.B. and Song, S.N. (1999). *Superconductivity*. Cambridge University Press, Cambridge.

Kittel, C. (1996). *Introduction to Solid State Physics* (7th edn). John Wiley & Sons, New York.

Klauder, J.R. and Skagerstam, B.-S. (1985), *Coherent States: Applications in Physics and Mathematical Physics*, World Scientific.

Lee, D.M. (1997). *Rev. Mod. Phys.* **69**, 645–666.

Leggett, A.J. (1975). *Rev. Mod. Phys.*, **47**, 332–414.

Leggett, A.J. (1980). *Suppl. Prog. Theor. Phys.*, **69**. 80–100.

Leggett, A.J. (2001). *Reviews of Modern Physics*, **73**, 307–356.

Leggett, A.J. (2002). *J. Phys. Condens. Matter*, **14**, R415.

Loudon, R. (1979). *The Quantum Theory of Light*, Oxford.

Ma, S.-K. (1974). *Modern Theory of Critical Phenomena*. Benjamin/Cummings.

Mackenzie, A.P. and Maeno, Y. (2003). *Rev. Mod. Phys.*, **75**, 657–712.

Mahklin, Y., Schön, G., and Shnirman, A. (2001). *Rev. Mod. Phys.*, **73**, 357.

Mandl, F. (1987). *Statistical Physics* (2nd edn). John Wiley, Chicester.

Matthews, J. and Walker, R.L. (1970). *Mathematical Methods of Physics* (2nd edn). Addison Wesley.

Mermin, D. (1990). *Boojums All the Way through: Communicating Science in a Prosaic Age*. Cambridge University Press, Cambridge.

Mineev, V.P. and Samokhin, K.V. (1999). *Introduction to Unconventional Superconductivity*. Gordon and Breach Science Publishers, Amsterdam.

Nakamura, Y., Pashkin, Y.A., and Tsai, J.S. (1999). *Science*, **398**, 786–788.

Osheroff, D.D. (1997). *Rev. Mod. Phys.*, **69**, 667–682.

Overend N., Howson M.A., and Lawrie I. D. (1994). *Phys. Rev. Lett.*, **72**, 3238–3241.

Pethick, C.J. and Smith, H. (2001). *Bose-Einstein Condensation in Dilute Gases*, Cambridge University Press, Cambridge.

Phillips, W.D. (1998). *Rev. Mod. Phys.*, **70**, 721–741.

Pines, D. (1961). *The Many Body Problem.* Benjamin Cummings, Reading, MA.

Pitaevskii, L.P. and Stringari, S. (2003). *Bose-Einstein Condensation*, International Series of Monographs in Physics, Clarendon Press, Oxford.

Poole, C.P. (2000). *Handbook of Superconductivity*. Academic Press.

Ramakrishnan, T.V. and Rao, C.N.R. (1992). *Superconductivity Today*. Wiley Eastern, New Delhi.

Raman, C., Abo-Shaeer, J.R., Vogels, J.M., Xu, K. and Ketterle, W. (2001). Phys. Rev. Lett. **87**, 210402.

Rickayzen, G. (1980). *Green's Functions and Condensed Matter*. Academic Press, London.

Salomaa, M.M. and Volovik, G.E. (1987). *Rev. Mod. Phys.*, **59**, 533–613.

Sauls, J.A. (1994). *Advances in Physics*, **43**, 113–141.

Silver, R.N. and Sokol, P.E. (eds.) (1989). *Momentum Distributions*. Plenum, New York.

Schrieffer, J.R. (1964). *Theory of Superconductivity*. Benjamin/Cummings, Reading, MA.

Schneider, T. and Singer, J.M. (2000). *Phase Transition Approach to High Temperature Superconductors*. Imperial College Press, UK.

Singleton, J. (2001). *Band Theory and Electronic Properties of Solids*. Oxford University Press, Oxford.

Tilley, D.R. and Tilley, J. (1990). *Superfluidity and Superconductivity* (3rd edn). Adam Hilger and IOP Publishing, Bristol.

Tinkham, M. (1996). *Introduction to Superconductivity* (2nd edn). McGraw-Hill, New York.

van der Wal, C. *et al.* (2000). *Science*, **290**, 773.

van der Wal (2001). *Quantum Superpositions of Persistent Josephson Currents*. PhD thesis, Delft University Press, Delft.

Tsuei, C.C. and Kirtley, J.R. (2000). *Rev. Mod. Phys.*, **72**, 969–1016.

Van Harlingen, D.J. (1995). *Rev. Mod. Phys.*, **67**, 515.

Vollhardt, D. and Wölfle, P. (1990). *The Superfluid Phases of Helium 3*. Taylor and Francis, London.

Volovik G.E. (1992). *Exotic Properties of Superfluid ^{3}He*. World Scientific, Singapore.

Waldram, J.R. (1996). *Superconductivity of Metals and Cuprates*. Institute of Physics, Bristol.

Wheatley, J.C. (1975). *Rev. Mod. Phys.*, **47**, 415–470.

Wollman, D.A. *et al.* (1993). *Phys. Rev. Lett.*, **71**, 2134.

Yeshurun, Y., Malozemoff, A. P., and Shaulov, A. (1996). *Rev. Mod. Phys.*, **68**, 911–949.

Ziman, J.M. (1979). *Principles of the Theory of Solids* (2nd edn). Cambridge University Press.

Index

The manufacturer's authorised representative in the EU for product
safety is Oxford University Press España S.A. of El Parque Empresarial
San Fernando de Henares, Avenida de Castilla, 2 - 28830 Madrid
(www.oup.es/en or product.safety@oup.com). OUP España S.A. also acts
as importer into Spain of products made by the manufacturer.
Printed and bound by CPI Group (UK) Ltd, Croydon, CR0 4YY
29/03/2025
01838460-0001